BIOMATERIALS

BIOMATERIALS

Novel Materials from Biological Sources

Editor
David Byrom

stockton
press

Published in the United States and Canada by
STOCKTON PRESS, 1991
257 Park Avenue South, New York, N.Y. 10010, USA

ISBN 978-1-349-11169-5   ISBN 978-1-349-11167-1 (eBook)
DOI 10.1007/978-1-349-11167-1

First published in the United Kingdom by
MACMILLAN PUBLISHERS LTD, 1991
Distributed by Globe Book Services Ltd
Brunel Road, Houndmills, Basingstoke, Hants RG21 2XS

British Library Cataloguing in Publication Data

Byrom, David
    Biomaterials.
    I. Title
    660

    ISBN 978-0-333-51175-6

Contents

Preface

Biomaterials are the very stuff of life. They provide the structure, form, function, protection and storage products of all organisms on earth. Man has just started to take advantage of the properties of some of this vast array of natural polymers and to find applications in medicine, food, packaging, consumer products and other fields.

In describing biomaterials as I have above, I have chosen to define those of commercial interest as polymeric substances produced biologically either directly, for example, by fermentation, or as products extracted from biological sources, as is the case with alginate. This book describes the state of research in a number of biomaterials of different types and with different applications.

The term biomaterials is commonly applied to chemically derived polymeric materials that are used in medical applications such as prostheses, bone plates and other implantable devices. This field of application is a vast area which has received much attention and will not be considered here except where it impinges on the use of biologically produced polymers.

The academic and industrial interest in biopolymers as materials results from a number of factors. They represent materials which have a range of specific properties which may not easily be emulated by synthetic polymers. They can be synthesised from renewable resources and are often biodegradable, which is important in some applications.

Although there is a fascination with biopolymers, there are a limited number of these materials in industrial production. This reflects their cost relative to that of chemically derived products coupled with, in some cases, limitations in mechanical and physical properties in areas where existing polymeric materials are used. It has therefore been necessary to find niche or speciality applications for biomaterials which have been developed to support their cost.

Biomaterials research is a multidisciplinary activity both in the sense that a range of expertise is necessary to produce, characterise and bring to market any novel material and in the sense that it has attracted scientists from the biological, chemical and polymer science communities. It is my intention that this book should reflect this fact and I believe that this has been illustrated in most of the examples included.

The biomaterials covered in this book also reflect the range of techniques which are used to make these biopolymers. Some are produced directly by microbial fermentation; some are extracted from natural sources and used directly; others are modified, fabricated or derived from natural polymers. Perhaps slightly paradoxically, one material included here is made by chemical synthesis using a biologically derived monomer. Some products at an early stage of development have been described in a chapter which gives an indication of the most speculative directions of biomaterials research.

Although the synthesis, modification, derivatisation and application of biomaterials may be technically sophisticated, biomaterials research is still in its infancy since most products in this class are single polymers or copolymers. Biotechnology is developing rapidly and novel methods for the synthesis of materials will arise as a result. Molecular genetics has already been used to produce a material − mollusc bioadhesive − which would be extremely difficult to manufacture were this technology not available.

Nature is adept in the synthesis of composite materials such as bone, horn, wood and shells, which have extremely interesting mechanical and physical properties. At the moment, man can only make use of this phenomenal production capability by utilisation of the materials once they are made. Gaining an understanding of the mechanism of synthesis and the fabrication of natural composites *in vivo* may allow the possibility of intervention in these processes using the techniques of molecular genetics. The capability to manufacture novel biomaterials with new properties, for example, in plants or by fermentation will be the end result of this work. Clearly the future holds exciting prospects for the field of biomaterials research.

David Byrom
November 1991

1   Silks

D. L. Kaplan, S. J. Lombardi, W. S. Muller &
S. A. Fossey

Biotechnology Division, U.S. Army Natick Research, Development
& Engineering Center, Natick, Massachusetts 01760-5020, USA

Silks: Chemistry, Properties and Genetics

Introduction

A number of reviews have been published detailing the chemical and physical properties of silkworm silks[1-6] and the genetics of silk production in the silkworm[7-9]. Here we overview some of the key chemical and physical properties of both silkworm and spider silks, and then review the genetic aspects of silk production in these organisms.

Silks represent a broad class of polymers which can be loosely defined as externally spun fibrous protein secretions[10]. Of all the natural fibres, silks are the only spun ones. Fibrous proteins, including silks, collagen, elastin and keratin, are distinctive owing to their repetitive primary structure, encoded by repetitive gene sequences under tight regulation. These protein products, containing repetitive domains that influence higher-order conformations, result in fibres and materials with unusual physical and mechanical properties[11]. In addition to the unusual mechanical properties, silks exhibit interesting interference patterns within the electromagnetic spectrum[12], unusual viscometric patterns related to processing[13], and piezoelectric properties[12,14,15]. Silks differ in composition, properties and morphology depending on their source. Differences are found between species as well as between different silks produced by the same organism, as is the case for many orb-weaving spiders. In general, silks are formed in sets of specialized glands by insects in the class Insecta and spiders in the class Arachnida. Silks are also produced by pseudoscorpions, mites[16] and the aquatic larvae of the midge, *Chironomus tentans* (Diptera), which produce a family of secretory proteins with unusual repeats including periodic cysteine residues[17]. There are thousands of silk-spinning insects and spiders, yet only a few have been investigated in detail.

Sericulture involves the growth of the domesticated silkworm *Bombyx mori* and related species for silk production for textiles. This art has been practised for thousands of years, originating in China. World silk production has declined since the 1930s because of competition from less expensive synthetic fibres[2]. Silkworms produce silk during one stage in their lifecycle. Silkworm silk (*B. mori* unless otherwise specified) consists of two fibroin protein filaments adhered together

3

with a sericin protein gum resulting in a single thread about 10 to 25 μm in diameter[2]. Threads of silks from wild silkworms often have larger diameters up to 65 μm[6]. Silk threads from different organisms can have different morphologies in cross-section, including triangular, crescent, rectangular and wedge.

The life cycle of *B. mori* lasts 55 to 60 days and passes through a series of developmental stages or molts. Cocoon formation occurs at around 26 days in the cycle, during the later stages of the fifth instar[18]. This stage is characterized by the extrusion of silk from a spinneret located in the head and drawing of the fibre by a characteristic figure-of-eight head movement. The silk is extruded into air and a change in conformation to a β-sheet occurs owing to shear, tension due to the movement of the head of the silkworm, and loss of water[6]. Rheology experiments with the soluble silk illustrate that crystallinity is positively correlated with shear rate and rate of drawing, and negatively correlated with the diameter of the spinneret. A critical extrusion rate of 500 mm per min was found by Magoshi *et al.*[13] to induce the conformational change to a β-sheet. During this process the sericin does not undergo a conformational change owing to its higher water content (86%); a shear rate several orders of magnitude higher than for the fibroin is required for the change to the β-sheet conformation. Commercially, silk fibres are harvested from silkworm cocoons, which are exposed to boiling-hot soapy water to remove the sericin before reeling the silk. About 300 to 1200 metres of usable thread are recovered per cocoon[3].

The sericin component of silkworm silk is represented by a family of proteins. Komatsu[19] distinguished sericin I, II, III and IV on the basis of solubility in hot water, and each of these proteins has a different amino-acid composition. Sericin exhibits primarily a random-coil conformation with some β-sheet structure. Conversion to the β-sheet occurs with moisture, heat or mechanical stretching, but the conversion is to a crystalline but not fully oriented form.

Spiders represent some of the most diverse and abundant terrestrial organisms on earth, with over 35,000 species having been identified[20,21]. Despite their presence for over 300 million years, spiders have never been commercialized like silkworms, although they have been of interest for medical reasons for hundreds of years[5]. The diversity of lifestyles among spiders is so great as to preclude further description here. In general, spiders are carnivorous and can be classified as either web-builders or hunters; the silks under review here are produced by web-builders. Unlike the silkworm, some orb-spinning spiders produce silks throughout their adult life and some spiders produce a variety of different silks, each synthesized in separate sets of specialized glands[22]. Female spiders are responsible for orb-web construction, and web recycling by partial resolubilization has been demonstrated for *Araneus cavaticus* using radioactive tracers[23].

Lucas *et al.*[5] referenced work from the nineteenth century where silk from the spider *Nephila madagascariensis* was spun into thread. This spider spins webs that, with a diameter of over a metre, can trap small birds. Historically, spider silks have also been used in eyepieces of optical instruments[5]. Devices for the controlled silking of spiders have been described[24,25]. Spider silks are diverse and serve in prey capture, reproduction and as vibration receptors. Emerging young spiders use silks for dispersion by wind and also to build miniature webs. Unusual visual displays on orb webs are characteristic of some spiders, such as the stabilimenta or zigzag pattern displayed on orb webs produced by the Argiope spider[10]. The function of this pattern is unknown, although it has been speculated that it stabilizes and strengthens the web, disguises the spider, absorbs water and warns off birds[10].

Gosline *et al.*[10] have studied structure–function relationships with fibrous natural polymers including spider silks. Spiders produce a variety of silks for different functions and are therefore useful organisms for the study of molecular design of natural structural polymers. The orb-weaving Aranaeid spider produces at least seven different silks, each of which is synthesized, processed and spun in a different set of glands[10,22,26–28]. Other species of spider synthesize only one type of silk[29]. The different glands and their associated silks include: major ampullate gland which produces structural silk for the orb frame, radii and dragline; flagelliform gland which produces the viscid silk for prey capture; aggregate gland which produces a glue-like silk; minor ampullate gland which produces support fibres for the frame and dragline silks; cylindrical gland for cocoon silk; aciniform gland for wrapping captured prey; and piriform gland for attachment silks which couple the frame and dragline silks to environmental substrates[10]. Each of these silks has a different characteristic amino-acid composition which reflects their functions. Primary DNA, RNA and amino-acid sequence data are beginning to be determined for these different silks to permit the detailed characterization of these relationships and of the genetic regulation involved in the synthesis of the various silks.

Silks play a major structural role in the survival of orb web-spinning spiders throughout the adult life cycle, unlike silkworm silk which functions in cocoon formation at one stage of the life cycle. Spider silk web fibres generally vary in diameter from 1 µm to 5 µm or larger, depending on the source and type of silk[10,22,25,30,31]. The diameters of the web fibres from *Araneus diadematus*, *A. sericatus* and *A. gemma* averaged 3 µm and ranged between 1.03 µm and 3.75 µm (ref. 31). Gosline *et al.*[10], in studies on *A. sericatus*, reported that a 75-mg spider uses 180 µg of protein to spin a web up to 100 cm^2, representing very efficient engineering and use of this material.

Anderson and Tillinghast[32] studied *Argiope aurantia* and *Argiope trifasciata* and identified soluble chemical compounds present on web fibres that may function in species identification, prey capture or in water retention to maintain the adhesive and elastomeric properties of the silk. Salts and neurotransmitters such as γ-aminobutyric acid were indentified on the web fibres. Tillinghast and Christenson[32] identified similar water-soluble compounds on the orb web of *N. clavipes*. Vollrath *et al.*[34] studied *A. diadematus*, an orb weaver, and also found webs covered with aqueous droplets containing salts and neurotransmitters.

Structure

Silks adopt a variety of secondary structures including α helices, β-sheets and cross-β-sheets. The α-helical conformations are characteristic of silks from bees, wasps and ants[35,36], β-sheets are characteristic of many silkworm silks and spider silks, and cross-β-sheets are characteristic of many insects, such as the green lace-wing fly *Chrysopa flava*[37]. The β-sheets incorporate the silk polymer chains parallel to the fibre axis, while in the cross-β conformation the polymer chains are positioned at right angles to the fibre axis and are often characterized by a higher content of serine residues. The secondary and tertiary structures of the silk β-sheet have been described[2,6,38]. The polymer chains run antiparallel, with hydrogen-bonds, roughly perpendicular to the chain axis, being formed between the carbonyl and amine groups from the peptide bonds that hold the chains together in sheets. The sheets assemble through hydrophobic interactions owing to their close packing density, which is attributable to the preponderance of short side-chain amino acids in the polymer chain[5].

The structure of cocoon silk fibroin from *B. mori* consists of antiparallel β-sheets first described by Marsh *et al.*[39]. The fibroin consists of both crystalline and amorphous domains with the amorphous regions characterized by the presence of amino acids with bulkier side chains. The crystalline domains are characterized by a high percentage of glycine, alanine and serine, in a 3:2:1 ratio, which contain short side chains to permit the close packing densities for overlying sheets.

The statistical coil form, β-sheet form (also called silk II or β-silk) and a second crystalline form called silk I have been reported for silkworm fibroin (Fig. 1). The nomenclature silk I and silk II was first proposed by Kratky *et al.*[40,41]. Silk I has also been called α-silk or water-soluble silk[42]. The use of the term α-silk should probably be avoided since it may lead to confusion with the α-helix of Pauling and Corey. The two crystalline forms of *B. mori* silk fibroin have been shown by Lotz *et al.*[4] to be isomorphous with the two crystalline forms of poly(L-alanine-glycine). Poly(L-alanine-glycine) has often been used as a model for the crystalline regions of silk, both experimentally[1,43] and computationally[44,45]. Unfortunately, the crystalline forms of silk and poly(L-alanine-glycine) were named independently of one another and the AGII form of poly(L-alanine-glycine) is isomorphous to silk I, and AGI isomorphous to silk II, which may lead to some confusion.

Warwicker[46] studied spider silks and silkworm silks with X-ray diffraction, including cocoon silk from the spider *Nephila senegalensis* and reeled silk from *N. madagascariensis*. Both silks exhibited antiparallel β-sheet structures but had different intersheet distances. Five different classes of fibroins were identified. The *a*-axis (dimension of the unit cell, or direction of the side chains or intersheet distance) went from 0.93 to 1.57 nm depending on the classification of the fibroin, while the *b*-axis (hydrogen-bonds of the antiparallel β-pleated sheets, or interchain distance) was 0.944 nm and the *c*-axis (fibre axis) was 0.695 nm. The 0.695-nm distance is characteristic of an antiparallel β-sheet. X-ray diffraction data on spider silk indicate 55% to 60% of the crystallinity of silkworm silk[10].

Infrared and X-ray diffraction data indicate a degree of cystallinity of between 62% and 66% for *B. mori* cocoon silk (silk II) fibroin, while comparable values for cocoon silk from wild-type silkworms ranged slightly lower, from 50 to 63%[47]. The crystallite size of silk II based on X-ray diffraction data for silkworm silks varies from 1.5 to 2.0 nm perpendicular to the plane on the sheets in the domesticated silk and from 4.7 to 6.8 nm in wild silks (summarized in ref. 47). Marsh *et al.*[39] reported that the unit cell for the crystalline domains included

Biomaterials

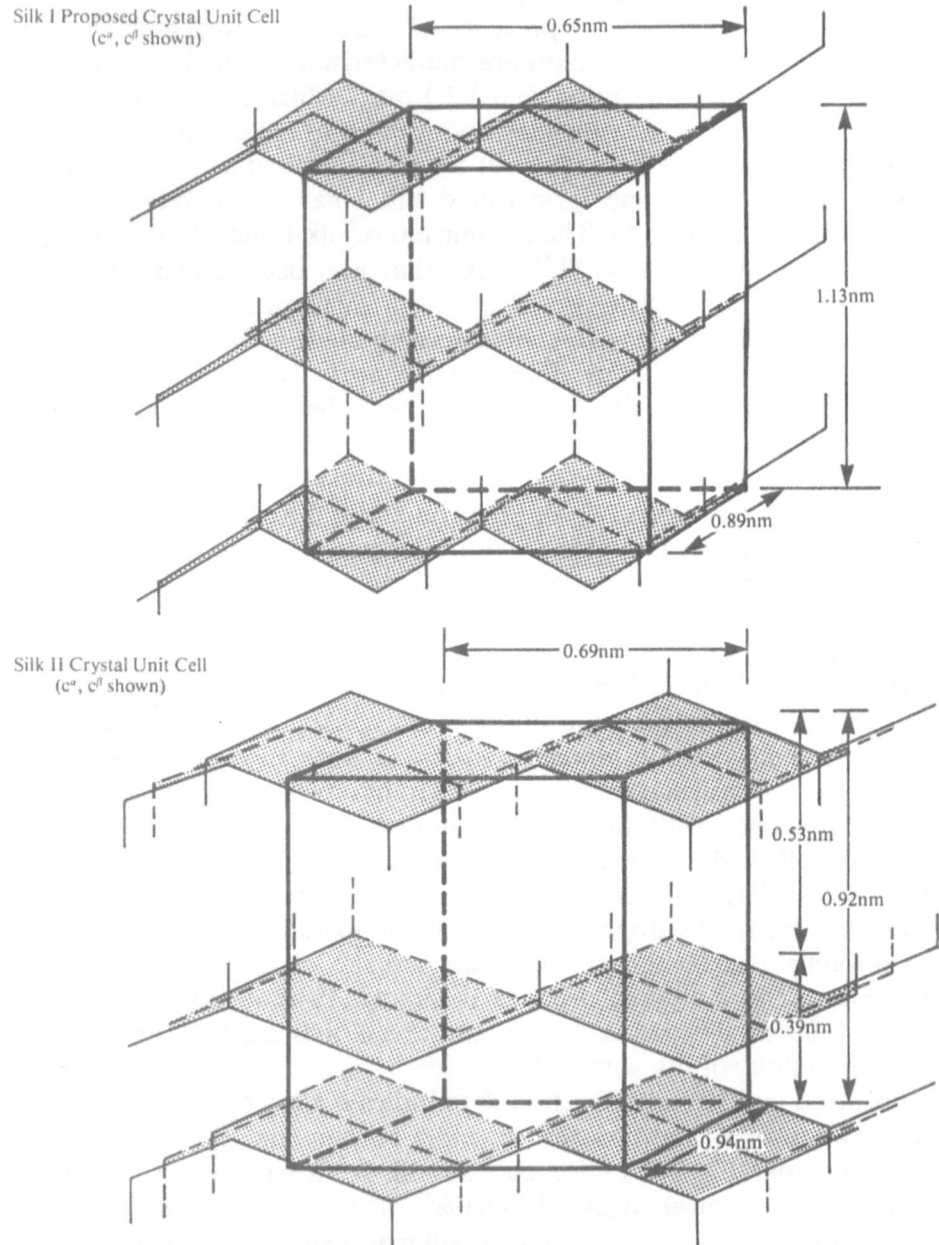

Silk I Proposed Crystal Unit Cell
(cᵃ, cᵝ shown)

Silk II Crystal Unit Cell
(cᵃ, cᵝ shown)

Fig. 1 Model structure of silk I and silk II, end view with hydrogen-bonds shown, illustrating orientation of amino-acid side chains.

8

an interchain distance of 0.94 nm (*a*), a fibre axis distance of 0.697 nm (*b*) and an intersheet distance of 0.92 nm (*c*). The smaller glycine side chains alternate with serine or alanine side chains. The *c* dimension of a unit cell has been related to the glycine content[2] and ranges from 21 glycine per 1000 monomers to 450 glycine per 1000 monomers, corresponding to *c* dimensions of 1.57 nm in *N. senegalensis* spider silk and 0.92 nm in *B. mori* silk, respectively[1,36]. In *B. mori* fibroin (silk II), the long glycine-X repeat within the crystal results in a structure in which the glycine side chains all project from the same side of the sheet and the alanine and serine side chains from the other side of the sheet. The sheets then stack with the glycine faces together and the alanine/serine faces together.

Chou *et al.*[48–50] have carried out energy-minimization studies on silk II structural motifs, including homopolymers of alanine, valine, isoleucine, serine, threonine, phenylalanine and tyrosine in parallel and antiparallel configurations for neighbouring polymer chains. They found a higher-energy less-stable structure for the parallel versus the antiparallel chain configuration in the case of the glycine, alanine, leucine and aminobutyric acid homopolymers, and the reverse for valine, isoleucine, serine, threonine, phenylalanine and tyrosine. Tsukada *et al.*[51] studied poly(Gly-L-Ala) and copolymers of (L-Ala-Gly and L-Ser-Gly) in 2:1 ratios using X-ray diffraction, infrared spectroscopy and circular dichroism. Introduction of the serine into the polymer backbone resulted in better crystallization into the antiparallel or cross-β forms. Colona-Cessari *et al.*[44] modelled the silk II form with conformational energy calculations on several sheets of poly(L-alanine-glycine). Oka *et al.*[52] have recently used Ala-Gly energy-minimization calculations of single strands to predict right-handed and left-handed helical configurations as models for silk I.

Either silk I or silk II forms can be obtained from the solvated form depending on crystallization conditions. The silk II form is, however, more stable than the silk I form and any mechanical agitation during drying will result in the conversion to the silk II form. Transformation from silk I to silk II in the solid state has also been effected by exposure to hydrophilic organic solvents such as methanol or acetone by an electric field, changes in temperature and by mechanical drawing[53–60]. Similar thermal transition will occur above 180 °C and on exposure to steam.

Because of the instability of silk I to mechanical strain, attempts to induce orientation in a sample for X-ray diffraction, NMR or other analytical methods cause a transition to the silk II form. Experimental studies of silk I have been only on samples of low orientation. Elucidation of the structure of silk I must therefore follow from model building and comparison of the model with experimental evidence. A number of models have been proposed for silk I on the basis of computer-generated predictions usually tied to experimental data. For example, Lotz and Keith[61] proposed a crankshaft model using poly(L-Ala-Gly) lamaellar crystals with resulting unit-cell dimensions of $a = 0.472$ nm (interchain), $b = 1.44$ nm (intersheet) and $c = 0.96$ nm, the fibre axis repeat being a multiple of 0.32 nm. Ilzuka[2] reported that fibroin contains three subunit chains, each of which folds back on itself many times to form rod-like structures with adjacent segments of each chain interacting through hydrogen-bonds; it was reported that the silk I form is a meta-stable crystalline structure in an orthorhombic cell containing two polymer chains. Recent work by Fossey *et al.*[45] based on conformational energy calculations with homopolymers of glycine or alanine and copolymers of Gly-Ala repeats, has resulted in a new proposed model for silk I (fig. 1). The unit cell structure proposed has a fibre axis repeat of 0.646 nm and an interchain spacing of 8.94 nm and an intersheet spacing of 1.13 nm. The structure is an alternating left-hand and right-hand 3_1 helix with alanine residues configured in a left-handed helix and glycine residues in a right-handed helix.

Physical properties

Silk fibroin represents one of the largest polypeptides in nature with reported relative molecular masses of 350K to 415K (refs 62,63). Initially, fibroin molecular masses from 33K up to 400K were reported using a variety of methods, until the use of denaturants such as urea or guanidine hydrochloride produced more consistent results. The difficulty in obtaining homogeneous preparations of the fibroin and complete solubilization without degradation is primarily responsible for these early discrepancies. Tashiro and Otsuki[64] and Tashiro *et al.*[65] had previously reported molecular masses of 300K to 420K for fibroin extracted from the posterior region of the silk gland using ultracentrifugation and sedimentation equilibrium analyses. Molecular masses reported for sericin protein have ranged from 20K to 310K (refs 66,67). Candelas and Cintron[68] reported a molecular mass of 320K for silk collected from the major ampullate gland from the spider *N. clavipes*. Lombardi and Kaplan[69] have reported a mass of 350K from this species.

Silkworm fibroin proteins vary in composition depending on source, extraction method and analysis. In general, the principal component is a high relative molecular mass protein of around 350K, also called the heavy (H) chain[66,69]. Data supporting the presence of another fibroin subunit, termed P25 or light (L) chain, of about 25K has been reported on the basis of direct extraction from the posterior region of the silk gland followed by reductive alkylation[65,66,70].

There is speculation that a disulphide linkage exists between the small and large chains, owing to the reductive alkylation and the cysteine content of the small chain. The two chains were separated by SDS-PAGE only after 2-mercaptoethanol treatment, and a 1:1 stoichiometry was found. Recent isolation of mRNA coding for the L-chain and cloning results have provided more definitive support for the existence of this second component of fibroin[71].

Unlike silkworm silks, which contain multiple proteins (L- and H-fibroin chains, sericins), the spider appears to produce only one fibroin-like protein per gland for incorporation into the silk fibre. No sericin proteins are present. But Work[72] has reported a protein sheath or skin core around major ampullate gland silk from the spider *A. diadematus*. Previous studies on proteins in the ampullate glands of *N. madagascariensis* found proteins of 300K with a 50K subunit[73]. This finding may have been due to the presence of a mixture of large and small ampullate gland silks or the failure to inhibit protease activity during the extraction process[68].

Silks are insoluble in dilute acids and dilute alkali, are resistant to most proteolytic enzymes[23,74,75] and are hydrolysed by concentrated sulphuric acid[3]. Lithium bromide (9.3 M) solubilizes fibroin, while specific mixtures of acids solubilize silk fibres without decreasing the molecular mass of the polymer[69,76]. Silk assembles into a compact conformation as shown by intrinsic viscosity measurements. For example, a solution containing fibroin of molecular size 300K exhibits a viscosity significantly below values obtained for comparably sized and smaller sized synthetic polymers[2].

Silk fibres are hygroscopic, with a moisture regain of 10 to 15%; they are translucent with a high lustre or sheen; they exhibit good abrasion resistance; an ability to bind dyes; printability; resistance to mildew; and undyed are susceptible to sunlight[3,6,77]. Some spider silks have been shown to be supercontracting[78–80] and also to exhibit unusual electromagnetic reflectance patterns particularly in the ultraviolet region[12]. Zemlin[25] reported specific gravities of 1.347 to 1.35 for

11

spider silks, depending on the species of spider, and 1.25 for silkworm silk.

Chemistry

Lucas *et al.*[81] and Lucas and Rudall[36] among others have reported on the amino-acid composition of silkworm fibroin (Tables 1 and 2). In general, a high content of glycine, alanine and serine totalling 80 to 85% of the amino acids present is reported and values as high as 96% have been found for other species. Acid side-chain groups predominate by two to three times over basic side-chain groups and aspartic acid is usually present in higher amounts than glutamic acid. A trace of cysteine has been reported. The sericin, which comprises 25% of the cocoon weight, has an amino-acid composition that is significantly different from that of fibroin, with a high percentage of glycine, serine and aspartic acid residues, which total over 60%. Sericin proteins exhibit an isoelectric point of 4.0, reflecting the acidic side chains of its component amino acids, and contain a higher cysteine content than fibroin. The hydrophilic nature of sericin permits its separation and removal from the fibroin during the processing of silkworm cocoons in boiling water and also results in its susceptibility to proteolytic enzymes. The L-chain of fibroin has an amino-acid composition of 15% aspartate, 14% alanine, 11% glycine and 11% serine.

Nadiger and Halliyal[47] summarized much of the previous work on the composition of silkworm silks. The mulberry silk, which is the domesticated silk from *B. mori*, contains 72 to 81% fibroin, 19 to 28% sericin and traces of lipid, pigment and carbohydrate. The carbohydrate portion includes glucosamine. The trace components of mulberry *B. mori* cocoon silk comprise between 2.5 to 3.7% of the total material. The domestic silkworm converts approximately 65% of the amino acids in mulberry leaves to silk fibroin and sericin[2]. The alanine content in silks from wild-type silkworms is significantly higher than in the domesticated mulberry silkworm silk, with evidence for extended polyalanine domains[82].

Numerous authors have reported amino-acid sequences for *B. mori* fibroin, with 60% of the fibroin consisting of repeat units characterized as crystalline domains containing only four different amino acids. A hexapeptide repeat, Ser-Gly-Ala-Gly-Ala-Gly (SGAGAG), was first reported as the main structural element, and a 59-mer repeat as representative of the crystalline domain of *B. mori* silk[36,81,83].

Table 1 Amino-acid composition of silkworm silks

Silkworm source	Ala	Arg	Asx	Cys	Glx	Gly	His	Ile	Leu	Lys	Met	Phe	Pro	Ser	Thr	Tyr	Trp	Val
Total fibroin																		
Bombyx mori[5]	29.40	0.47	1.30	0.20	1.02	44.60	0.14	0.66	0.53	0.32	0.10	0.63	0.63	12.10	0.91	5.17	0.11	2.20
Antheracea pernyi[36]	44.10	2.60	4.70	—	0.80	26.50	0.80	—	0.80	0.10	—	0.60	0.30	11.80	0.10	4.90	1.10	0.70
Antheracea moloneyi[36]	53.00	0.10	0.50	—	0.30	42.40	0.20	0.10	0.20	0.10	—	0.10	0.20	0.30	0.20	0.10	—	2.10
Sericin																		
Bombyx mori[149]	5.97	3.10	16.71	0.15	4.12	13.49	1.30	0.72	1.14	3.30	0.04	0.53	0.68	33.43	9.74	2.61	0.21	2.75
Light chain																		
Bombyx mori[71]	14.20	4.50	14.80	1.40	9.20	9.20	2.30	7.80	7.50	1.20	0.40	2.70	3.20	9.00	3.00	2.80	—	6.40

Amino-acid residues per 100 total residue (mole %).

Table 2 Amino-acid side chains of silkworm silks

Source	Type of amino-acid side chain								
	Small	Polar	Acidic/amide	Basic	Cyclicimino	Aromatic	Sulphur	Aliphatic	Hydroxyl
Total fibroin									
Bombyx mori[5]	86.10	31.43	2.32	0.93	0.36	5.80	0.30	31.90	13.01
Antheracea pernyi[36]	82.40	25.80	5.50	3.50	0.30	5.50	—	45.60	11.90
Antheracea moloneyi[36]	95.70	1.80	0.80	0.40	0.20	0.20	—	55.40	0.50
Sericin									
Bombyx mori[149]	52.89	74.31	20.84	7.70	0.68	3.14	0.19	10.58	43.17
Light chain									
Bombyx mori[71]	32.40	46.80	24.00	8.00	3.20	5.50	1.80	35.90	12.00

Amino acids expressed as residues per 100 total residues (mole %). Side chains: small, Gly + Ala + Ser; polar, Asx + Thr + Ser + Glx + Tyr + Lys + His + Arg; acidic/amide, Asx + Glx; basic, Lys + His + Arg; cyclicimino, Pro; aromatic, Phe + Tyr; sulphur, Met + Cys; aliphatic, Ala + Val + Leu + Ile; hydroxyl, Ser + Thr.

Although the sequence of the 59-mer repeat is not conclusively known, it is thought to be Gly-Ala-Gly-Ala-Gly-[(Ser-Gly-(Ala-Gly)$_n$]$_8$-Ser-Gly-Ala-Ala-Gly-Tyr (GAGAG[SG(AG)$_n$]$_8$SGAAGY), where n is usually 2. In subsequent studies by Strydom *et al.*[84] using automated Edman degradation, a related sequence was proposed as more representative of the core repeat in the crystalline domains: Gly-Ala-Gly-Ala-Gly-Ser-Gly-Ala-Ala-Gly-[Ser-Gly-(Ala-Gly)$_n$]$_8$-Tyr (GAGAGSGAAG[SG(AG)$_n$]$_8$Y), where n is usually 2. There have been comparable studies to characterize fully the sequence of amino acids in the amorphous domains, although they would be expected to contain the bulkier side-chain amino acids. For example, Lucas *et al.*[83] and Lucas and Rudall[36], among others, reported peptide sequence data for some soluble (amorphous) fractions, which comprise approximately 40% of the total protein, and included the following motifs: Gly-Ala-Gly-Ala-Gly-Ala-Gly-Tyr (GAGAGAGY), Gly-(Gly$_3$-Ala$_2$-Val)-Tyr [G(G$_3$A$_2$V)Y], Gly-Ala-Gly-Tyr (GAGY), Gly-(Gly$_2$-Ala-Asp)-Tyr [G(G$_2$AD)Y], Gly-Val-Gly-Try (GVGY), Ser-Gly-Tyr (SGY) and Gly-Pro-Tyr (GPY).

Work and Young[85], Zemlin[25], Lucas *et al.*[81], Peakall[86,87], Andersen[27], Lombardi and Kaplan[76] and others have reported the amino-acid composition of spider silks (Tables 3 and 4). Many similarities exist between *B. mori* and *N. clavipes* in terms of gross morphology of the silk gland and silk protein structure. As in the silkworm, alanine, glycine and serine predominate, but in spider silk there is also a higher percentage of larger side-chain amino acids such as glutamine, tyrosine, leucine and valine. Peakall[86–88] determined the amino-acid composition of different spider silks, including frame, cocoon and swathing silks from the spider *A. diadematus*, and reported different compositions depending on the type of silk. Andersen[27] reported on the amino-acid composition of different spider silks from *A. diadematus* collected directly from the seven different glands, and correlated this information with the functional properties of the different silks. Some of the conclusions from this study were that glands producing long structural fibres produced silks containing over 50% short side-chain amino acids, compared with 30 to 40% from the other glands; silks with high elasticity, which are produced in the flagelliform gland, contain about 20% proline; and the sticky glue-like silks contained a much higher percentage of basic amino acids. Work and Young[85] generated additional data, and compiled much of the previously published data on amino-acid composition of spider silks. They discussed

Table 3 Amino-acid composition of spider silks

Source	Ala	Arg	Asx	Cys	Glx	Gly	His	Ile	Leu	Lys	Met	Phe	Pro	Ser	Thr	Tyr	Trp	Val
Major ampullate gland																		
Argiope aurantia[33]	22.80	1.40	1.10	—	12.30	37.80	0.20	0.50	1.80	0.50	0.60	1.20	10.30	4.60	0.60	—	3.40	1.40
Araneus diadematus[85]	18.30	0.49	1.08	0.16	11.86	41.30	0.68	0.53	1.76	0.28	0.00	3.50	9.55	4.74	0.46	—	4.38	0.90
Araneus marmoreus[85]	13.10	0.66	0.91	0.06	17.99	44.23	0.64	0.41	1.31	0.20	0.35	0.33	11.70	4.17	0.21	—	2.82	0.91
Eriophora fuliginea[85]	13.11	0.39	1.08	0.03	8.90	51.05	0.40	0.44	1.52	0.14	0.00	0.24	11.29	7.06	0.10	—	3.50	0.87
Eriophora ravilla[85]	14.71	0.50	0.90	0.03	11.34	49.08	0.20	0.55	1.56	0.07	0.00	0.18	12.00	4.24	0.07	—	3.82	0.75
Metazygia wittfeldae[85]	20.53	0.69	1.13	0.18	9.19	43.09	0.40	0.61	0.74	0.08	0.06	0.20	11.86	6.52	0.11	—	3.65	0.96
Neoscona hentzii[85]	21.01	0.40	1.09	0.39	11.30	43.11	1.16	0.58	1.44	0.32	0.03	0.43	9.22	5.13	0.79	0.00	2.40	1.60
Argiope argentata[85]	19.28	1.25	0.97	0.23	11.16	45.48	1.07	0.30	2.11	0.16	0.00	0.49	8.32	5.56	0.34	—	2.52	0.66
Argipe aurantia[85]	22.26	0.97	0.90	0.24	12.20	42.04	0.79	0.54	2.64	0.50	0.22	1.49	8.39	3.67	0.29	0.07	2.64	0.24
Leucauge argentea[85]	15.86	1.00	1.05	0.93	10.31	46.89	1.19	0.07	4.00	0.20	0.03	0.86	6.73	5.98	0.38	—	4.00	0.93
Nephila clavipes[85]	24.85	1.94	0.95	0.06	10.49	48.69	0.41	0.13	2.62	0.21	0.06	0.24	2.15	2.11	0.40	0.00	2.62	0.06
Nephilengys cruentata[76]	26.63	0.65	0.93	0.04	9.92	47.92	2.05	0.25	2.85	0.41	0.16	0.17	2.50	2.60	0.47	0.00	2.85	0.04
Nephila clavipes[76]	21.10	7.60	2.50	0.10	9.20	37.10	0.50	0.90	3.80	0.50	0.40	0.70	4.30	4.50	1.70	—	2.90	1.80
Argiope aurantia[76]	22.20	2.90	1.60	0.30	11.10	34.70	0.50	0.80	4.20	0.50	0.30	0.40	6.40	5.10	0.80	—	3.80	1.50
Neoscona domiciliorum[76]	18.00	0.60	0.60	0.70	10.00	38.00	0.30	0.50	1.20	0.20	0.20	0.80	11.20	6.80	0.90	—	3.70	0.70
Minor ampullate gland																		
Argiope aurantia[74]	31.30	2.20	2.80	—	2.20	41.20	0.30	1.00	1.80	1.00	0.40	0.90	0.90	5.20	2.00	—	4.80	2.30
Araneus diadematus[85]	29.91	1.51	1.79	0.42	1.94	49.28	1.15	0.30	0.98	0.16	0.00	0.67	0.48	4.38	1.57	—	3.99	1.52
Eriophora fuliginea[85]	28.42	0.72	0.97	0.03	0.97	54.34	0.21	1.00	1.04	0.07	0.09	0.33	0.37	3.89	1.76	—	4.74	1.05
Eriophora ravilla[85]	28.37	0.77	1.00	0.03	1.01	55.75	0.25	0.78	1.08	0.04	0.06	0.25	0.20	3.62	1.52	—	4.36	0.92
Neoscona hentzii[85]	29.51	0.00	1.19	0.25	0.75	58.33	0.68	0.91	0.62	0.08	0.00	0.17	0.00	2.29	0.72	—	3.36	1.14
Argiope argentata[85]	31.88	2.17	1.85	0.12	2.99	41.98	0.00	1.24	1.77	1.08	0.00	0.77	1.12	5.27	1.70	0.00	4.68	1.38
Argiope aurantia[85]	35.49	2.22	1.82	0.08	1.46	40.95	0.00	1.15	1.93	0.71	0.00	0.65	0.00	4.52	1.87	0.00	4.72	1.69
Nephila clavipes[85]	27.63	1.82	2.32	0.13	3.60	45.86	0.80	1.34	1.78	0.29	0.04	0.58	0.62	4.46	2.12	—	3.77	2.84
Nephilengys cruentata[85]	31.78	1.55	1.37	0.02	3.12	46.38	1.55	0.77	1.00	0.95	0.22	0.54	0.71	2.61	1.10	—	4.82	1.51

Amino-acid residues per 100 total residues (mole %).

Table 4 Amino-acid side chains in both major and minor ampullate gland silks from different spider species

Source		Type of amino-acid side chain							
	Small	Polar	Acidic/amide	Basic	Cyclicimino	Aromatic	Sulphur	Aliphatic	Hydroxyl
Major ampullate gland									
Argiope aurantia[33]	65.20	20.70	13.70	2.10	10.30	1.20	0.60	26.50	5.20
Araneus diadematus[85]	64.34	19.59	12.35	1.45	9.55	3.50	0.16	21.49	5.20
Araneus marmoreus[85]	61.50	24.78	18.65	1.50	11.70	0.33	0.41	15.73	4.38
Eriophora fuliginea[85]	71.22	18.07	9.29	0.93	11.29	0.24	0.03	15.94	7.16
Eriophora ravilla[85]	68.03	17.32	11.84	0.77	12.00	0.18	0.03	17.57	4.31
Metazygia wittfeldae[85]	70.14	18.12	9.88	1.17	11.86	0.20	0.24	22.84	6.63
Neoscona hentzii[85]	69.25	20.19	11.70	1.88	9.22	0.43	0.42	24.63	5.92
Argiope argentata[85]	70.32	20.51	12.41	2.48	8.32	0.49	0.39	22.35	5.90
Argiope aurantia[85]	67.97	19.39	13.17	2.26	8.39	1.56	0.46	25.68	3.96
Leucauge argenta[85]	68.73	20.11	11.31	2.39	6.73	0.86	0.96	20.86	6.36
Nephila clavipes[85]	75.65	16.51	12.43	2.56	2.15	0.24	0.12	27.66	2.51
Nephilengys cruentata[85]	77.15	17.03	10.57	3.11	2.50	0.17	0.20	29.77	3.07
Nephila clavipes[76]	62.70	26.50	16.80	8.60	4.30	0.70	0.50	27.60	6.20
Argiope aurantia[76]	62.00	22.50	14.00	3.90	6.40	0.40	0.60	28.70	5.90
Neoscona domiciliorum[76]	62.80	19.40	10.60	1.10	11.20	0.80	0.90	20.40	7.70
Minor ampullate gland									
Argiope aurantia[74]	77.70	15.70	4.10	3.50	0.90	0.90	0.40	36.40	7.20
Araneus diadematus[85]	83.57	12.50	3.45	2.82	0.48	0.67	0.42	32.71	5.95
Eriophora fuliginea[85]	86.65	8.59	1.69	1.00	0.37	0.33	0.12	31.51	5.65
Eriophora ravilla[85]	87.74	8.21	1.78	1.06	0.20	0.25	0.09	31.15	5.28
Neoscona hentzii[85]	90.13	5.71	0.75	0.76	0.00	0.17	0.25	32.18	4.50
Argiope argentata[85]	79.13	15.06	5.16	3.25	1.12	0.77	0.12	36.27	6.06
Argiope aurantia[85]	80.96	12.60	3.68	2.93	0.00	0.65	0.08	40.26	6.41
Nephila clavipes[85]	77.95	15.41	5.42	2.91	0.62	0.58	0.17	33.59	6.61
Nephilengys cruentata[85]	80.77	12.25	4.67	4.05	0.71	0.54	0.24	35.06	6.33

Amino acids expressed as residues per 100 total residues (mole %). Side chains: small, Gly + Ala + Ser; polar, Asx + Thr + Ser + Glx + Tyr + Lys + His + Arg; acidic/amide, Asx + Glx; basic, Lys + His + Arg; cyclicimino, Pro; aromatic, Phe + Tyr; sulphur, Met + Cys; aliphatic, Ala + Val + Leu + Ile; hydroxyl, Ser + Thr.

the variability found in the published data and related this to possible inherent variability in silk composition. Lombardi and Kaplan[76] recently published a detailed study on the amino-acid composition of dragline silk from *N. clavipes* and found consistent results between individuals and within individual spiders.

Partial acid hydrolysis of spider silks produced the following short sequences for the most common peptides from *N. clavipes* and *Araneus gemmoides* dragline silk[69,178]: Gly-Gln-Gly-Ala-Gly (GQGAG) and Gly-Tyr-Gly-Gly-Leu-Gly (GYGGLG). Related data from solubilized *N. clavipes* dragline indicated a 54-mer crystalline domain sequence similar to the 59-mer found in *B. mori*, and a variety of sequences correlating with amorphous domains based on the presence of amino acids with bulkier side chains, including arginine, isoleucine, threonine, glutamine, proline, phenylalanine and leucine[69].

Mechanical properties

Silks exhibit an unusual combination of strength and toughness which set them apart from other natural and synthetic fibres. A summary of some of the mechanical properties of silkworm and spider silks is included in Table 5.

Zemlin[25] completed one of the most comprehensive studies on the mechanical properties of spider silks, and related functional properties to amino-acid composition. The study included several species of spiders: *N. clavipes*, *A. arrantia*, *Nephila cruentata* and *Parawixla audax*. He reported initial modulus values for some spider silks of up to 288.2 g.p.d. (= 3.4×10^{10} N/m^2) and elongation values of up to 39.3%. *N. clavipes* had the highest average tenacity (7.93 g.d.p. = 9×10^8 N/m^2) of the five spider species studied, and the highest average initial modulus (98.6 g.d.p. = 1.2×10^{10} N/m^2). In general, spider silks are comparable to silkworm silks in terms of extensibility, but higher in modulus and strength.

Silk fibres represent an exception to most materials where strength and modulus increase with increased loading rates, while elongation decreases. With silks, increasing rates of loading result in increased elongation, so that work to rupture increases at higher rates of loading[90]. At low loading rates, silkworm silk has a higher modulus than nylon (4154 mN/tex versus 3540 mN/tex), but lower strength (440 mN/tex versus 795 mN/tex), while at high loading rates silk out-performs

17

Table 5 Mechanical properties of silks and other fibres

Sample	Elongation (%)	Modulus (N/m^2)	Strength (N/m^2)	Energy to break (J/kg)
Spider silks: dragline				
Nephila clavipes[10,25]	9.8–32.1	1–30×10^9	3–18×10^8	3–10×10^4
Argiope aurantia[25]	18.3–21.5	6–24×10^9	5–13×10^8	2×10^4
Nephila creuentata[25]	10.1–39.3	2–21×10^9	2–18×10^8	1×10^4
Parawixla audax[25]	20.5–35.5	2–5×10^9	2–9×10^8	4×10^4
Ariope argentata[25]	10.1–24.3	3–6×10^9	2–8×10^8	3×10^4
Araneus sericatus[10]	30.0	1×10^{10}	1×10^9	1×10^5
Silkworm silks: cocoon				
Bombyx mori[10,47,90,177]	15–35	5×10^9	6×10^8	7×10^4
Antheraea mylitta[49]	35.0	2×10^9	5×10^8	6×10^4
Anaphe moloneyi[49]	12.5	4×10^9	6×10^8	3×10^4
Byssus pina nobilias[49]	50.0	2×10^9	1×10^8	4×10^4
Other fibres[10,90,179,180]				
Nylon	18.0–26.0	3×10^9	5×10^8	8×10^4
Cotton	5.6–7.1	6–11×10^9	3–7×10^8	5–15×10^3
Kevlar	4	1×10^{11}	4×10^9	3×10^4
Steel	8.0	2×10^{11}	2×10^9	2×10^3

nylon[90]. Single-fibre tests on silkworm silk indicated that at about 270 m/s loading, silk had a greater tensile strain velocity than nylon 6,6, a slightly lower breaking tenacity (4.3 versus 5.5 g.p.d.), and a lower breaking extension (7.9 versus 12.2). In general, because of these unusual properties, silks have an excellent ability to absorb energy at high rates of loading[90].

Ilzuka[91] studied the mechanical properties of six different races of *B. mori*. The silks ranged from 1.5 to 3.8 deniers in size, and when extended after reeling from the cocoon, from 200 to 1400 m in length. Three different layers of the cocoons were studied to investigate changes in the silk diameter and mechanical properties associated with cocoon formation. The elastic modulus increased as the size of the silk thread decreased, in accordance with the theory that the thinner thread located at the exterior of the cocoon has a more complex fibrous structure[91]. The theoretical value of the longitudinal elastic modulus for the crystal of silk fibroin was about 1.6×10^{12} dyn/cm^2, assuming a density of the crystal of 1.36 g/cm^3 (ref. 91). Cocoons were found to vary in size, and therefore the length of the silk provided also varies from cocoon to cocoon and within and among races.

Lucas *et al.*[5,81] and Lucas and Rudall[36] correlated mechanical properties of silkworm silk with the ratio of long side-chain amino acids to short side-chain amino acids (glycine, alanine, serine, threonine) in the fibroin, and found some correlation of high ratios with higher elongation and lower tenacity. In addition, Iizuka[91] showed a linear correlation between crystallinity and denier, and elastic modulus and denier; the finer denier have higher tenacity and lower elongation.

The mechanical properties of spider silks are influenced by temperature, water content, and deformation rate. These silks perform a key natural function as structural fibres in absorbing impact energy from flying insects without breaking. They serve to dissipate energy over a broad area and must balance stiffness, strength and extensibility. About 70% of the energy of impact resulting in extension of the spider orb-web silk is lost as heat through viscoelastic processes and therefore is not recoverable as elastic-recoil energy. This property is important in terms of the natural function of these materials, since the mechanical energy of impact is therefore neither available to break the threads nor to eject the impinging insect from the web. The stickier silks also serve to bind the insect into the web once contact is made[10,26]. Denny[22,28] correlated the mechanical properties of viscid and frame silks of orb webs produced by *A. sericatus* with the natural function of the web as a shock absorber and aerial filter, functioning with minimal energy costs and minimal material from the spider. The viscid silk (catching spiral) is covered by glue droplets. This silk has low stiffness and is very elastic, over 200%, and therefore also contributes to the energy absorption and dissipation properties of the web[10].

Work[30,31,78,79], Work and Morosoff[80] and Fornes *et al.*[92] studied the physical/mechanical properties of spider silks including supercontraction behaviour. They compared web fibres to mechanically silked fibres. Unrestrained fibres were reported to supercontract to about one-half of their original length on exposure to water. Unlike silk fibres from the major ampullate gland, minor ampullate gland silk does not supercontract in water when axially unrestrained[31]. Silkworm silks do not display this property of supercontraction unless exposed to harsh chemical treatments or high temperatures. Therefore, Gosline[10] postulated that spider silk consists of a composite of crystalline domains in an amorphous network. The contracted silk is still birefringent but less so than the dry material. Viscid silk is birefringent only if stretched. Spider silks contain a lower degree of crystallinity than silkworm silk, which is also reflected in the lower percentage of short side-chain amino acids. The amorphous regions in the spider silk were postulated to be about 1.4kb in size between crystalline domains, although they probably are larger and extend through a number of crystalline and amorphous regions[10]. The crystalline and aligned polymer network contribute to the strength and toughness of spider silk.

Lucas and Rudall[36] described the α to β transformations of silkworm silk. Gosline[10,26] proposed rubber-like mechanisms or entropy driven processes to explain the mechanical properties of spider silk. Dong *et al.*[93] have studied changes in silk structure during mechanical deformation using FTIR spectroscopy. They studied *N. clavipes* and *A. gemmoides* to determine the molecular mechanisms of spider silk elasticity using major and minor ampullate gland silks. With applied tension, helical conformations appeared, possibly deriving from alanine-rich sequences found in the primary structure. This force-induced formation of ordered structure has been used to explain the mechanical properties of spider silk.

Processing

FIBRES

The unusual mechanical properties of silk fibres result from the primary and higher-order structures of the protein as well as the processing conditions used by the organism to spin fibres. The processing conditions function to produce molecular orientation in the protein chains. This occurs at ambient temperatures in an aqueous solution, unlike the conditions employed in synthetic fibre processing.

Silkworm fibroin is produced in a single cell type lining a pair of long tube-like glands located on each side of the silkworm. The gland consists of three relatively distinct regions, the posterior section where the fibroin is synthesized by epithelial cells, the middle gland where the fibroin product is stored and the sericin is synthesized but separated from the fibroin, and the anterior section where mixing of fibroin and sericin occurs and the two glands join to produce a single duct leading to the spinneret (Fig. 2). The concentration of silk fibroin in the posterior gland is 12 to 15% and in the middle portion of the gland the fibroin and sericin concentration is approximately 30%. The diameter of the silk thread decreases towards its proximal end during

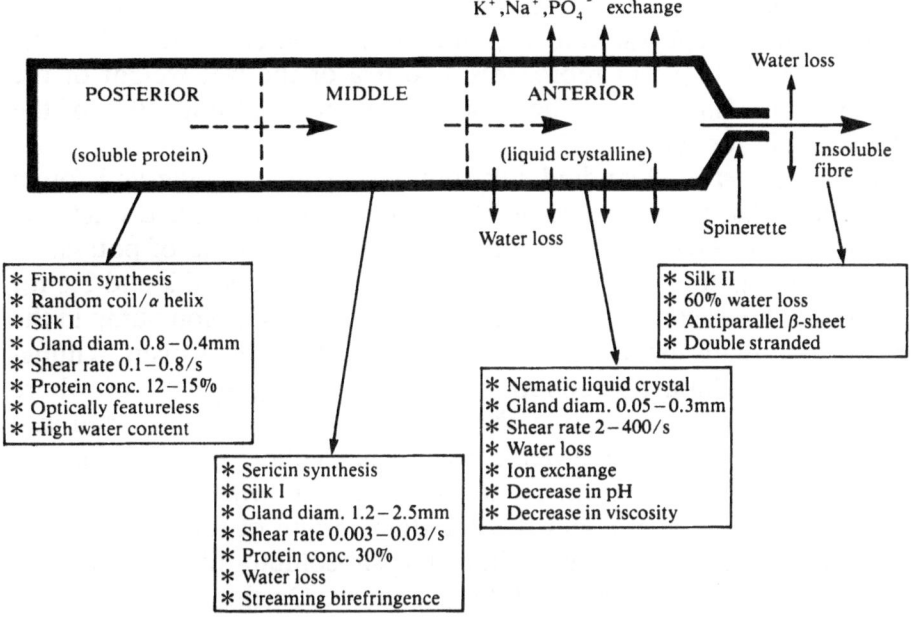

Fig. 2 Schematic representation of silk processing in silkworms and spiders (compiled from refs 13, 91, 96, 98–100, 162).

cocoon spinning, and the spinning rate increases as the size of the silk thread decreases during the formation of the cocoon[91]. Initial fibroin synthesis results in a protein solution which is progressively modified during passage through the gland and finally spun into an insoluble crystalline polymer characterized by the stable β-sheet structure. The two glands join together immediately before the spinneret and two fibres of fibroin are coated by a layer of sericin and the fibre is spun into air.

Comparable solution behaviour studies are underway with spider silks, which demonstrate a nematic liquid crystalline phase for partially dried extracts from the major ampullate glands of *N. clavipes*[94]. Polarized light microscopy is being used, and with shear or fibre drawing, banded microstructure is observed.

In contrast to silkworm silk, which is produced at one stage in the life cycle, spiders can produce silks throughout their adult life cycle. Multiple forms of silk are produced, each emanating from a different set of glands. As in the silkworm, there are three regions in the major-ampullate gland, the tail where over 90% of fibroin synthesis occurs, the storage sac and the duct where secretion occurs[95]. The major ampullate gland represents 4 to 6% of the wet weight of the spider in *A. diadematus*. The protein content represents 10% of the gland weight for each web[86].

Tillinghast *et al.*[96] studied silk spinning by *A. aurantia* and found water loss due to extraction in the major ampullate gland duct and potassium ion exchange for sodium ions, since the ratio of potassium ions to sodium ions was 0.6 for major ampullate gland silk and 2.6 for web silk. The existence of a sodium-ion and potassium-ion pump in the major ampullate gland duct was postulated but not proven. Tillinghast[97] further characterized ion-transport in the aggregate duct, where potassium ions and inorganic phosphate are actively transported into the aggregate gland during web construction. Tillinghast[74] fractionated orb-web components from *A. aurantia* and *A. trifasciata* to define further the components and the glandular origin of the components. Orb webs were partitioned into trypsin-soluble fibroin, trypsin-insoluble fibroin and water-soluble fractions. On the basis of amino-acid composition, the trypsin-insoluble fibroin originates in the major ampullate gland, whereas the trypsin-soluble fraction originates in multiple glands.

The anterior division of the silkworm gland is very narrow, 0.05 to 0.3 mm in diameter, which results in a high shear rate of 2 to 400/s at 1.0 cm/s spinning rate[91]. The soluble silk in the anterior gland of *B. mori* undergoes a conformational change owing to mechanical shear, drawing at the spinneret, and loss of water (Fig. 2)[91]. The conversion of the random coil/helix soluble protein to the β form has been demonstrated *in vivo* and *in vitro* through the use of a variety of techniques including X-ray crystallography, infrared spectroscopy and NMR spectroscopy. Magoshi *et al.*[13] summarized previous work on processing liquid silk to the fibre form in the silkworm. The processing involves shear and elongation stress during cocoon formation, with fibroin orientation parallel to the drawing direction during spinning. The spinneret functions both as a press and as a fulcrum for drawing the silk by the silkworm[13].

Magoshi *et al.*[98,99] first studied liquid crystalline behaviour of silkworm silk and reported that fibroin from the anterior portion of the silk gland of *B. mori* exhibited liquid-crystalline properties. Conformations of fibroin in the mid-gland and the anterior region of the gland had been previously described as principally random coil with some

α-helices[98,99], and on drying, the liquid silk fibroin assumes the silk I (α) form. The fibroin isolated from the anterior region of the silk gland does not cause birefringence of polarized light; however, this behaviour is exhibited by cells lining the walls of the gland even after the liquid silk was removed from the gland. The same liquid silk, if allowed to dry within the anterior portion of the excised gland, does exhibit orientation and birefringence directional with, but not perpendicular to, the gland[98,99]. A nematic liquid-crystalline phase was presumed by the authors to be present. The liquid silk from the gland can be drawn by hand at 7 mm/s, similar to the approximate rate of cocoon spinning, which results in an orientated β-sheet conformation[98,99]. It was concluded that there is approximately a 60% loss of water due to evaporation to the air during the processing of the soluble silk into the final insoluble silk fibre during cocoon spinning[13,98].

Magoshi *et al.*[13] reported stress–strain curves for liquid silk collected from the gland over the range of extension rates from 10 to 1000 mm/ min. They found that as the extension rate increased, the stress required to stretch the liquid silk increased. The yield point increased linearly over the range of extension rates of 50 to 450 mm/min and remained constant above this rate. Water loss occurred during the drawing process. X-ray diffraction patterns reflected the conformational changes, with characteristic peaks for the β form appearing above 500 mm/min, while none of these peaks were observed when the solution was drawn at lower rates. Birefringence patterns also changed above about 450 mm/min. In general, in comparison to spinning synthetic fibres, the spinning rate and pressure of the silkworm process is much lower, which was cited as evidence for liquid crystallinity[13]. The critical extension rate is dependent on drawing temperature, being 400 mm/min at 3 °C compared to well above 500 mm/min at 35 °C.

Magoshi *et al.*[13] also studied the viscosity of liquid silk at different locations within the silk gland and found a decrease in viscosity as the silk progressed towards the anterior division of the gland, suggesting an ordering of structure. The pH of fibroin solution decreased to 4.0 as it approached the anterior division of the gland, which correlates with the formation of β-form crystals at pH 4.5, and crystallization to the α form above pH 9.0 from aqueous solution[13].

Ilzuka[91] reported on shear rates required to convert the soluble silk to the β form in the anterior portion of the silk gland leading to the spinneret, and the acceleration of this effect by calcium ions. On the basis of the solution properties and evidence of ordered structure within the silk gland, it was speculated that the silkworm simulates a liquid-crystal spinning process during spinning silk fibre for the co-coon. Both native silk and regenerated fibroin from *B. mori* were studied. Regenerated fibroin exhibited a large increase in shear stress, which suddenly occurred at a specific shear rate, indicating the form-ation of some ordering in solution and evidenced by the formation of crystallite-like centres with a structure similar to silk thread. The critical shear rate was dependent on concentration, and on the basis of these effects a model was proposed for fibroin as a loosely defined structure where the transformation from α to β forms involves unfolding.

For the reconstituted fibroin at high concentration, the viscosity was very low (0.9 poise at 10.1 g/100 ml at 14 °C) for a fibroin of relative molecular mass about 300K. This was cited as additional indirect evidence that the concentrated solution is liquid crystalline[91]. Critical shear rates were more difficult to quantify in the native fibroin extrac-ted from the posterior portion of the middle silk gland owing to the presence of some three-dimensional structure in the native fibroin solution. This structure, present without any applied shear stress, was thought to be due to calcium ions and magnesium ions. Although change in the critical shear rate (50 per s) was found over the range of fibroin concentration from 0.5 to 4 g/100 ml, work from other labora-tories was cited where the critical shear rate of silk fibroin from a wild silkworm (*Antheracea pernyi*) did decrease from 900 to 2000 per s over increasing fibroin concentration from 0.4 to 5.0 g/100 ml (summarized in ref. 91).

Ilkzuka[91] calculated shear rates within the silk gland at an assumed spinning rate of 1.0 cm/s. From native and regenerated fibroins, the author chose a critical shear rate of aqueous silk to be 10^2 to 10^3 per s for polymer concentrations between 15 and 30% (wt). Microfibrils formed owing to α to β transformation, with increasing shear rate as the solution proceeded towards the spinneret. Additional evidence for a liquid crystalline state was derived from the calculations on draw ratio required to produce silk fibres. Aqueous silk would have to be lengthened more than ten-fold at a drawing ratio of 10 cm/s or more to achieve the native cross-sectional area of silk fibre. However, on the basis of the cross-sectional area of the spinneret, only a three-fold increase was calculated. Therefore, the silk solution presumably has already acquired some ordered structure within the anterior silk gland, as also reflected by the lowered viscosity. Sericin may serve to improve uniform denaturation and also aid passage through the spinneret after crystallization. It has also been postulated that the sericin may provide a source of divalent cations and an acceptor of water derived from the fibroin during the extrusion process[91]. The spinneret itself is only very slightly tapered.

Li and Yu[100] also investigated the liquid-crystalline state for silk-worm silk fibroin and reported evidence for this behaviour in the middle portion of the silk gland; thus shear stress would not be a factor as is potentially the case in the anterior portion of the gland. The authors used polarizing microscopy and small-angle X-ray scattering (SAXS) to deduce the presence of liquid crystallinity. Birefringence was only detected parallel to the gland and was observed in the anterior and middle, but not the posterior gland regions. A nematic liquid-crystalline phase was postulated for the silk fibroin in the middle portion of the gland, characterized by rigid rod-like molecules. SAXS data illustrate an anisotropic distribution of particle sizes in the middle region of the silk gland, with sizes in the direction perpendicular to the spinning direction smaller than in the parallel direction, suggesting rod-like forms.

FILMS

Conformations of silkworm fibroin films cast from aqueous solutions of silk are influenced by concentration and temperature[13]. Concentrations of fibroin above about 4% (wt) generally result in α forms below 40°C, and β forms above this temperature. Below 40°C, random-coil conformations occur at lower concentrations depending on temperature. The β form is also induced by freezing an aqueous solution of fibroin before drying; the effect is not dependent on the concentration of fibroin when quenching temperatures are −2°C to −20°C. Spherulites of silk fibroin are formed and the optical direction is dependent on drying temperature, drying rate and quenching temperature. Solvent-induced crystallization has also been reported, with β forms induced in polar hydrophilic solvents such as methanol, and α forms induced in hydrophobic solvents. A glass transition occurs at 173°C on heating soluble amorphous fibroin film, as water is lost up to 100°C; hydrogen-bonds are broken in the 150 to 180°C range; and transformation to the β form with reformation of hydrogen-bonds occurs above 180°C. Yamamura *et al.*[101] formed very thin films (450 to 1050 Å in thickness) and fine fibres (0.2 μm in diameter) from regenerated silk fibroin by flow-induced crystallization.

On the basis of the behaviour of silk in solution and in film form, Asakura *et al.*[102–105], Asakura and Demura[106], Demura *et al.*[107,108] and Yoshimizu and Asakura[109] have extended the work on conformational changes in silk films induced by a number of different mechanisms. These changes have been extensively characterized by ^{13}C-NMR and circular dichroism. Membranes prepared from cocoon silk from *B. mori* have found application in the immobilization of enzymes for biosensors. The membranes were manufactured by removal of the serioin in soap solution at 100°C for 0.5h, followed by washing in distilled water, solubilization in 9M LiBr at 40°C, dialysis for 4 days, centrifugation and film casting[105]. Silk fibroin was also reported as a suitable medium for the immobilization of enzymes by Grasset *et al.*[110].

In the film-forming process, the solution of silkworm silk is cast on acrylic resin plates at 20°C at a relative humidity of 60%. Polyethylene glycol can be added to the solution before casting to form a more porous matrix. Enzymes such as peroxidase, glucose oxidase[103,107,108] and invertase[109] have been co-cast with the solubilized fibroin for immobilized membrane formation. Once the film is formed, it can be immersed in 80% methanol aqueous solution to form an insoluble membrane in water[53,113]. Other techniques such as mechanical stretching also result in the formation of an insoluble matrix[53]. This approach provides membranes containing entrapped enzymes without the need for cross-linking reagents. X-ray diffraction revealed the presence of silk II after treatment with methanol[113].

Asakura *et al.*[102] studied the solution structures and conformational changes in silks from both *B. mori* and the wild-type silkworm *Philosamia cynthis ricini*. The silk fibroin from *P. c. ricini* in aqueous solution contains an α-helical conformation derived from the presence of polyalanine regions. Up to 22-mers of polyalanine were identified which normally are not present in *B. mori* silk. In *P. c. ricini*, alanine and glycine account for approximately 80% of the amino acids present in the fibroin; however, the percentage of alanine is higher than glycine in this silkworm, whereas the opposite is the case in *B. mori*. The α-helical structure can be transformed to the random coil by a temperature increase or an increase in urea content in aqueous solution. The helicity of the fibroin from *P. c. ricini* was found to be 15% at 25°C and 26% at 0°C[102]. *B. mori* silk fibroin, with a 30.0mol% L-alanine content, assumed only a random-coil conformation, whereas *P. c. ricini* silk, with up to 48.8mol% L-alanine, locally assumes an α-helical conformation in aqueous solution[111]. ^{13}C-NMR was used to characterize the conformation of the silk fibroin both in solution and while stored within the silkworm silk gland[102], as well as conformations in the solid state[112].

Silkworm genetics

TRANSCRIPTION OF THE FIBROIN GENE

Expression of the fibroin gene is localized to the posterior region of the silk gland and is controlled in a precise temporal and spatial manner[114–116]. Control is exerted at the level of transcription. Activation of fibroin gene transcription starts from the anterior part of the posterior silk gland at the beginning of the fifth instar and gradually spreads towards the posterior end of the gland[117]. A summary picture of the fibroin gene structure is shown in Fig. 3. In 1972, Suzuki and Brown[114] first characterized mRNA for silk fibroin from the posterior region of the silk gland. The isolation of fibroin mRNA by sucrose-gradient centrifugation was based on a predicted codon use derived from the amino-acid composition of fibroin. From fibroin, with a high percentage of glycine (45%), alanine and serine residues, it was estimated that the mRNA would have a minimum guanine plus cytosine content of 57%, with a guanine content of 40%. The high guanine plus cytosine content permits separation of fibroin mRNA by density gradient centrifugation because of its high buoyant density compared with other RNAs. The value of 57% for guanine plus cytosine content of fibroin mRNA is significantly higher than the value for the DNA (39%) or remaining RNA (50%) in the organism. Fibroin mRNA was relatively stable compared with heterogeneous nuclear RNA

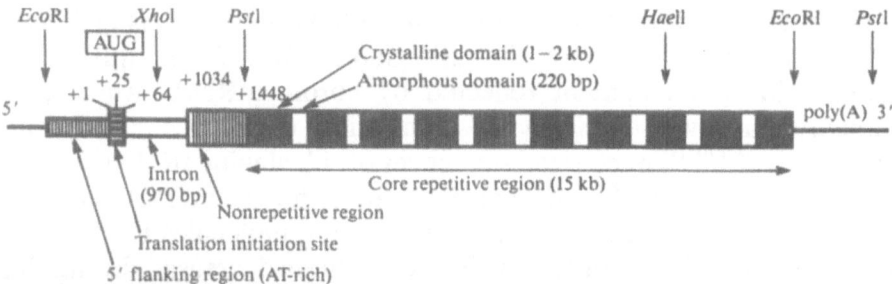

Fig. 3 Summary representation of the structure of the fibroin gene in *B. mori* (compiled from refs 7, 126, 128, 132, 133).

(hnRNA) and precursor ribosomal RNA (rRNA) and represented about 0.8 to 1.4% of total RNA in the posterior region of the silk gland at the end of larval development. This mRNA was found only in the posterior region of the silk gland where fibroin is synthesized. It was not present in the middle region of the gland or the remaining portions of the silkworm. The fibroin mRNA was digested with RNase T_1 and pancreatic ribonuclease to produce predicted oligonucleotide patterns and for the assignment of the major codons for the three principle amino acids (GGU and GGA for glycine, GCU for alanine, UCA for serine).

Suzuki et al.[118] reported that the amounts of DNA complementary to fibroin mRNA in the carcass, middle region of the silk gland and posterior region of the silk gland were similar, whereas fibroin is only synthesized in the posterior region of the silk gland. The fibroin mRNA in the cells of the posterior region of the silk gland is encoded by only 0.0022% of the total DNA in these cells, yet an extraordinarily high translational activity with highly stable fibroin mRNA (-10 kcal/mol) is evident from the synthesis of about 300 µg of fibroin per cell[18] in the last 4 days of the fifth instar. Each polyploid cell contains about 10^6 fibroin genes responsible for the synthesis of 10^{15} fibroin molecules[118]. Each gene produces 10^4 mRNA molecules, each of which can produce about 10^5 protein molecules in 4 days[118]. Specific amplification of fibroin genes was not responsible for this level of protein production. The synthesis of very stable mRNA and strong translational controls appear to be responsible[118].

Gage and Manning[121] using Southern-blot hybridization determined that a single copy of the fibroin gene exists per haploid genome. Each mature cell in the posterior region of the silk gland contains 10^5 more DNA than the diploid cell, which represents replication of the entire genome and not selective amplification of fibroin DNA[122].

Garel et al.[119] reported the functional adaptation of transfer RNAs in the posterior region of the silk gland. They found qualitative changes in tRNA content reflected by fibroin mRNA, which presumably improves efficiency of translation of mRNA to fibroin. Sprague et al.[120] characterized sequences of silk alanine tRNAs.

Suzuki and Giza[115] found a constant rate of fibroin mRNA synthesis throughout the fifth larval instar, seven to ten molecules per minute, while rRNA and heterogeneous RNA synthesis decreased during this period. Therefore, fibroin mRNA represented an increasing percentage of total RNA during pulse labelling using intracoelomic injection of tritiated uridine. Fibroin mRNA accounts for up to 7.5% of the RNA synthesized *in vivo* in 30 min and as much as 4.4% of total cellular RNA by the end of the fifth instar. This rate represents an extremely high level for the transcription of a single mRNA. The regulation and expression of the fibroin gene is illustrated by the steady-rate of fibroin synthesis during the fifth instar but the absence of synthesis during the fourth instar. The stability of the mRNA is reflected in its long half life, which was reported to be several days[114,115]. The finding that isolated nuclei from the fifth instar retain an ability to reinitiate RNA synthesis provides a well-defined system for the *in vitro* study of fibroin gene regulation. Suzuki and Giza[115] found a 5' flanking sequence upstream from the TATA box in genomic clones that is involved in the initiation of transcription.

Lizardi[125] used affinity chromatography to isolate fibroin mRNA from total hnRNA. He found about 19% of the hnRNA from 40S rRNA was fibroin mRNA with one predominant product of about 5.8kb. He reported that the fibroin mRNA is processed into a mature mRNA by the excision of a single intron. The *B. mori* genome is small in size, whereas the fibroin gene is relatively large, comprising 0.004% of the total genome[24].

STRUCTURE OF THE FIBROIN GENE

Manning and Gage[60] used Southern blots of genomic DNAs, restriction mapping and partial sequence analysis to determine the conserved nature of the fibroin gene. Most of the silk gene plus flanking regions at the 5' end where mRNA transcription is initiated was contained on an *Eco*RI fragment. To develop the restriction map, genomic DNA was digested with restriction enzymes and fragments were hybridized with fibroin ^{125}I-labelled mRNA. The gene was about 20 kilobases (kb), the protein-coding region consisting of about 11kb. The DNA flanking region at the 5' end was 3kb and the DNA flanking region at the 3' end was 6.5kb. The coding region was resistant to digestion by many of the commonly used restriction enzymes, including *Hae*III, which recognizes the sequence GGCC which would be expected to occur at a high frequency within DNA of 60% guanine plus cytosine content. The full number of *Hae*II fragments expected on the basis of mRNA sequences reported by Suzuki and Brown[114] were not detected. *Hae*II yielded only a few large gene fragments yet should cleave Gly-Ala codon pairs. Since less than 0.1% of the DNA is 5-methyl cytosine[122], it was determined that less than 1% of the glycines preceding alanine are encoded by GGC, representing a strong selection against this codon. Only three other enzymes cut within the fibroin gene, *Hha*I (GCGC) which produced ten fragments, *Alu*I (AGCT) which cleaved the gene into very small fragments, and *Hpa*II (CCGG).

A restriction map of DNA encoding silk fibroin was developed with restriction digests of total silk gland DNA. The orientation of fibroin mRNA was also determined[60]. Hybridization of ^{125}I-labelled mRNA is specific for the fibroin gene[121] owing to the purity of mRNA and interval repeats in the fibroin gene coding sequence. Therefore this approach would not detect nonrepetitive domains (amorphous regions) which are estimated to constitute less than 15% of the mRNA[114,116,121]. No large additions or deletions in the DNA adjacent to the fibroin gene were found during the polyploidization process or during the transition from inactive to active fibroin gene expression[60]. No differences were found in the physical map of DNA adjacent to the fibroin gene in the posterior region of the silk gland where the gene is active, within the middle region of the silk gland where the gene is inactive, or in carcass cells where the gene is also inactive and of low ploidy[60]. Manning and Gage[60] concluded that there was little evidence for large crystalline and amorphous domains and little evidence for noncoding regions within the fibroin gene. Some recombination within the *Eco*RI fragment of the major coding region was also reported.

CLONING OF THE FIBROIN GENE

Suzuki and Ohshima[123] and Ohshima and Suzuki[89] were the first to clone the *B. mori* silk fibroin gene fragment. Clones were generated in two ways, both by ligation of *Eco*RI-digested *B. mori* DNA with plasmid and by poly(dA)·poly(dT) annealing of sheared *B. mori* DNA with plasmid vector. *B. mori* DNA was enriched for fibroin genes 15- to 1000-fold by ultracentrifugation in CsCl gradients based on the GC-rich fibroin. Colonies of transformed *E. coli* HB101 were immobilized on filters and hybridized with ^{125}I-labelled fibroin mRNA to identify positive clones.

Twelve fibroin gene clones were identified using this technique[24,123]. One clone, pFb29, was constructed from the ligation of *Eco*R1-digested DNA with *Eco*R1-digested pMB9. Other clones included pFb19, which contains the 5' end of the gene and a 5' flanking sequence, and pFb10, which contains the 3' end of the gene and a 3' flanking sequence. The authors noted some instability of clones pFb29 and pFb19 and heterogeneity in the size of intragenic sequences, possibly due to the recombination of the repetitive fibroin gene resulting in deletions or rearrangements. Flanking sequences were more stable and pFb10 was the most stable, presumably owing to the smaller size of the insert. The fibroin gene identified consisted of a coding region of more than 14 kb, was highly repetitive (85% of the repetitive region consists of three amino acids, alanine, glycine, serine), and exhibited a preference in codon assignment owing to natural selection. A variety of other clones from the fibroin gene have been generated. These include pBF36, which contains about 1300 base pairs of complementary DNA at the 3' end to the poly(A) region; pBF41, which contains a 14-kb DNA fragment of the fibroin gene with a 7-kb deletion in the core region and 5.6 kb of the 5' flanking region; and pBF48, which contains an 11.6-kb DNA fragment outside the 3' end of the fibroin gene and about 4 kb within the 3' end of the gene[180].

Tsujimoto and Suzuki[126] presented a detailed restriction map from genomic clones of the fibroin gene and sequenced about 700 base pairs (bp) to begin to understand regulation of fibroin gene expression. Differential expression was presumed to be regulated at the transcriptional level. No difference in the structure of the fibroin gene was found in comparing translated (active) versus nontranslated (silent) cells during this study. Tsujimoto and Suzuki[127] first described the sequence of the 5' end of the fibroin gene clone including the large intervening and flanking (promoter) sequences using S1 mapping and the pFb29 clone. They sequenced over 2 kb of the fibroin gene cDNA, including the 5' flanking region, the intervening sequence and part of the fibroin coding region. The sequence confirmed an intervening sequence of 970 bases and identified more specific sites for the 5' coding–intervening junction at position +64 to +66 and the 3' intervening–coding junction at position +1034 to +1036. The repetitious Gly-Ala region began at position +1448. Tsujimoto and Suzuki[126,127] identified a 5' cap site (+1), a TATA box at position −30 to −24 and a sequence from position −85 to −71 homologous to many prokaryote promoters. An *in vitro* transcription initiation site was located at the 5' end of the mRNA with the coding sequence ATCAG[128]. The detailed sequence and structural analysis of the intervening sequence was studied to understand the control of post-transcriptional RNA splicing and to understand the physiological and evolutionary relevance of this sequence. The 5' end was mapped using restriction enzymes and was found to be AT-rich.

Gage and Manning[132,133] detailed the structure of the repetitive region of the fibroin gene which had been difficult to characterize owing to its size (16 kb) and instability when cloned[126,127]. They presented detailed studies on the organization of the fibroin gene using restriction endonuclease mapping. They reported on sequences corresponding to the repetitious crystalline domains and the amorphous domains, focusing on the conserved repetitive regions and the polymorphism in the nonrepetitive region. The restriction endonucleases (*Mbo*II, *Ava*II, *Hin*fI) cleaved sequences representative of the amorphous regions. They found a 15-kb repetitive gene core containing ten large crystalline domains interspersed by smaller amorphous domains. The amorphous regions consisted of about 220 base pairs interrupting crystalline domains of 1 to 2 kb. They also found that the glycine codon GGA is not used before serine codons but is often used before alanine codons.

The conserved repetitive region in the core, the alternating crystalline and amorphous domains, and the specific codon usage were described in detail. The authors felt that the alternating domains were a key in processing and solubility, which would be influenced by the size and spacing of the amorphous domains to prevent premature crystallization within the gland. They cited supporting evidence of Lucas *et al.*[83] that chymotrypsin treatment results in the precipitation of crystalline polypeptides. Comparisons of fibroins in 22 variants of *B. mori* resulted in the identification of 19 alleles which differed in length and internal sequence organization. Differences were found in the lengths of fibroin-encoding regions owing to variable lengths of crystalline domains resulting from recombination between these highly repetitive domains. A high degree of polymorphism was found in the length and organization in the fibroin gene locus and the encoded proteins varied in size by as much as 15%.

Because of the repetitive nucleotide sequence of the large 16-kb fibroin gene, many restriction enzymes cleaved only near the 5' and 3' ends, including *Hind*III, *Hae*III, *Pst*I and *Bgl*I[126,129]. Deletions from this gene-core repetitive region were found in plasmid pFb29 DNA when grown in *Escherichia coli*. About 60% of the fibroin mRNA had a complete methylated cap (m[7]GpppAmUmCXG).

Sprague *et al.*[63] were the first to speculate on the possibility of crossover recombination in the core region, which was later confirmed to occur. They used restriction endonuclease mapping to investigate fibroin genes encoding different-sized fibroin proteins, and concluded that the size variation was due to unequal rearrangements or crossover within the highly repetitive part of the gene, resulting in variant alleles. The authors concluded that considerable rearrangement may be manageable because the resulting encoded proteins do not significantly alter the β-sheet structure and, in turn, the function of the protein.

Lizardi[129] also studied silk fibroin polymorphism and correlated mRNA size with protein size. Fingerprints were developed of translational pauses which occur in the *in vitro* translation of fibroin mRNA. Different silk fibroins were used from a number of *B. mori* strains with polypeptide phenotypes from 365K to 415K, as measured by SDS-PAGE. He found that the size of the mRNAs corresponded with polypeptide length, ranging from 5.5 to 6.3 kb.

CONTROL OF FIBROIN SYNTHESIS

The regulatory controls over the synthesis of this large protein have resulted in intensive interest at both DNA and mRNA levels. The first cell-free synthesis of fibroin was demonstrated by Ikariyama *et al.*[130] using polyribosomes isolated from the posterior region of the silk gland that were studied free or adsorbed onto DEAE-Sephadex. The addition of salts, sucrose, dithiothreitol, enzymes, ATP, GTP and amino acids in ratios reflecting their concentration in fibroin was required for the synthesis. SDS-PAGE was used to monitor fibroin synthesis, and proteins from 300K to 400K were formed, in part due to the very stable mRNA. More than 1 ng of protein was synthesized in a 4-hour reaction.

Lizardi[129] also demonstrated *in vitro* translation of full-length fibroin using a heterologous cell-free system from rabbit reticulocyte lysates which required supplementation with specific insect tRNAs. He found that translation was discontinuous both in the *in vitro* cell-free system and in silk glands maintained in culture. This effect was not due to the secondary structure of mRNA but reflected the different times required by the ribosome for the recognition-binding reactions at each codon. Since glycine, alanine and serine constitute about 85% of the amino acids in fibroin, about 70% of the total tRNA present corresponds to that required for those amino acids. The discontinuous chain elongation may be due to suboptimal concentrations of specific tRNAs, despite supplementation in the *in vitro* experiments and in intact silk gland cells. The authors speculated that this could be a control mechanism used in the cell to regulate elongation rates, to obtain specific levels of polysome loading or to optimize residence time for nascent chains that may be modified by specific enzymes in a post-translational mode, for example by glycosylation or by proteases. Other authors have also studied cell-free synthesis of fibroin[131].

Young *et al.*[142] studied transcriptional controls of gene expression. The RNA distribution in the silk glands becomes enriched for tRNA species corresponding to the predominant amino acids in fibroin (40% glycine, 20% alanine, 10% serine and 10% tyrosine). For the tRNA specific for alanine, most of this increase is due to the accumulation of a novel tRNA unique to the silk gland. Therefore, there are two types of tRNAs specific for alanine—constitutive tRNA in all silkworm cells, and fibroin-specific tRNA in only the posterior portion of the silk gland. Up to 70% of this tRNA during active fibroin synthesis is the fibroin-specific type.

In a further study of the regulatory controls of fibroin gene expression, Tsujimoto *et al.*[134] identified an enhancer region and an essential region for expression of the fibroin gene at the 5' end. *In vitro* mutagenesis of the cloned gene was used to determine the function of the 5' flanking region in transcriptional control. Okamoto *et al.*[135] found that the 5' flanking sequences of fibroin and sericin genes were 70% homologous. They also identified a CAAT box-like sequence at position −93 to −83. Tsuda and Suzuki[124] developed an *in vitro* replication system to study genes of isolated fibroin cells. Using this cell-free system they found a 5' upstream flanking region up to position −238 that was required in the control of transcription. Tsuda and Suzuki[136] further identified upstream 5'-flanking modulating sequence elements from the TATA box located within a 200-bp region between position −238 and −116 and position −73 and −53. Suzuki *et al.*[8] concluded that a *cis* element is present in this upstream flanking region that functions specifically in cells located in the posterior region of the silk gland.

To elucidate transcriptional and translational controls, fibroin genes purified directly from cells isolated from the posterior region of the silk gland (producers of fibroin) and cells from the middle region of the silk gland (nonproducers of fibroin) have been compared. Tsujimoto and Suzuki[137] found no differences in DNA sequence around the 5' end (position −171 to +104) including the promoter, and no differences in hybridization patterns between nucleotides of position −650 to +326. They concluded that transcriptional controls are not a result of nucleotide base changes in these regions. No differences in methylation patterns were found when comparing control regions of active and inactive fibroin-producing cells.

Suzuki and Adachi[138] sequenced 1910 bases in the fibroin-producing and fibroin-nonproducing genes. They found identical sequences from position −297 to +333 and position +746 to +1261 where transcriptional controls are located (−29 to +6 for the promoter function, −238 to −116 and −73 to −53 for transcriptional enhancement signals, and +66 to +1037 for exon/intron boundaries where the noncoding region ends and the repetitive sequence begins). The results from cell-free transcription studies also indicated no distinguishable differences between the fibroin DNA from producing and nonproducing cells. The nucleotide differences in the region from position −649 to +1261 apparently do not influence transcriptional controls. Sequence polymorphism in the upstream 5′-flanking region was also studied by Ueda et al.[139] and polymorphism in the 5′ flanking region and nonrepetitive region near the 5′ end of the gene was found. Base sequence changes in the 2-kb to 4-kb upstream region from the 5′ end were noted. Small differences in gel patterns were found from restriction digests of introns and of the 5′ flanking regions within 1.1 kb of the 5′ end.

Studies on transcriptional and translational controls by Hirose et al.[140] using in vitro transcription of the fibroin gene by posterior silk gland extracts identified factors present in the extracts that supercoil covalently closed circular DNA. This increased the efficiency of the fibroin gene promoter, regardless of the upstream enhancer sequence. There was a 3–10-fold increase in transcription when compared to nicked-circular or linear DNA. Adams et al.[141] characterized small nuclear RNAs from the posterior region of the silk gland. These low relative molecular mass nonribosomal- and nontransfer-RNAs function in a variety of gene and protein expression controls and transport. Some highly conserved sequences with vertebrate U-RNAs were noted.

Suzuki et al.[8,143] studied cell-free systems to further elucidate transcriptional enhancement of the B. mori fibroin gene. The upstream region from position −72 to −32 contained an enhancing signal stimulated by extracts from a number of different silk gland regions and other cells. The other enhancement region, from position −238 to −73, was tissue-specific for extracts from posterior silk gland versus middle silk gland. They also summarized mutagenicity studies used to elucidate signals and factors involved in fibroin gene transcription, using six in vitro transcription systems and one in vivo system. Fibroin transcription required an intact region from position −29 to +6, including the TATA box and cap site, and is the putative promoter[139].

The surrounding bases, particularly the second and fourth, are critical in directing the site of initiation. The mutagenesis studies also identified a region around position −20 that was required for efficient transcription *in vitro*. The distance between this site and the TATA box is roughly one turn of the DNA double helix[7].

Koga-Ban and Suzuki[144] studied differences in DNAase I sensitivity of fibroin and sericin genes during developmental stages in the posterior region of the silk gland. The fibroin gene was more sensitive to the enzyme during larval feeding stages, which are periods of active growth, than during larval molting stages when the gene is inactive. The authors concluded that this related to changes in chromatin structure. It was suggested that fibroin gene expression was controlled at the levels of chromatin structure and DNA conformation, since no differences were found between the producing and nonproducing genes. Ohta *et al.*[117] used RNAase mapping to evaluate transcription in *B. mori*, and found that fibroin gene transcription is functional for the first time once embryonic silk gland development is complete. They concluded that this would be a good point in the developmental cycle to look for factor(s) influencing transcriptional control. DNA supercoiling factors were recently identified in extracts from the posterior region of the silk gland[145].

Kusuda *et al.*[146] cloned the fibroin gene from the mulberry wild silkworm *Bombyx mandarina*, and determined the structure of the 5′-end and upstream elements. No significant differences were found in the genes of this species and the *B. mori* silkworm except for some mismatches, insertions or deletions in intron regions or upstream of the 5′ end. Even with the large differences in silk production between these two species, no significant differences in fibroin gene structure were found. Tamura[19] studied *Antheracea yamamai*, where fibroin synthesis and storage is in the posterior region of the silk gland. The protein had a relative molecular mass of 450kd by SDS-PAGE, and when treated with reducing agents, dissociated into two proteins of 220kd to 250kd instead of the L- and H-chains found in *B. mori* fibroin. This fibroin gene was cloned and its structure determined. The 5′ end contained a 69-bp exon and a 140-bp exon, and the core region begins 800bp from the start site. A cell-free system was also studied.

Couble *et al.*[147,148] found the synthesis of fibroin H- and L-chains are synchronized with the presence of molar equivalents of the corresponding mRNAs. Kimura *et al.*[149] cloned the L-chain cDNA.

A cDNA was generated from L-chain mRNA isolated from the posterior region of the silk gland and inserted into plasmid pBR322 using oligo(dC)·oligo(dG) tailing. *Pst*I restriction sites were used to clone-in the insert consisting of about 800 bp.

Mita *et al.*[150] characterized the codon preferences in the repetitive core region of the fibroin mRNA from *B. mori*. Two components were identified in this core region, a repetitive component and a joining component. They had different codon preferences, attributed to restrictions imparted by the secondary structure of the mRNA. The authors speculated that the two domains are required for the proper folding of the silk fibroin mRNA into a fully functional conformation, and would also correlate with the translational pauses previously described[129].

Yamaguchi *et al.*[71] characterized the L-chain and obtained a cDNA clone containing the putative full-length L-chain nucleotide sequence. The genes for the H- and L-chains are located on different chromosomes, although they are possibly coordinately regulated and translated in membrane-bound polysomes. The H- and L-chain assembly was required for efficient intracellular transport and secretion of fibroin[151,152]. Yamaguchi *et al.*[71] found a relative molecular mass of 25.8K for the L-chain with three cysteine residues, two of which are involved in an intramolecular disulphide linkage and the third located towards the C-terminal end potentially available for binding with the H-chain. The authors speculated on the binding of the H- and L-chains through this disulphide bond, as well as on other hydrophobic interactions, which would interfere with secondary structure formation that would otherwise result in premature precipitation of the protein. Therefore this L- and H-chain interaction may ensure solubilization of fibroin during transport and secretion until the protein is spun into a fibre by the silkworm.

Hui and Suzuki[153] and Hui *et al.*[154] identified homeodomains in binding sites in the 5′ flanking region of *B. mori* silk fibroin L- and H-chain genes. They found multiple AT-rich sequences located near the 5′ flanking region of the fibroin L-chain gene that are homologous to fibroin and sericin-1 genes and that can also bind nonsilkworm proteins. One of the silk gland-specific factors may play a role in the expression of silk protein genes, since it binds to the 5′ flanking region of the *B. mori* fibroin gene, sericin-1 gene, and two additional silk protein genes of *B. mori* and *A. yamamai*, a wild-type silkworm.

CONTROL OF SERICIN SYNTHESIS

Sericin consists of many different proteins. Okamoto *et al.*[135] identified two genes, from the middle region of the silk gland, that code for at least six different sericin proteins. Differences in splicing in the two genes were responsible for their encoding different proteins. For example, one gene results in mRNAs of 10.5kb, 9.0kb, 4.0kb and 2.8kb, and the other gene results in mRNAs of 5.4kb and 3.1kb. The 5' flanking region is strongly homologous to that of the fibroin gene. Gammo[155] has mapped one of the sericin genes. Ishikawa *et al.*[156] have characterized sericin gene expression during the development of the silkworm.

SUMMARY

Research on silkworm genetics began with a burst of activity in the 1970s which resulted in the successful preliminary mapping of the fibroin gene structure. The complete crystalline or core region and the details at the 3'-end and 3'-flanking regions have not been reported. The focus since the late 1970s has been on the 5'-end and 5'-flanking regions to develop a better understanding of the regulatory controls of the gene. It appears that specific factors, not yet identified, control gene activity, while chromatin structure may also be important. The full understanding of these control mechanisms will be important, since this represents a highly specialized, productive and controlled system for protein expression at a specific developmental stage. There are also no reports in the literature of any of the *B. mori* fibroin clones being expressed in viral, bacterial, fungal or other host system for the production of recombinant silk protein.

Spider silk synthesis and genetics

The study of spider silk expression has resulted in the following observations: a full-length silk protein can be produced in the excised gland; protein synthesis can be initiated at any point in the adult life cycle; and many different silks are able to originate from specialized glands. These attributes provide some important considerations in gene regulation for comparison with the silkworm system, where only one silk is produced at only one specific stage in the life cycle[157]. Additional interest in spider silks derives from their unusual physical and mechanical properties.

PROTEIN SYNTHESIS

Candelas and Cintron[68] isolated the major ampullate glands from the orb-spinning spider *N. clavipes*, which are similar to the glands of *B. mori* and represent highly specialized and differentiated structures for the production of specific proteins. They found a relative molecular mass of 320K for the soluble silk protein in the gland. Excised glands maintained protein-synthetic activity for up to 5 hours and were capable of synthesizing full-length protein with incorporation of tritiated alanine and glycine. The spiders were mechanically silked just before gland removal to stimulate translational activity.

Candelas and Lopez[158], Candelas *et al.*[159] and Candelas *et al.*[157] reported the initiation of both transcription and translation in excised major ampullate glands by depleating the supply of stored silk by mechanical silking. Peakall[86,160] also reported on the stimulation of protein synthesis by the administration of acetylcholine, which causes secretion of stored fibroin followed by the onset of protein synthesis. Approximately 1 hour after mechanical silking, protein synthesis peaks[158]. This corresponds with the average delay found in eukaryotic systems for a primary transcript to be translated after processing and transport. An array or ladder of peptides is found in the epithelium of the gland, similar to that reported by Lizardi *et al.*[129] for *B. mori* during cell-free and whole-gland fibroin synthesis, and is thought to be due to translational pauses. Protease inhibitors were used during the excision process, so it was concluded that the ladder was not due to protease activity[159]. Nonstimulated glands did not incorporate tritiated-labelled amino acids into the protein.

Candelas *et al.*[161] isolated tubiliform glands of *N. clavipes* and compared the protein to the large ampullate gland protein. Three pairs of tubiliform glands secrete fibroin for egg-case formation. They found a protein of 380kd by SDS-PAGE and also found that the glands were active after their removal. Discontinuous extension of the fibroin chain was again noted although the response to the stimulus of mechanical silking was less than that found for the major ampullate gland. Tillinghast and Townley[162] also reported the stimulation of protein synthesis by mechanical silking of the spider *A. cavaticus*, both *in vitro* and *in vivo*. They also reported independent control of protein synthesis in the two different major ampullate glands.

CLONING OF THE FIBROIN GENE

Lombardi and Kaplan[69,163,164] and Lombardi *et al.*[165] have cloned a full-length fibroin cDNA from *N. clavipes*. Total cellular RNA was isolated and purified from the excised major ampullate glands, and mRNA was isolated by passing the sample over an oligo(dT) column. The mRNA was size-selected and large transcripts (>9 kb) were used to synthesize cDNA. The cDNA was cloned using unique site-directional primer-adapter systems. In-frame transcription/translation was obtained in viral and plasmid vectors used to transfect *E. coli*. The recombinant cDNA library was screened with oligonucleotide probes generated from dragline silk peptide sequences labelled with [32]P. The plaque purified cDNA library was expressed in *E. coli* as a fusion protein. The recombinant silk protein was purified; it had a relative molecular mass of 350K as determined by SDS-PAGE[69]. The recombinant silk protein was concentrated and spun into a short fibre. Subsequently, the amino-acid composition of the recombinant silk was comparable with reeled dragline and glandular silk from *N. clavipes*. Preliminary results show minimal differences between the three sources of silk. The structure of the gene is being determined.

A genomic library was constructed using DNA isolated and purified from whole abdomens of mature, adult female spiders[69,163]. Following further purification, the high relative molecular mass DNA (>50 kb) was partially digested to yield fragments of 25 kb for ligation into a λ phage vector. Oligonucleotide and nick-translated probes corresponding to silk peptide sequences and a 1.3-kb fragment of the *B. mori* 5' protein-encoding domain were used to screen the library. Colony plaque hybridizations confirmed the presence of fibroin gene sequences in *N. clavipes* DNA. Southern-blot hybridizations were performed to determine the copy number of the major ampullate gland gene. Preliminary data indicate that the dragline silk gene is a single-copy gene like *B. mori*, regulated in part by tissue-specific factors. The full-length cDNA clone was determined to be about 10–11 kb, similar to the results reported by Lizardi *et al.*[125] for *B. mori* fibroin. Additional sequencing of both cDNA nested deletion clones and genomic clones is continuing. So far the 5' noncoding regions, 5' cap structure, and a portion of the protein-encoding region have been sequenced from the major ampullate gland cDNA library.

Dong *et al.*[165] and Xu and Lewis[167] used oligonucleotide probes designed from partial acid hydrolysates of dragline silk to screen a cDNA library constructed from *N. clavipes* major ampullate gland silk and cocoon silk mRNAs. Xu and Lewis[167] identified more than 20 positive clones using these probes; the inserts ranged from 800 bp to 2.4 kb. Partial sequencing of the largest clone showed a repetitive sequence and high G+C content. This content may cause some inserts to delete and religate. A 34-mer crystalline repeat (not highly conserved) for the core region was identified. The published data represent the 3′ end of the mature silk message, including the beginning of the poly(A) tail, polyadenylation site and a portion of the 3′ protein-encoding region.

Studies on spider silk genetics and regulation are in their infancy in comparison to work on silkworm silks. Recent work on cloning and expression of spider silk genes has resulted from new interest in these systems and their protein products, which should stimulate increased research. The expression of the multiplicity of silks in spiders, and the regulation and control mechanisms involved, will complement strongly the ongoing studies on the silkworm system and provide new insights into transcriptional and translational controls.

Synthetic silk genetics

A number of groups have attempted the construction of synthetic silk-like genes for the production of synthetic silks, with a primary focus on the crystalline region of the silkworm silk. For example, Amgen (California, USA) used rDNA technology to produce synthetic silks in recombinant yeast and *E. coli* based on the *B. mori* crystalline region. Silk-like polymers were produced ranging from 5K to 50K (personal communication). PA Technologies (UK) have performed similar work with the crystalline domain of silkworm silk combined with an amorphous domain based on a small region of the enzyme β-galactosidase fusion protein[168,169]. There were problems with instability of the repetitive domain, codon use, tRNA ratios and proper mRNA secondary structure. The crystalline domain used in these constructs was the same as cited earlier for the *B. mori* core 59-mer repeat. This domain was alternated with the amorphous domains (β-galactosidase) in an attempt to mimic the native silk core fibroin structure. A 3.3-kb synthetic gene was constructed, ligated with vector pPA1 and used to transform *E. coli*. No details were provided on the expression of the synthetic gene, recovery of protein, or yields from the host cells studied.

Syntro and Protein Polymer Technolgies (California, USA) have also produced silk-like proteins using synthetic gene technology[170–174]. Crystalline domains have been coupled to elastin-like repeats to mimic the amorphous domains in the core silkworm fibroin gene. The work also included general methods for the cloning of repetitive synthetic structural genes. Monomeric coding units, with variations in codon usage, self-ligate to form multimers which are screened and subcloned to obtain the desired size of the synthetic gene. Monomer sequences are designed, synthesized and assembled into homopolymers and block copolymers. The polymers produced are essentially monodisperse. Synthesis is regulated by a promoter specific for the synthetic gene and not recognized by the host RNA polymerase. This initiation factor is inducible for the insert. This inducible transcription initiator allows for the growth of the host cell to high density before the initiation of synthetic protein expression. Crystalline domains were based on the repeat hexamer Gly-Ala-Gly-Ala-Gly-Ser (SGAGAG), and the amorphous domains consisted of the elastin-like pentamer repeat Val-Pro-Gly-Val-Gly (VPGVG). The elastin-like blocks function to reduce crystallinity and increase flexibility in the polymer chain. The synthetic genes ranged in size from 60 to 500 b. Multimers ranging from 1 to 5 kb can be formed into peptides from 30K to 150K. The peptides with greater than 90% purity have been produced in gram or larger quantities by fermentation.

E. coli strain HB101 stably maintains the gene, presumably by the variable codon usage in the synthetic gene construct. The copolymers are soluble in water to 1%, and films have been formed from the homopolymer on glass slides. No data on fibres generated from these silk-like materials have been published. The homopolymers were purified by extraction with concentrated LiBr solution and dialysed. Polyamide homoblock polymers and alternating copolymers with silk-like to elastin-like ratios from 1:4 to 2:1 have been produced. A primary sequence from human fibronectin has also been incorporated into the copolymer to develop biological activity related to cell attachment. This modified polymer can be reversibly denatured, is stable to 120°C and is resistant to proteolytic attack.

McGrath *et al.*[175,176], Creel *et al.*[177] and Modler *et al.*[176] constructed synthetic silk-like repeats modelled after the crystalline domain of *B. mori* fibroin to accomplish the synthesis of new peptides with controlled crystal structure and surface chemistry. Examples of repeats studied include [(Ala-Gly)$_3$-Pro-Glu-Gly]$_n$ and [(Ala-Gly)$_4$-Pro-Glu-

Gly]$_n$, which were predicted to assume a β-sheet crystalline structure with carboxylic acid surfaces on the basis of glutamic acid side chains located at the predetermined β-turns along with the antiparallel β-sheets. Similar technology as described above for Protein Polymer Technolgies was used to accomplish the synthesis and expression of these peptides. The oligomers were synthesized and then self-ligated to multimers, linkers were attached, the insert directionally ligated into a host plasmid and *E. coli* transformed and grown with ^{3}H-labelled glycine. Protein expression was monitored by the *in vivo* incorporation of ^{3}H-labelled glycine. Protein production was induced by addition of IPTG (isopropyl-β-D-thiogalactopyranoside) in some studies. The solid state conformations of these peptides have been partially determined. In the case of the trimer core repeat, the predicted crystalline structure is not obtained, while with the tetramer repeat a closer approximation of the predicted crystallity was found by X-ray diffraction analysis. A comparison was made with the chemical synthesis of one of these peptides [(Ala-Gly)$_3$-Pro-Glu-Gly)$_n$. The chemical approach required an 18-step procedure and resulted in the formation of a polydisperse product[177].

Acknowledgements

We thank Scott Stockwell, Janet Ward, John Song and Marvin Greenberger (Natick Research, Development & Engineering Center) for help and suggestions; Marion Goldsmith (University of Rhode Island) for generous supplies of silkworms and helpful discussions; Christopher Viney (University of Washington) for insights and collaboration on solution properties; Bob Work (North Carolina State University) for guidance and suggestions on controlled silking; Randy Lewis (University of Wyoming) for helpful discussions; Tsunehiro Mukai (National Cardiovascular Center Research Institute, Osaka) for the generous donation of plasmid pBF41; and David Adams (Worcester Polytechnic Institute) for early guidance. Very special thanks to Ted Thaunhauser (Cornell University), George Nemethy (Mount Sinai School of Medicine) and Harold Scheraga (Cornell University) for guidance and support in the modelling and analytical work. Additional thanks go to Tetsuo Asakura (Tokyo University of Agriculture and Technology), Takuma Gamo (National Institute of Agrobiological Resources, Tsukuba City, Ibaraki) and Morio Ikehara (Protein Engineering Research Institute, Osaka) for helpful discussions.

References

1. Fraser, R.D.B. & MacRae, T.P. (1973) Conformation in fibrous proteins and related synthetic polypeptides. In *Conformation of Fibrous Proteins* (Academic, New York).
2. Ilzuka, E. (1985) Silk: an overview. *J. Appl. Poly. Sci.* **41**: 163–171.
3. Livengood, C.D. (1990) Silk. In *Polymers-Fibers and Textiles: A Compendium* (ed. Kroschowitz, J.I.) Encyclopedia Reprint Series, pp. 789–797. Wiley, New York.
4. Lotz, B. & Colonna-Cesari, F. (1979) The chemical structure and the crystalline structures of *Bombyx mori* silk fibroin. *Biochimie* **61**: 205–214.
5. Lucas, F., Shaw, J.T.B. & Smith, S.G. (1960) Comparative studies of fibroins I. The amino acid composition of various fibroins and its significance in relation to their crystal structure and taxonomy. *J. Mol. Biol.* **2**: 339–349.
6. Robson, R.M. (1985) Silk composition, structure, and properties. In *Fiber Chemistry Handbook of Science and Technology* (eds Lewin, M. & Pearce, E.), Vol. IV, pp. 647–700. Marcel Dekker, New York.
7. Goldsmith, M.R. & Kafatos, F.C. (1984) Developmentally regulated genes in silkmoths. *Ann. Rev. Genet.* **18**: 443–487.
8. Suzuki, Y., Takiya, S., Hara, W., Obara, T., Suzuki, T. & Hui, C.-C. (1987) Developmental regulation of the tissue-specific genes and the homeotic genes in *Bombyx mori*. In *Gunma Symp. Endocrinol.* (ed. Iwai, K.) Vol. 24, 13–26. Center for Academic Publications Japan, Tokyo VNU Science Press, Utrecht.
9. Tamura, T. (1989) Silk synthesis and genetic engineering. *Sen'i Gakkaishi* **45(8)**: 345–349.
10. Gosline, J.M., DeMont, M.E. & Denny, M.W. (1986) The structure and properties of spider silk. *Endeavour* **10(1)**: 37–43.
11. Parry, D.A.D. (1979) Determination of structural information from the amino acid sequences of fibrous proteins. In *Fibrous Proteins: Scientific, Industrial and Medical Aspects* (eds Parry, D.A.D. & Creamer, L.K.). Academic, London.
12. Craig, C.L. & Bernard, G.D. (1989) Insect attraction to ultraviolet-reflecting spider webs and web decorations. *Ecology* **71(2)**: 616–623.
13. Magoshi, J., Magoshi, Y. & Nakamura, S. (1985) Crystallization, liquid crystal, and fiber formation of silk fibroin. *J. Appl. Poly. Sci.* **41**: 187–204.
14. Ando, Y., Okano, R., Nishida, K., Miyata, S. & Fukada, E. (1980) Piezoelectric and related properties of hydrated silk fibroin. *Rep. Prog. Polymer Physics Japan* **23**: 775–778.
15. Fukada, E. (1956) On the piezoelectric effect of silk fibers. *J. Phys. Soc. Japan* **12**: 1301.
16. Hazan, A., Gertler, A., Tahori, A.S. & Gerson, U. (1975) Spider mite webbing. III. Solubilization and amino acid composition of the silk protein. *Comp. Biochem. Physiol.* **51**B: 457–462.
17. Case, S. (1986) Correlated changes in steady-state levels of Balbiani ring mRNAs and secretory polypeptides in salivary glands of *Chironomus tentans*. *Chromosoma* **94**: 483–491.
18. Tashiro, Y., Morimoto, T., Matsura, S. & Nagata, S. (1968) Studies on the posterior silk gland of the silkworm, *Bombyx mori*. I. Growth of posterior silk gland cells and biosynthesis of fibroin during the fifth larval instar. *J. Cell Biol.* **38**: 574–588.
19. Komatsu, K. (1985) Chemical and structural studies on silk sericin. *Proc. 7th Int. Wool Tex. Res. Conf.* **1**: 373–382.
20. Foelix, R.F. (1982) *Biology of Spiders*. Harvard University Press, Cambridge, Massachusetts.

21. Nentwig, W. (1987) *Ecophysiology of Spiders*. Springer-Verlag, New York.

22. Denny, M.W. (1980) Silks – their properties and functions. In *Mechanical Properties of Biological Materials* (eds Vincent, J.F.V. & Currey, J.D.), pp. 247–272. Cambridge University Press.

23. Townley, M.A. & Tillinghast, E.K. (1988) Orb web recycling in *Araneus cavaticus* (Araneae, Araneidae) with an emphasis on the adhesive spiral component, gabamide. *J. Arachnol.* **16**: 303–319.

24. Work, R.W. & Emerson, P.D. (1982) An apparatus and technique for the forcible silking of spiders. *J. Arachnol.* **10**: 1–10.

25. Zemlin, J.C. (1968). A study of the mechanical behavior of spider silks. *Technical Report 69-29-CM (AD 684333)*. U.S. Army Natick Laboratories, Natick, Massachusetts.

26. Gosline, J.M., Denny, M.W. & DeMont, M.E. (1984) Spider silk as rubber. *Nature* **309**: 551–552.

27. Andersen, S.O. (1970) Amino acid composition of spider silks. *Comp. Biochem. Physiol.* **35**: 705–711.

28. Denny, M.W. (1976) The physical properties of spider's silk and their role in the design of orb-webs. *J. Exp. Biol.* **65**: 483–506.

29. Palmer, J.M., Coyle, F.A. & Harrison, F.W. (1982) Structure and cytochemistry of silk glands of the mygalomorph spider *Antrodiaetus unicolor* (Araneae, Antrodiaetidae). *J. Morph.* **174**: 269–274.

30. Work, R.W. (1976) The force-elongation behavior of web fibers and silks forcibly obtained from orb-web-spinning spiders. *Tex. Res. J.* **46**: 485–492.

31. Work, R.W. (1977) Dimensions, birefringences, and force-elongation behavior of major and minor ampullate silk fibers from orb-web-spinning spiders – the effects of wetting on these properties. *Tex. Res. J.* **47**: 650–662.

32. Anderson, C.M. & Tillinghast, E.K. (1980) GABA and taurine derivatives on the adhesive spiral of the orb web of *Argiope* spiders, and their possible behavioural significance. *Physiol. Entomol.* **5**: 101–106.

33. Tillinghast, E.K. & Christenson, T. (1984) Observations on the chemical composition of the web of *Nephila clavipes* (Araneae, Araneidae). *J. Arachnol.* **12**: 69–74.

34. Vollrath, F., Fairbrother, W.J., Williams, R.J.P., Tillinghast, E.K., Bernstein, D.T., Gallagher, K.S. & Townley, M.A. (1990) Compounds in the droplets of the orb spider's viscid spiral. *Nature* **345**: 526–528.

35. Atkins, E.D.T. (1967) A four-stranded coiled-coil model for some insect fibrous proteins. *J. Mol. Biol.* **24**: 139–141.

36. Lucas, F., Shaw, J.T.B. & Smith, S.G. (1958) The silk fibroins. In *Advances in Protein Chemistry* (eds Anfinsen, C.B., Anson, M.L., Bailey, K. & Edsall, J.T.), Vol. 13, pp. 107–242. Academic, New York.

37. Geddes, A.J., Parker, K.D., Atkins, E.D.T. & Beighton, E. (1968) 'Cross-B' conformation in proteins. *J. Mol. Biol.* **32**: 343–358.

38. Manning, R.F. & Gage, L.P. (1980) Internal structure of the silk fibroin gene of *Bombyx mori*. II. Remarkable polymorphism of the organization of crystalline and amorphous coding sequences. *J. Biol. Chem.* **255(19)**: 9451–9457.

39. Marsh, R.E., Corey, R.B. & Pauling, L. (1955) An investigation of the structure of silk fibroin. *Biochim. Biophys. Acta* **16**: 1–34.

40. Kratky, O. (1956) An X-ray investigation of silk fibroin. *Farraday Soc.* **52(1)**: 558–570.

41. Kratky, O., Schauenstein, E. & Sekora, A. (1950) An unstable lattice in silk fibroin. *Nature* **165**: 319–320.

42. Ambrose, E.J., Bamford, C.M., Elliot, A. & Hanby, W.E. (1951) Water

soluble silk: An alpha-protein. *Nature* **167**: 264–265.

43. Lotz, B., Brack, A. & Spach, G. (1974) β-structure of periodic copolypeptides of L-alanine and glycine. *J. Mol. Biol.* **87**: 193–203.

44. Colonna-Cesari, F., Premilat, S. & Lotz, B. (1975) Conformational analysis of the beta sheet structure of poly-L-alanine and poly(L-alanine-glycine). *J. Mol. Biol.* **95**: 71–82.

45. Fossey, S.A., Nemethy, G., Gibson, K.D. & Scheraga, H.A. (1990) Conformational energy studies of β-sheets of model silk fibroin peptides. I. Sheets of alanine and glycine. *Biopolymers* (in press).

46. Warwicker, J.O. (1960) Comparative studies of fibroins. II. The crystal structures of various fibroins. *J. Mol. Biol.* **2**: 350–362.

47. Nadiger, G.S. & Halliyal, V.G. (1984) Relation between structure and properties of natural silk. *Colourage* **31(20)**: 23–32.

48. Chou, K.-C., Nemethy, G. & Scheraga, H.A. (1983) Effect of amino acid composition on the twist and the relative stability of parallel and antiparallel β-sheets. *Biochemistry* **22**: 6213–6221.

49. Chou, K.-C., Nemethy, G. & Scheraga, H.A. (1983) Role of interchain interactions in the stabilization of the right-handed twist of β-sheets. *J. Mol. Biol.* **168**: 389–407.

50. Chou, K.-C., Pottle, M., Nemethy, G., Ueda, Y. & Scheraga, H.A. (1982) Structure of β-sheets. Origin of the right-handed twist and of the increased stability of antiparallel over parallel sheets. *J. Mol. Biol.* **162**: 89–112.

51. Tsukada, M., Bertholon, G. & Hirabayashi, K. (1982) Structure of polypeptides model related to silk fibroin. *J. Soc. Fiber Technol. Japan* **38(11)**: 451–455.

52. Oka, M., Baba, Y., Kagemoto, A. & Nakajima, A. (1990) Theoretical conformational analysis on silk fibroin model polypeptide with Ala-Gly repeated sequence. *Poly. J.* **22(5)**: 416–425.

53. Ishida, M., Asakura, T., Yokoi, M. & Saito, H. (1990) Solvent- and mechanical-treatment-induced conformational transition of silk fibroins studied by high resolution solid-state ^{13}C NMR spectroscopy. *Macromolecule* **23**: 88–94.

54. Magoshi, J. (1972) Studies on physical properties and structure of silk. I. Crystallization of silk fibroin. *Rep. Prog. Polymer Physics Japan* **15**: 599–604.

55. Magoshi, J. (1974) Comformational changes of silk fibroin treated by various temperatures. *Kobunshi Ronbunshu, Eng. Ed.* **3**: 1668–1670.

56. Magoshi, J. (1974) Thermal properties of silk fibroin films treated by organic solvents. *Kobunshi Ronbunshu, Eng. Ed.* **3**: 1908–1915.

57. Magoshi, J. (1976) Studies on physical properties and structure of silk. VII. Conformational changes of silk fibroin. *Rep. Prog. Polymer Physics Japan* **29**: 661–664.

58. Magoshi, J. & Magoshi, Y. (1977) Physical properties and structure of silk. V. Thermal behavior of silk fibroin in the random-coil conformation. *J. Polymer Sci.* **15**: 1675–1683.

59. Magoshi, J., Nobutami, K. & Kakudo, M. (1973) Studies on the crystallization and alpha-beta transition of silk fibroin. *Kobunshi Kagaku, Eng. Ed.* **2**: 936–941.

60. Manning, R.F. & Gage, L.P. (1978) Physical map of the *Bombyx mori* DNA containing the gene for silk fibroin. *J. Biol. Chem.* **253(6)**: 2044–2052.

61. Lotz, B. & Keith, H.D. (1971) Crystal structure of poly(L-Ala-Gly)II. *J. Mol. Biol.* **61**: 201–215.

62. Lizardi, P.M. (1979) Genetic polymorphism of silk fibroin studied by two-dimensional translation pause fingerprints. *Cell* **18**: 581–589.

63. Sprague, K.U., Roth, M.B., Manning, R.F. & Gage, L.P. (1979) Alleles of the fibroin gene coding for proteins of different lengths. *Cell* **17**: 407–413.

64. Tashiro, Y. & Otsuki, E. (1970) Studies on the posterior silk gland of the silkworm *Bombyx mori*. IV. Ultracentrifugal analyses of native silk proteins, especially fibroin extracted from the middle gland of the mature silkworm. *J. Cell Biol.* **46**: 1–16.

65. Tashiro, Y., Otsuki, E. & Shimadzu, T. (1972) Sedimentation analyses of native silk fibroin in urea and guanidine HCl. *Biochim. Biophys. Acta* **257**: 198–209.

66. Gamo, T., Inokuchi, T. & Laufer, H. (1977) Polypeptides of fibroin and sericin secreted from the different sections of the silk glands in *Bombyx mori*. *Insect Biochem.* **7**: 285–295.

67. Sprague, K.U. (1975) The *Bombyx mori* silk proteins. Characterization of large polypeptides. *Biochemistry* **14**: 925–931.

68. Candelas, G.C. & Cintron, J. (1981) A spider fibroin and its synthesis. *J. Expt. Zool.* **216**: 1–6.

69. Lombardi, S.J. & Kaplan, D.L. (1990) Recombinant spider silk proteins through genetic engineering. *U.S. Patent Appl.*

70. Shimura, K., Kikuchi, A., Ohtomo, K., Katagata, Y. & Hyodo, A. (1976) *Studies on silk fibroin of Bombyx mori.* I. Fractionation of fibroin prepared from posterior silk gland. *J. Biochem.* **80**: 693–702.

71. Yamaguchi, K., Kikuchi, Y., Takagi, T., Kikuchi, A., Oyama, F., Shimura, K. & Mizuno, S. (1989) Primary structure of the silk fibroin light chain determined by cDNA sequencing and peptide analysis. *J. Mol. Biol.* **210**: 127–139.

72. Work, R.W. (1984) Duality in major ampullate silk and precursive material from orb-web-building spiders (Araneae). *Trans. Am. Microsc. Soc.* **103(2)**: 113–121.

73. Braunitzer, V.G. & Wolff, D. (1955) Vergleichende chemische unterschungen über die fibroine von *Bombyx mori* und *Nephila madagascariensis*. *Z. Naturforscb.* **10(b)**: 404–412.

74. Tillinghast, E.K. (1984) The chemical fractionation of the orb web of *Argiope* spiders. *Insect Biochem.* **14(1)**: 115–120.

75. Tillinghast, E.K. & Kavanagh, E.J. (1977) The alkaline proteases of *Argiope* and their possible role in web digestion. *J. Exp. Zool.* **202**: 213–222.

76. Lombardi, S.J. & Kaplan, D.L. (1990) The amino acid composition of major ampullate gland silk (dragline) of *Nephila clavipes* (Araneae, Tetragnathidae). *J. Arachnol.* **18(3)**: 297–306.

77. Morimoto, S. (1970) Silk-like fiber K-6 (chinon). *Industr. Eng. Chem.* **62(3)**: 23–32.

78. Work, R.W. (1981) A comparative study of the supercontraction of major ampullate silk fibers of orb-web-building spiders (Araneae). *J. Arachnol.* **9**: 299–308.

79. Work, R.W. (1985) Viscoelastic behaviour and wet supercontraction of major ampullate silk fibres of certain orb-web-building spiders (Araneae). *J. Exp. Biol.* **118**: 379–404.

80. Work, R.W. & Morosoff, N. (1982) A physico-chemical study of the super-contraction of spider major ampullate silk fibers. *Tex. Res. J.* **52**: 349–356.

81. Lucas, F., Shaw, J.T.B. & Smith, S.G. (1962) Some amino acid sequences in the amorphous fraction of the fibroin of *Bombyx mori*. *Biochem. J.* **83**: 164–171.

82. Nadiger, G.S., Bhat, N.V. & Padhye, M.R. (1985) Investigation of amino acid composition in the crystalline region of silk fibroin. *J. Appl. Poly. Sci.* **30**: 221–225.

83. Lucas, F. & Rudall, K.M. (1968) Extracellular fibrous proteins: the silks. In *Comprehensive Biochemistry* (eds Florkin, M. & Stotz, F.H.), Ch. 7, 126(B), pp. 475–558. Elsevier, Amsterdam.

84. Strydom, D.J., Haylett, T. & Stead, R.H. (1977) The amino-terminal sequence of silk fibroin peptide Cp – a reinvestigation. *Biochem. Biophys. Res.*

Commun. **79(3)**: 932–938.

85. Work, R.W. & Young, C.T. (1987) The amino acid compositions of major and minor ampullate silks of certain orb-web-building spiders (Araneae, Araneidae). *J. Arachnol.* **15**: 65–80.

86. Peakall, D.B. (1964) Composition, function and glandular origin of the silk fibroin of the spider *Araneus diadematus* Cl. *J. Exp. Zool.* **156**: 345–352.

87. Peakall, D.B. (1968) The silk glands. In *A Spider's Web* (eds Witt, P.N., Reedy, C.F. & Peakall, D.B.). Springer, Berlin.

88. Peakall, D.B. (1966) Regulation of protein production in the silk glands of spiders. *Comp. Biochem. Physiol.* **19**: 253–258.

89. Ohshima, Y. & Suzuki, Y. (1977) Cloning of the silk fibroin gene and its flanking sequences. *Proc. Natl. Acad. Sci. U.S.A.* **74(12)**: 5363–5367.

90. Laible, R.C. (1980) Fibrous armour. In *Ballistic Materials and Penetration Mechanics* (ed. Laible, R.C.), pp. 73–115. Elsevier, Amsterdam.

91. Ilzuka, E. (1985) Silk thread: mechanism of spinning and its mechanical properties. *J. Appl. Poly. Sci.* **41**: 173–185.

92. Fornes, R.E., Work, R.W. & Morosoff, N. (1983) Molecular orientation of spider silks in the natural and supercontracted states. *J. Poly. Sci.* **21**: 1163–1172.

93. Dong, Z., Lewis, R.V. & Middaugh, C.R. (1991) Molecular mechanism of spider silk elasticity. *Arch. Biochem. Biophys.* **284**: 53–57.

94. Kerkam, K., Viney, C., Kaplan, D.L. & Lombardi, S.J. (1991) Liquid crystalline characteristics of natural silk secretions. *Nature* **349**: 596–598.

95. Bell, L.A. & Peakall, D. (1969) Changes in fine structure during silk protein production in the ampullate gland of the spider *Araneus serricatus*. *J. Cell. Biol.* **42**: 285–295.

96. Tillinghast, E.K., Chase, S.F. & Townley, M.A. (1984) Water extraction by the major ampullate duct during silk formation in the spider *Argiope aurantia* Lucas. *J. Insect Physiol.* **30(7)**: 591–596.

97. Tillinghast, E.K. (1986) The gut, stercoral pocket, and hemolymph as potential storage sites of K^+ and phosphate for the spider's orb web. *Comp. Biochem. Physiol.* **84A(2)**: 331–334.

98. Magoshi, J., Magoshi, Y. & Nakamura, S. (1985) Physical properties and structure of silk: 9. Liquid crystal formation of silk fibroin. *Poly. Commun.* **26**: 60–61.

99. Magoshi, J., Magoshi, Y. & Nakamura, S. (1985) Physical properties and structure of silk: 10. The mechanism of fibre formation from liquid silk of silkworm *Bombyx mori*. *Polymer Comm.* **26(10)**: 309–311.

100. Li, G. & Yu, T. (1989) Investigation of the liquid-crystal state in silk fibroin. *Makromol. Chem. Rapid Commun.* **10**: 387–389.

101. Yamura, K., Okumura, Y., Ozaki, A. & Matsuzawa, S. (1985) Flow-induced crystallization of *Bombyx mori* L. silk fibroin from regenerated aqueous solution and spinnability of its solution. *J. Appl. Polym. Sci.* **41**: 205–220.

102. Asakura, T., Kashiba, H. & Yoshimizu, H. (1988) NMR of silk fibroin. 8. ^{13}C NMR analysis of the conformation and the conformational transition of *Philosamia cynthia ricini* silk fibroin protein on the basis of Bixon-Scheraga-Lifson theory. *Macromolecule* **21**: 644–648.

103. Asakura, T., Yoshimizu, H. & Kakizaki, M. (1990) An ESR study of spin-labeled silk fibroin membranes and spin-labeled glucose oxidase immobilized in silk fibroin membranes. *Biotech. Bioeng.* **35**: 511–517.

104. Asakura, T., Demura, M. & Tsutsumi, M. (1988) ^{23}Na and ^{27}Al NMR studies of the interaction between *Bombyx mori* silk fibroin and metal ions trapped in the porous silk fibroin membrane. *Makromol. Chem. Rapid Commun.* **9**: 835–839.

105. Asakura, T., Watanabe, Y., Uchita, A. & Minagawa, H. (1984) NMR of silk fibroin. 2. ^{13}C NMR study of the chain dynamics and solution structure of *Bombyx mori* silk fibroin. *Macromolecule* **17**: 1075–1081.

106. Asakura, T. & Demura, M. (1988) NMR of silk fibroin. 11. ^{1}H NMR analysis of water orientation in porous silk fibroin membrane. *Transaction* **44(11)**: 535–540.

107. Demura, M., Asakura, T. & Kuroo, T. (1989) Immobilization of biocatalysts with *Bombyx mori* silk fibroin by several kinds of physical treatment and its application to glucose sensors. *Biosensors* **4**: 361–372.

108. Demura, M., Asakura, T., Nakamura, E. & Tamura, H. (1989) Immobilization of peroxidase with a *Bombyx mori* silk fibroin membrane and its application to biophotosensors. *J. Biotechnol.* **10**: 113–120.

109. Yoshimizu, H. & Asakura, T. (1990) Preparation and characterization of silk fibroin powder and its application to enzyme immobilization. *J. Appl. Poly. Sci.* **40**: 127–134.

110. Grasset, L., Cordier, D. & Ville, A. (1977) Woven silk as a carrier for the immobilization of enzymes. *Biotech. Bioeng.* **19**: 611–618.

111. Kashiba, H., Asakura, T. & Komoto, T. (1989) Nuclear magnetic resonance study on the relationship between the helicity and L-alanine content of L-alanine/B-alanine random copolypeptides in connection with silk fibroin conformation. *Makromol. Chem.* **190**: 811–818.

112. Saito, H., Ishida, M., Yokoi, M. & Asakura, T. (1990) Dynamic features of side chains in tyrosine and serine residues of some polypeptides and fibroins in the solid as studied by high-resolution solid-state ^{13}C NMR spectroscopy. *Macromolecule* **23**: 83–88.

113. Minoura, N., Tsukada, M. & Nagura, M. (1990) Fine structure and oxygen permeability of silk fibroin membrane treated with methanol. *Polymer* **31**: 265–269.

114. Suzuki, Y. & Brown, D.D. (1972) Isolation and identification of the messenger RNA for silk fibroin from *Bombyx mori*. *J. Mol. Biol.* **63**: 409–429.

115. Suzuki, Y. & Giza, P.E. (1976) Accentuated expression of silk fibroin genes *in vivo* and *in vitro*. *J. Mol. Biol.* **107**: 183–206.

116. Suzuki, Y. & Suzuki, T. (1974) Quantitative measurements of fibroin messenger RNA synthesis in the posterior silk gland of normal and mutant *Bombyx mori*. *J. Mol. Biol.* **88**: 393–407.

117. Ohta, S., Suzuki, Y., Hara, W., Takiya, S. & Suzuki, T. (1988) Fibroin gene transcription in the embryonic stages of the silkworm, *Bombyx mori*. *Dev. Growth Diff.* **30(3)**: 293–299.

118. Suzuki, Y., Gage, L.P. & Brown, D.D. (1972) The genes for silk fibroin in *Bombyx mori*. *J. Mol. Biol.* **70**: 637–649.

119. Garel, J.P., Hentzen, D. & Daillie, J. (1974) Codon responses of tRNA[Ala], tRNA[Gly] and tRNA[Ser] from the posterior part of the silk gland of *Bombyx mori* L. *FEBS Lett.* **39(3)**: 359–363.

120. Sprague, K.U., Hagenbuchle, O. & Zuniga, M.C. (1977) The nucleotide sequence of two silk gland alanine tRNAs: implications for fibroin synthesis and for initiator tRNA structure. *Cell* **11**: 561–570.

121. Gage, L.P. & Manning, R.F. (1976) Determination of the multiplicity of the silk fibroin gene and detection of fibroin gene-related DNA in the genome of *Bombyx mori*. *J. Mol. Biol.* **101**: 327–348.

122. Gage, L.P. (1974) Polyploidization of the silk gland of *Bombyx mori*. *J. Mol. Biol.* **86**: 97–108.

123. Suzuki, Y. & Ohshima, Y. (1977) Isolation and characterization of the silk fibroin gene with its flanking sequences. *Cold Spring Harbor Symp. Quant. Biol.* **42**: 947–957.

124. Tsuda, M. & Suzuki, Y. (1981) Faithful transcription intitiation of fibroin gene in a homologous cell-free system reveals an enhancing effect of 5' flanking sequence far upstream. *Cell* **27**: 175–182.

125. Lizardi, P.M. (1976) The size of pulse-labeled fibroin messenger RNA. *Cell* **7**: 239–245.

126. Tsujimoto, Y. & Suzuki, Y. (1979) Structural analysis of the fibroin gene at the 5' end and its surrounding regions. *Cell* **16**: 425–436.

127. Tsujimoto, Y. & Suzuki, Y. (1979) The DNA sequence of *Bombyx mori* fibroin gene including the 5' flanking, mRNA coding, entire intervening and fibroin protein coding regions. *Cell* **18**: 591–600.

128. Tsuda, M., Ohshima, Y. & Suzuki, Y. (1979) Assumed initiation site of fibroin gene transcription. *Proc. Natl. Acad. Sci. U.S.A.* **76**: 4872–4876.

129. Lizardi, P.M. (1979) Discontinuous translation of silk fibroin in a reticulocyte cell-free system and in intact silk gland cells. *Proc. Natl. Acad. Sci. U.S.A.* **76(12)**: 6211–6215.

130. Ikariyama, Y., Aizawa, M. & Suzuki, S. (1979) Cell-free protein synthesis by posterior silk gland polyribosome. *J. Solid-Phase Biochem.* **4(4)**: 279–288.

131. Greene, R.A., Morgan, M., Shatkin, A.J. & Gage, L.P. (1975) Translation of silk fibroin messenger RNA in an ehrlich ascites cell-free extract. *J. Biol. Chem.* **250(13)**: 5114–5121.

132. Gage, L.P. & Manning, R.F. (1980) Internal structure of the silk fibroin gene of *Bombyx mori*. I. The fibroin gene consists of a monogeneous alternating array of repetitious crystalline and amorphous coding sequences. *J. Biol. Chem.* **255(19)**: 9444–9450.

133. Gage, L.P. & Manning, R.F. (1980) Internal structure of the silk fibroin gene of *Bombyx mori*. II. Remarkable polymorphism of the organization of crystalline and amorphous coding sequences. *J. Biol. Chem.* **255(19)**: 9451–9457.

134. Tsujimoto, Y., Hirose, S., Tsuda, M. & Suzuki, Y. (1981) Promoter sequence of fibroin gene assigned by *in vitro* transcription system. *Proc. Natl. Acad. Sci. U.S.A.* **78**: 4838–4852.

135. Okamoto, H., Ishikawa, E. & Suzuki, Y. (1982) Structural analysis of sericin genes: Homologies with fibroin gene in the 5' flanking nucleotide sequences. *J. Biol. Chem.* **257(24)**: 15192–15199.

136. Tsuda, M. & Suzuki, Y. (1983) Translation modulation *in vitro* of the fibroin gene exerted by a 200-base-pair region upstream from the 'TATA' box. *Proc. Natl. Acad. Sci. U.S.A.* **80**: 7442–7446.

137. Tsujimoto, Y. & Suzuki, Y. (1984) Natural fibroin genes purified without using cloning procedures from fibroin-producing and nonproducing tissues reveal indistinguishable structure and function. *Proc. Natl. Acad. Sci. U.S.A.* **81**: 1644–1648.

138. Suzuki, Y. & Adachi, S. (1984) Signal sequences associated with fibroin gene expression are identical in fibroin-producer and -nonproducer tissues. *Dev. Growth Diff.* **26(2)**: 139–147.

139. Ueda, H., Hyodo, A., Takei, F., Sasaki, H., Ohshima, Y. & Shimura, K. (1984) Sequence polymorphisms in the 5'-upstream region of the fibroin H-chain gene in the silkworm *Bombyx mori*. *Gene* **28**: 241–248.

140. Hirose, S., Tsuda, M. & Suzuki, Y. (1985) Enhanced transcription of fibroin gene *in vitro* on covalently closed circular templates. *J. Biol. Chem.* **260(19)**: 10557–10562.

141. Adams, D.S., Herrera, R.J., Luhrmann, R. & Lizardi, P.L. (1985) Isolation and partial characterization of U1–U16 small RNAs from *Bombyx mori*. *Biochemistry* **24**: 117–125.

142. Young, L.S., Takahashi, N. & Sprague, K.U. (1986) Upstream sequences

confer distinctive transcriptional properties on genes encoding silkgland-specific tRNA^Ala. *Proc. Natl. Acad. Sci. U.S.A.* **83**: 374–378.

143. Suzuki, Y., Tsuda, M., Takiya, S., Hirose, S., Suzuki, E., Kameda, M. & Ninaki, O. (1986) Tissue-specific transcription enhancement of the fibroin gene characterized by cell-free systems. *Proc. Natl. Acad. Sci. U.S.A.* **83**: 9522–9526.

144. Koga-Ban, Y. & Suzuki, Y. (1987) Changes of DNaseI sensitivity of the fibroin and sericin genes during the silkworm development. *Dev. Growth Diff.* **29(4)**: 363–372.

145. Ohta, T. & Hirose, S. (1990) Purification of a DNA supercoiling factor from the posterior silk gland of *Bombyx mori. Proc. Natl. Acad. Sci. U.S.A.* **87**: 5307–5311.

146. Kusuda, J., Tazima, Y., Onimaru, K., Ninaki, O & Suzuki, Y. (1986) The sequence around the 5' end of the fibroin gene from the wild silkworm *Bombyx mandarina*, and comparison with that of the domesticated species, *B. mori. Molec. Gen. Genet.* **203**: 359–364.

147. Couble, P., Garel, A. & Prudhomme, J.-C. (1981) Complexity and diversity of polyadenylated mRNA in the silk gland of *Bombyx mori*: Changes related to fibroin production. *Dev. Biol.* **82**: 139–149.

148. Couble, P.M., Chvillard, M., Moine, A., Ravel-Chapuis, P. & Prudhomme, J.-C. (1985) Structural organization on the P$_{25}$ gene of *Bombyx mori* and comparative analysis of its 5' flanking DNA with that of the fibroin gene. *Nucleic Acid Res.* **13**: 1801–1814.

149. Kimura, K., Oyama, F., Ueda, H., Mizuno, S. & Shimura, K. (1985) Molecular cloning of the fibroin light chain complementary DNA and its use in the study of the expression of the light chain gene in the posterior silk gland of *Bombyx mori. Experientia* **41**: 1167–1171.

150. Mita, K., Ichimura, S., Zama, M. & James, T.C. (1988) Specific codon usage pattern and its implications on the secondary structure of silk fibroin mRNA. *J. Mol. Biol.* **203**: 917–925.

151. Takei, F., Oyama, F., Kimura, K., Hyodo, A., Mizuno, S. & Shimura, K. (1984) Reduced level of secretion and absence of subunit combination of the fibroin synthesized by mutant silkworm, No. 2. *J. Cell Biol.* **99**: 2005–2010.

152. Takei, F., Kikuchi, Y., Kikuchi, A., Mizuno, S. & Shimura, K. (1987) Further evidence for importance of the subunit combination of silk fibroin in its efficient secretion from the posterior silk gland cells. *J. Cell Biol.* **105**: 175–180.

153. Hui, C.-C. & Suzuki, Y. (1990) Homeodomain binding sites in the 5' flanking region of the *Bombyx mori* silk fibroin light-chain gene. *J. Mol. Biol.* **213**: 395–398.

154. Hui, C.-C., Matsuno, K. & Suzuki, Y. (1990) Fibroin gene promotor contains a cluster of homeodomain binding sites that interact with three silk gland factors. *J. Mol. Biol.* **213**: 651–670.

155. Gamo, T. (1982) Genetic variants of the *Bombyx mori* silkworm encoding sericin proteins of different length. *Biochem. Genet.* **20**: 165–177.

156. Ishikawa, E. & Suzuki, Y. (1985) Tissue- and stage-specific expression of sericin genes in the middle gland of *Bombyx mori. Dev. Growth Diff.* **27(1)**: 73–82.

157. Candelas, G.C., Ortiz, N., Ortiz, A., Candelas, T.M. & Rodriguez, O.M. (1984) A novel method of translation in fibroin. In *Growth, Cancer and the Cell Cycle* (eds Shehan & Friedman), pp. 99–106. Humana Press, New Jersey.

158. Candelas, G.C. & Lopez, F. (1983) Synthesis of fibroin in the cultured glands of *Nephila clavipes. Comp. Biochem. Physiol.* **74B(3)**: 637–641.

159. Candelas, G.C., Candelas, T., Ortiz, A. & Rodriguez, O. (1983) Translational pauses during a spider fibroin synthesis. *Biochem. Biophys. Res. Commun.* **116(3)**: 1033–1038.

160. Peakall, D.B. (1964) Effects of cholinergic and anticholinergic drugs on the

synthesis of silk fibroins of spiders. *Comp. Biochem. Physiol.* **12**: 465–470.

162. Candelas, G.C., Ortiz, A. & Molina, C. (1986) The cylindrical or tubiliform glands of *Nephila clavipes. J. Exp. Zool.* **237**: 281–285.

163. Tillinghast, E.K. & Townley, M.A. (1986) The independent regulation of protein synthesis in the major ampullate glands of *Araneus cavaticus* (Keyserling). *J. Insect. Physiol.* **32(2)**: 117–123.

164. Lombardi, S.J. & Kaplan, D.L. (1990) Isolation, cloning, and physiochemical characterization of spider silk from the golden orb-weaver, *Nephila clavipes. Poly. Preprints, Div. Poly. Chem., Amer. Chem. Soc.* **31(1)**: 195–196.

165. Lombardi, S.J. & Kaplan, D.L. (1990) The *Nephila clavipes* major ampullate gland silk protein, amino acid composition and detection of silk gene-related nucleic acids in the genome. *Acta. Zool. Fennica* **190**: (in the press).

164. Lombardi, S.J., Fossey, S. & Kaplan, D.L. (1990) Recombinant spider silk proteins for composite fibers. In *Proc. Am. Soc. Composites 5th Tech. Conf.*, pp. 184–187. Technomic Pub. Co., Lancaster, Pennsylvania.

166. Dong, Z., Xu, M., Middaugh, C.R. & Lewis, R. (1990) Spider silk proteins.

167. Xu, M. & Lewis, R.V. (1990) Structure of a protein superfiber: spider dragline silk. *Proc. Nat. Acad. Sci. U.S.A.* **87(18)**: 7120–7124.
Poly. Preprints, Div. Poly. Chem., Amer. Chem. Soc. **31(1)**: 197–198.

168. Edwards, R.M., Light, J.A. & Nicholson, K. (1988) Improvements in or relating to structural proteins. European Patent Appl. 0 294 979.

169. Petty-Saphon, S. & Light, J.A. (1986) Improvements in or relating to production of silk. UK Patent Application GB 2 162 190 A.

170. Cappello, J. & Ferrari, F.A. (1990) Functionally recombinantly prepared synthetic protein polymer. PCT Patent Appl. WO 9005177.

171. Cappello, J. & Crissman, J.W. (1990) The design and production of bioactive protein polymers for biomedical applications. *Poly. Preprints, Div. Poly. Chem., Amer. Chem. Soc.* **31(1)**: 193–194.

172. Cappello, J., Crissman, J., Dorman, M., Mikolajczak, M., Textor, G., Marquet, M. & Ferrari, F.A. (1990) Genetic engineering of structural protein polymers. *Biotech. Prog.* **6**: 198–202.

173. Cappello, J., Crissman, J., Dorman, M., Mikolajczak, M., Textor, G., Marquet, M. & Ferrari, F.A. (1990) The genetic production of synthetic crystalline protein polymers. In *Materials Synthesis Utilizing Biological Processes* (eds Rieke, R.C., Calvert, P.D. & Alper, M.), Vol. 174, pp. 267–276. Materials Research Society, Pittsburgh, Pennsylvania.

174. Ferrari, F.A., Richardson, C., Chambers, J., Causey, S.C. & Pollock, T.J. (1988) Construction of synthetic DNA and its use in large polypeptide synthesis. International Publication Number WO88/03533.

175. McGrath, K., Tirrel, D.A., Kawai, M., Mason, T.L. & Fournier, M.J. (1990) Chemical and biosynthetic approaches to the production of novel polypeptide materials. *Biotech. Prog.* **6**: 188–192.

176. McGrath, K.P., Fournier, M.J., Mason, T.L. & Tirrell, D.A. (1990) Synthesis and expression of an artificial gene encoding a novel sequential polypeptide. *Poly. Preprints, Div. Poly. Chem., Amer. Chem. Soc.* **31(1)**: 190–191.

177. Creel, H.S., Fournier, M.J., Mason, T.L. & Tirrell, D.A. (1990) Synthesis and expression of artificial genes encoding proteins with repeating -(AlaGly)$_4$ProGluGly- elements. *Poly. Preprints, Div. Poly. Chem., Amer. Chem. Soc.* **31(2)**: 165.

178. Modler, H., Fukushima, Y., McGrath, K.P., Tirrell, D. & Yashima, E. (1990) Synthesis and structural characterization of novel sequential polypeptides. *Poly. Preprints, Div. Poly. Chem., Amer. Chem. Soc.* **31(1)**: 192.

179. Calvert, P.D. (1988) Biological systems: Materials. In *Encyclopedia Materials Science and Engineering, Biological Macromolecules*, pp. 334–339. Pergamon Press, Oxford.

180. Morton, W.E. & Hearle, J.W.S. (1975) Physical properties of textile fibers. The Textile Institute, Heinemann, London.

181. Pearson, W.R., Mukai, T. & Morrow, J.F. (1981) Repeated DNA sequences near the 5'-end of the silk fibroin gene. *J. Biol. Chem.* **256(8)**: 4033–4041.

2 Collagen

Stephen D. Gorham

Johnson and Johnson Medical Biopolymers Group, Scottish Metropolitan Alpha Centre, Stirling University Innovation Park, Stirling FK9 4NF

Collagen as a Biomaterial

Biochemistry of collagen

STRUCTURE

Collagen constitutes more than a quarter of the total protein in animals[1]. It is a fibrous protein conferring many important characteristics on the connective tissues of the body, in particular mechanical properties such as tensile strength and biological properties such as the activation of the blood-clotting cascade.

Collagen derives its name from two Greek words: *kolla* meaning 'glue', and *gennan* meaning 'to produce'. Hence collagen – 'glue producer' or 'former'. Collagen was first recognized as the component of tissue that when boiled produced glue, and was known to the Romans as early as 50 AD when Pliny wrote that 'glue is cooked from the hide of bulls'[2].

Although most of the body 'scaffolding' in mammals is composed of collagen, the properties of the various connective tissues of the body are not determined by collagen alone. While collagen is indeed a major component of the extracellular matrix, the function of the connective tissues depends on the number, nature, relative proportions and interactions of all of their constituent macromolecules, including elastin, proteoglycans, glycosaminoglycans and cell-binding glycoproteins, such as fibronectin and laminin, as well as collagen itself.

Similarly, collagen cannot be considered simply as being a single component; 12 different types have been isolated so far, each one being designated by a Roman numeral[3-7]. However, here the discussion will be limited to the major fibre-forming interstitial collagens, types I, II and III, of which type I is by far the most abundant. Indeed most prosthetic materials for biomedical applications have type I collagen as their major component.

The structure of the major interstitial collagens has been described in several reviews[8-11]. The main feature of the fibrous collagens, types I, II and III, is their stiff triple-helical structure, shown in Fig. 1. Three collagen polypeptides, or α-chains, are wound around each other in a regular helix to generate a rope-like structure approximately 280 nm in

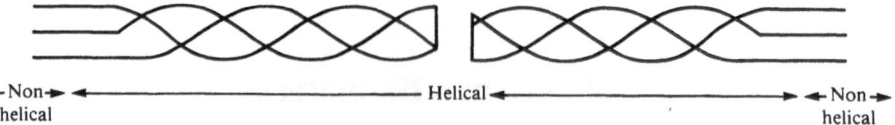

Fig. 1 Triple helical structure of fibre-forming collagen types I, II and III. Each individual α-chain is a left-handed helix, three of which are wound together to form a right-handed triple helix, 280 nm in length and 1.4 nm in diameter. Each α-chain consists of more than 1000 amino acids, with about 9–50 (according to the collagen type and species) being found in the nonhelical terminal regions. Collagen is characterized by its unique amino-acid sequence, a repeat unit of $(Gly-X-Y)_n$ in the triple-helical region. Glycine is the only amino acid small enough to fit into the centre of the triple helix.

length and 1.4 nm in diameter. Each α-chain is a left-handed helix and the three wound together produce a right-handed triple-helical structure. At each end of the collagen molecule is a short nonhelical region known as a telopeptide. Types II and III collagen consist of three identical α-chains known as α1(II) and α1(III), respectively. However, two different α-chains, α1(I) and α2(I), are found in type I collagen. The composition and characteristics of different types of collagen are summarized in Table 1.

Each α-chain consists of more than 1000 amino acids. It is characterized by a unique triple helix-forming sequence in which every third residue is glycine, that is, which is made up of repeat units of the form (Gly-X-Y), where X and Y are any other amino acids. This repeat is not present in the telopeptide regions of the α-chain. Roughly 35% of the nonglycine positions X and Y in the triple helical region are occupied by the alicyclic amino acids proline and 4-hydroxyproline. The structures of these two imino acids are shown in Fig. 2. Proline occurs almost exclusively at the X position and accounts for about 130 residues per 1000, whereas 4-hydroxyproline occurs at the Y position, accounting for about 90 residues per 1000.

These amino acids stabilize the collagen molecule, hydroxyproline having a particularly important role in the hydrogen-bonded structure of the triple helix. Also, because of their alicyclic nature, they 'stiffen' the α-chain where they occur by preventing rotation around the C–N bond (Fig. 3), again conferring unique properties on the molecule.

The simple triple-helical collagen molecule was originally known as tropocollagen, although this term is not now generally used.

Table 1   Different collagen types

Collagen type	Chain composition	Tissue distribution
I	$[\alpha1(I)]_2\alpha2(I)$	Bone, tendon, skin, vessel walls, heart valve, cornea
II	$[\alpha1(II)]_3$	Cartilage
III	$[\alpha1(III)]_3$	Muscle, vessel walls, skin, gut, heart valve (usually coexists with type I except in bone, tendon, cornea)
IV	$[\alpha1(IV)]_2\alpha2(IV)$	Basement membranes
V	$\alpha1(V)\alpha2(V)\alpha3(V)$ or $[\alpha1(V)]_2\alpha2(V)$ or $[\alpha1(V)]_3$	Most interstitial tissues
VI	$\alpha1(VI)\alpha2(VI)\alpha3(VI)$	Most interstitial tissues
VII	$[\alpha1(VII)]_3$	Anchoring fibrils
VIII	$\alpha1(VIII)\alpha2(VIII)$? (chain organization of helix unknown)	Some endothelial cells
IX	$\alpha1(IX)\alpha2(IX)\alpha3(IX)$	Catilage
X	$[\alpha1(X)]_3$	Hypertrophic and mineralizing cartilage
XI	$1\alpha2\alpha3\alpha_1$ or $\alpha1(XI)\alpha2(XI)\alpha1(II)$	Cartilage

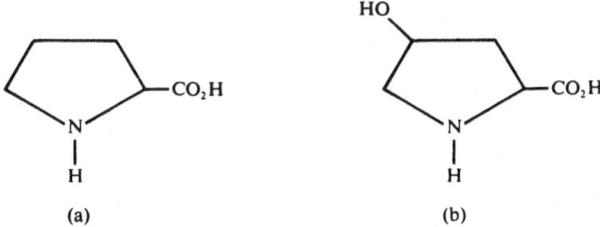

Fig. 2   Structures of (a) proline and (b) 4-hydroxyproline. Approximately 35% of the nonglycine positions in the triple helix are occupied by these two imino acids. Proline is found in position X of the $(Gly-X-Y)_n$ repeat and accounts for about 129 residues per 1000, whereas 4-hydroxyproline is found exclusively in position Y, accounting for about 92 residues per 1000.

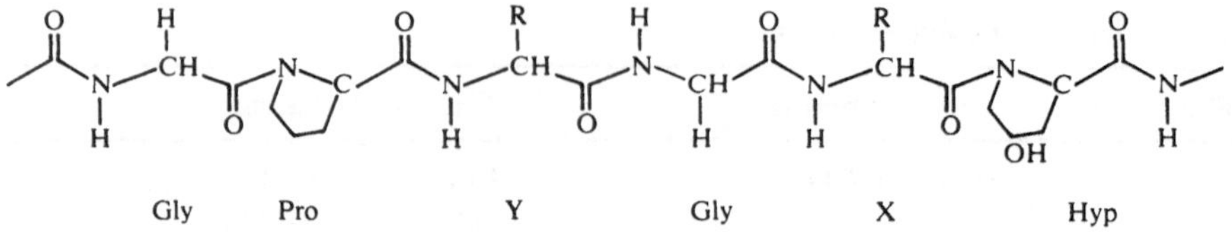

Biomaterials

![Chemical structure diagram showing collagen α-chain with labeled residues]

Gly   Pro   Y   Gly   X   Hyp

Fig. 3 Incorporation of proline and hydroxyproline into collagen α-chains (hydrogen-bonded structure and α-helix not shown).

Tropocollagen molecules can aggregate in a characteristic quarter-stagger arrangement (Fig. 4) to form long collagen fibrils 10–300 nm in diameter that are further stabilized by inter- and intramolecular cross-links[12]. Fibrils can associate to form fibres, which in turn can form much larger fibre bundles. The fibres and fibre bundles may also be stabilized by intermolecular crosslinks. Hence, through specific aggregation and crosslinking, types I, II and III collagen can form fibres of unusual strength and stability.

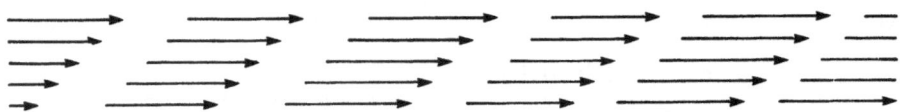

Fig. 4 Schematic representation of the arrangement of collagen molecules within fibrils. Individual tropocollagen molecules are arranged in an overlapping array with a stagger of 64 nm from one end of one molecule to that of the adjacent molecule.

BIOSYNTHESIS OF COLLAGEN

The biosynthesis of collagen has been extensively studied; here a brief account will suffice. For more detailed information refer to refs 13–15.

The biosynthesis is a complex multistep process (Fig. 5). Before secretion from the cell, collagen is synthesized (mainly in fibroblasts) in a precursor form known as procollagen, which contains both amino- and carboxy-terminal extension peptides[16–18].

During translation of the corresponding messenger-RNA to form the pro-α-chains, three important events take place in the rough endoplasmic reticulum[13]. First, many of the proline residues are hydroxylated by the enzyme 4-prolylhydroxylase to form 4-hydroxy-

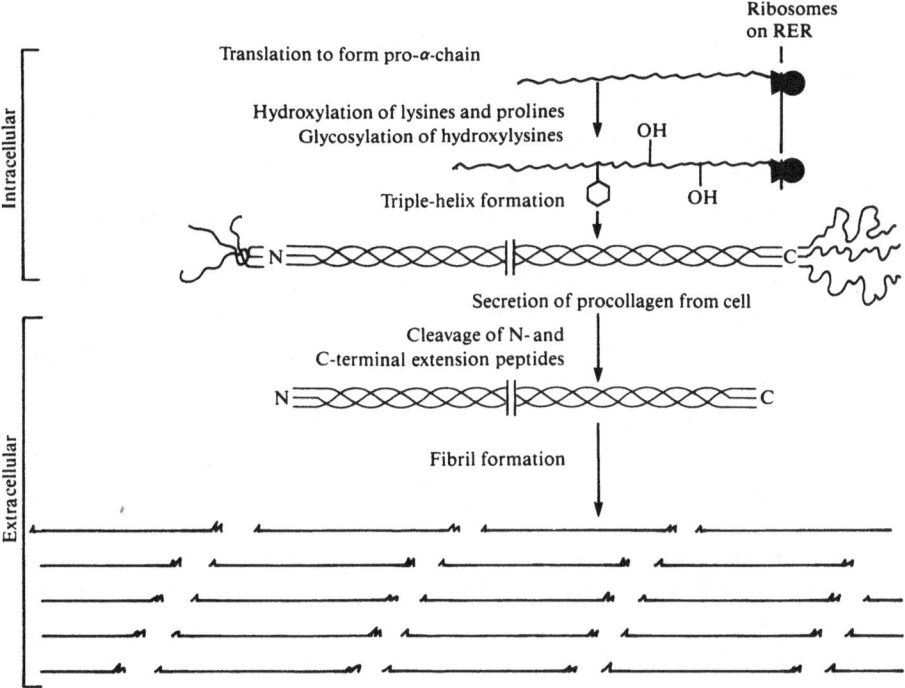

Fig. 5 Biosynthesis and processing of procollagen chains.

proline. Second, some of the lysine residues are hydroxylated by lysyl hydroxylase, with only a few of those at the Y position of the (Gly-X-Y)–sequence becoming so[9,19]. Third, some hydroxylated lysine residues are glycosylated in a process involving two glycosyltransferase enzymes. The first enzyme transfers a galactose residue to hydroxylysine, while the second can transfer a glucose residue onto the resulting galactosylhydroxylysine[13]. The second step is not always completed and hence hydroxylysine occurs in three forms, namely unglycosylated, glycosylated with one galactose residue, or glycosylated with the disaccharide to form 1-O-α-D-glycosyl-O-β-D-galactosyl hydroxylysine[20].

During translation of the pro-α-chains, mRNA, triple-helix formation occurs. Interchain disulphide bonds that form in the propeptide regions have been implicated in the correct alignment of the polypeptide chains before triple-helix formation[21-23].

The triple-helical procollagen is processed in the Golgi apparatus and then secreted from the cell by exocytosis[13]. Once outside the cell, procollagen is converted to tropocollagen by two specific procollagen peptidase enzymes that catalyse the cleavage of the extension peptides from the amino and carboxy termini of procollagen[13,18,24]. Recent evidence suggests that these enzymes might work on procollagen while it is still inside the secretory vacuole within the cell[25].

Individual collagen molecules then aggregate spontaneously in quarter-stagger arrangement in head-to-tail parallel bundles to form fibrils. The heads of the collagen molecules are staggered along the length of the fibres, accounting for the 64-nm spacing of the cross-striations in the regular banding pattern seen under the transmission electon microscope[13,26].

After fibril formation, crosslinking occurs (Fig. 6)[12]. First, hydroxy-lysines or lysines are oxidatively deaminated by the enzyme lysyl oxidase to form hydroxyallysine or allysine, respectively. The resulting aldehydic side chains can then react with another hydroxylysine residue to form either dehydrodihydroxylysinonorleucine (dehydro-DHLNL) or dehydrohydroxylysinonorleucine (dehydro-HLNL). Formation of hydroxylysino-5-ketonorleucine via an Amadori rearrangement of dehydro DHLNL can give further stabilization (Fig. 6).

BIOLOGICAL DEGRADATION OF COLLAGEN

The degradation of collagen *in vivo* can be considered from the viewpoint of several biological processes[27]. These are: (1) constructive morphogenesis as in embryonic development, that is, the degrading of old tissue to allow deposition of new (remodelling), or in change of function, such as metamorphosis in amphibians; (2) debridement in the repair or replacement of tissue, where old or damaged connective tissue must be removed in preparation for fresh replacement, such as in wound healing; (3) indiscriminate pathological tissue destruction such as lytic tumour invasion or lung emphysema. The degradation of implanted collagen following the application of prosthetic materials must also be considered (see below).

There are three main groups of enzymes responsible for the degradation of collagen, the metalloproteinases, neutral proteinases and lysosomal cathepsins[13,28]. The triple helix of collagen is particularly resistant to proteolytic attack. Indeed, the only mammalian enzymes shown to attack the intact triple helix are the specific collagenases, zinc-containing metalloproteinases requiring calcium as a cofactor[28,29]. These enzymes act at neutral pH and are therefore suited to the

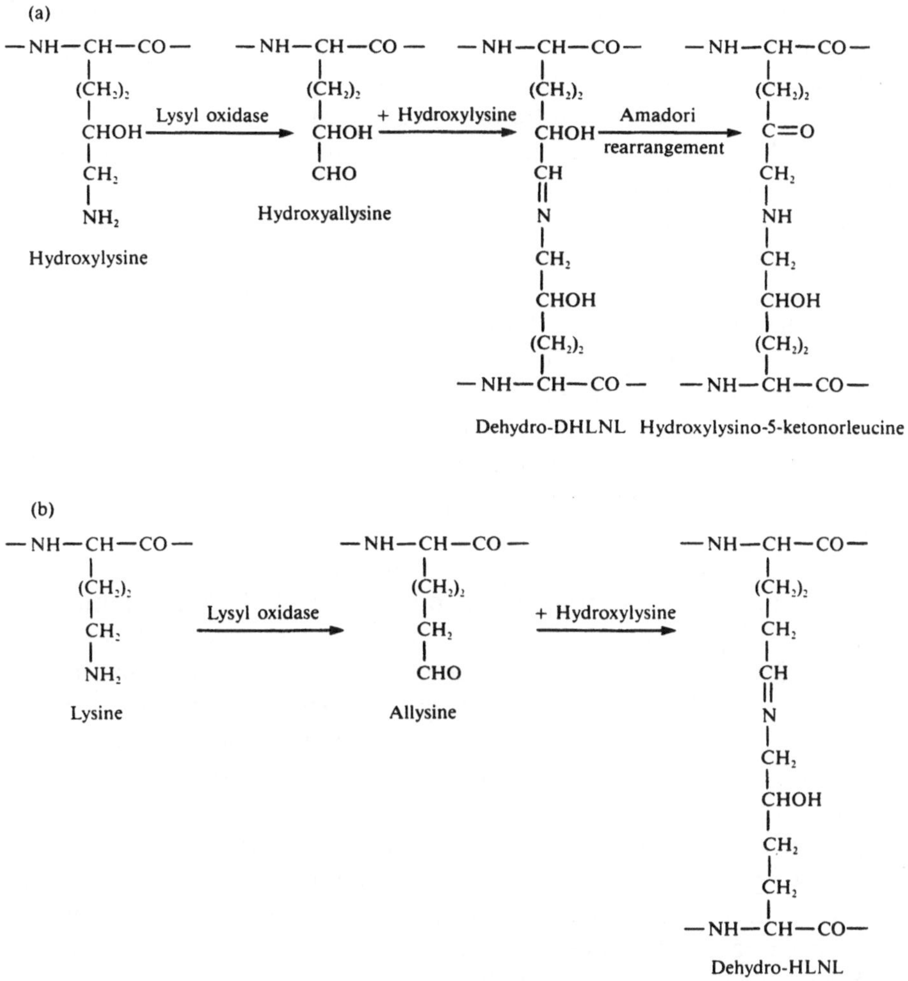

(a)

(b)

Fig. 6 Formation of the naturally occurring reducible crosslinks for (a) hydroxylysine and (b) lysine.

degradation of collagen in the extracellular matrix[28].

The first specific tissue collagenase was found in cultures of tadpole tail tissue[30], and mammalian collagenase activity was later demonstrated *in vivo* after the identification of a specific collagenase in the granules of human polymorphonuclear leukocytes[31,32]. Collagenases have since been shown to be secreted by a number of other cell types, such as fibroblasts and macrophages, as well as by cultures of connective tissues undergoing rapid resorption and pathological breakdown[13,27,28]. Collagenases have a unique specificity, cleaving the $\alpha1(I)$ chain of type I collagen at a specific glycine–isoleucine bond (position 772–773 in the amino-acid sequence), and the $\alpha2(I)$ chain at a glycine–leucine bond in the same position[27,30,33]. Hence the triple helix type I collagen is cleaved at the same locus in all three α-chains, giving two fragments known as TCA and TCB, which are ¾ and ¼ the length of the original molecule, respectively[13,28,29,33,34]. At physiological temperature (37°C), the fragments denature and become susceptible to another group of metalloproteinases, the gelatinases, as well as to nonspecific proteinases[13,28,35]. The breakdown of collagen at neutral pH is illustrated in Fig. 7.

Different collagen types show different susceptibilities to collagenases[28,29,33]. Type II collagen is more resistant to attack by a variety of collagenases than are types I and II. Type III collagen is susceptible to proteinases other than collagenase *in vitro*[10,28,36] and hence may be hydrolysed more rapidly *in vivo*.

Collagenases can be detected in media collected from mammalian connective tissue and cell cultures, but are rarely detectable in tissue extracts. Exceptions are tissues undergoing rapid resorption such as involuting uterus and the specific granule fraction of human polymorphonuclear leukocytes[37]. Both collagenase and gelatinase are secreted by neutrophils in response to soluble stimuli such as formyl-Met-Leu-Phe peptides, Ca^{2+} or ionophore A23187, and it has been suggested[38] that neutrophils may modulate their collagenolytic potential by selective release of collagenolytic protease. The latent forms of collagenase and gelatinase can be activated *in vitro* by trypsin, chaotropic agents or other reagents that can cause a thiol/disulphide exchange reaction[28]. The activation *in vivo* is more complex and takes place by proteolytic cleavage[27,33]. The regulation of the activity of collagenase *in vivo* occurs through mechanisms affecting its synthesis and latency, as well as through the action of naturally occurring inhibitors. These issues have been well reviewed[13,27,28,33,39,40] and will not be discussed further here.

Collagenases can cleave both insoluble and soluble collagen. Because of their triple-helix specificity, however, they cannot cleave the nonhelical telopeptide regions of insoluble collagen, that is the intra- and intermolecular crosslink regions. Thus the collagenases alone are unable to cause solubilization of collagen fibres. The process is therefore facilitated by enzymes that cleave the crosslinks in these regions[13,28] (Fig. 7). Enzymes that are particularly important in this role, acting extracellularly at neutral pH, are elastase and cathepsin G, which are found in the azurophilic granules of polymorphonuclear leukocytes[28]. Elastase solubilizes both insoluble type I and type III collagens, while cathepsin G is more active against type II than type I collagen.[28] Elastase can also catalyse the secondary cleavage of type I collagen in the triple-helical region[28]. There is also evidence that cathepsin G can cleave denatured collagen[41], as well as play a part in the activation of neutrophil collagenase[42].

The degree of crosslinking in collagen fibre affects its rate of degradation. Noncrosslinked reconstituted fibres are degraded more rapidly than native fibrils, while collagens from mature animals are degraded only slowly owing to multivalent crosslinks that accumulate with age[12,13].

In addition to enzymes acting extracellularly at neutral pH, collagen can be degraded by thiol proteinases (cathepsins) at acid pH[13,28,43-45].

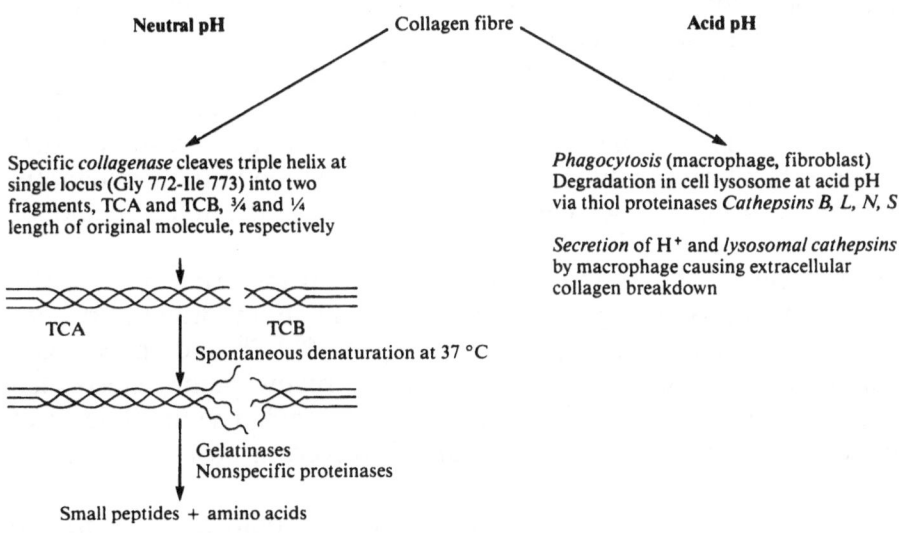

Fig. 7 Biodegradation of collagen.

Studies on implanted collagen sponge[46] indicated the importance of macrophages in the breakdown of the material with subsequent studies[47] showing that the pH of the microenvironment around the surface of activated macrophages could fall to below 5. This environment would favour collagen breakdown by proteinases acting at acidic rather than neutral pH. Further investigations using microelectrodes[43,48] demonstrated that osteoclasts as well as macrophages can create an acidic microenvironment between collagen films and attached cells. Furthermore, the lysosomal thiol proteinase, cathepsin B, can be excreted by the cell into this acidic medium[49], thus creating the ideal conditions for extracellular breakdown at acid pH. Acid proteases can also degrade collagen-fibre fragments or debris that has been phagocytosed and then incorporated into lysosomes[13]. The degradation of collagen at acid pH is shown in Fig. 7.

To date, cathepsins B, L, N and S have been shown to degrade collagen fibres at acid pH[43–45]. Cathepsins L and N solubilize collagen more effectively than do cathepsins B and S[44], collagenolysis being most rapid around pH 3.5[43]. Collagen fibres are densely packed at neutral pH but swell in acid, so at the lower pH range of the cathepsins' activity, the enzyme-sensitive bonds in the nonhelical domains become exposed and accessible to the enzymes[50]. Because the nonhelical terminal regions contain the intermolecular crosslinks that stabilize the fibres *in vivo*[12], the cathepsins function initially as depolymerases, causing further degradation only after the released monomers denature to gelatin.

Cathepsins B, L, N and S seem to have an important role in the biodegradation of collagen *in vivo*. They play an important part both in the degradation of implanted collagen[43,46,47,50] and in collagen remodelling[51] as well as in some cases of destructive pathogenesis such as the arthritic process[52–54] and tumour invasion[55]. The breakdown of connective tissue by cathepsins is also involved in postpartum involution[56].

Collagen breakdown *in vivo* is therefore a complex multienzyme process involving the metalloproteinases (collagenase and gelatinase) assisted by nonspecific proteinases and the serine proteinases elastase and cathepsin G acting extracellularly at neutral pH, as well as the thiol proteinases (cathepsins B, L, N, S) acting both intra- and extracellularly at acid pH. All classes of collagenolytic enzymes play a major role in the remodelling of collagen and in the healing process, as well as in the pathogenic destruction of connective tissues in such conditions as rheumatoid arthritis, lung emphysema and tumour invasion[28,53–55,57,58]. But during a healing or pathogenic process, it is not always clear which is the predominant or preferred mechanism of collagen degradation.

Collagen in Clinical Medicine

Introduction

Collagen has many features that make it an excellent naturally occurring biomaterial. Indeed, the use of collagen in medicine can be traced to the physician Galen in AD 175 who used absorbable catgut sutures[59]. Among the many useful features of collagen are its high tensile strength, controllable biodegradation, haemostatic properties[60], low antigenicity[60], low inflammatory and cytotoxic responses[61], and ability to promote cellular growth and attachment[62–68].

Collagen can be readily reconstituted into a number of forms such as sheets, tubes, sponges, powders and fleeces, all of which have found use in medical practice[1,60,69–72]. The fibrous protein has been used in haemostasis[60,73,74], nerve regeneration[75], tissue augmentation[76–77], burn and wound dressings[78–81], hernia repair[82,83], urinary tract surgery[84–87], drug delivery systems[60,88–92], occular surfaces[93], ophthalmology[90,94–97], a vaginal contraceptive barrier[98], bioprosthetic heart valves[99], vascular grafts[100,101], periodontology[91,92,102–105], thoracic surgery[60,106], repair of the duramater in neurosurgery[60,107–109], abdominal wall surgery[60,82] and surgical sutures[60]. A summary of the different physical forms of collagen and their clinical applications is given in Table 2.

Sources and preparations of collagen biomaterials

Although collagen is found almost everywhere in the mammalian body, those tissues particularly rich in fibrous collagen, such as skin and tendon (where type I collagen predominates), are generally used as starting materials for the preparation of reconstituted collagen for use in implants and dressings. Gut mucosa, however, is frequently used in the manufacture of surgical sutures[60].

The several physical forms of collagen all have real or potential uses in clinical medicine. Of these, films, sponges and injectables tend to be

Table 2 Physical forms of collagen and their clinical applications[60]

Form of collagen	Examples of application
Solution	Vehicle for drug delivery
	Injectable
Gel	Vitreous body
	Injectable
	Drug delivery
Powder	Haemostat
Film	Corneal replacement, corneal shield
	Dialysis membrane
	Wound dressing
	Patches (aneurysm, bladder)
	Abdominal wall
	Periodontal treatment
	Drug delivery
Sponge	Wound dressing
	Burn dressing
	Drug delivery
	Haemostat
Tube	Vessel prostheses
	Reconstruction of hollow organs (oesophagus, trachea)

the most frequently used. The preparation of the various forms of collagen such as porous sponges and porous sponge composites[60,66,69,78,110], films[111–113] and injectables[77,114–117], is well documented.

Porous collagen sponges are generally prepared from an acid-swollen collagen suspension which can be blended into a slurry and freeze-dried. The porosity of the sponge can be altered by varying the collagen concentration within the slurry. Composite sponges containing other materials such as glycosaminoglycans are usually prepared by adding a solution of the polysaccharide to the slurry and then by mixing further before freeze-drying[110].

Collagen films and composite films are most readily prepared in a similar way to sponges except that the slurries are allowed to dry in air (as opposed to being freeze-dried) to form a film. Alkali- as well as acid-swelling of the collagen can also be used[112].

Collagen as an injectable material is prepared from bovine corium by pepsin digestion to remove the terminal nonhelical peptides. The resulting material is further purified by precipitation, and supplied as a sterile suspension in buffered saline at concentrations of $35\pm5\,\mathrm{mg\,ml^{-1}}$ and $65\pm5\,\mathrm{mg\,ml^{-1}}$ (ref. 117).

Fate and degradation of implanted collagen

Having discussed the various mechanisms by which collagen fibres can be degraded *in vivo*, we now examine the fate and breakdown of implanted collagen. Clearly, this is related to the various cell types and the mechanisms by which they penetrate and subsequently degrade the prosthesis following implantation.

When collagen is implanted as a gel, the various inflammatory cell types migrate differently into its matrix. For example, fibroblasts adhere to the collagen fibres and cause considerable contraction of the gel[118,119], a factor which must be taken into account when using this form of collagen to correct facial defects. Certain types of epithelial cells can also cause contraction but to a much lesser extent[118], whereas little or no contraction is found with macrophages[118]. Polymorpho-nuclear leukocytes (neutrophils) adhere to collagen only poorly and their invasion of the gel is largely independent of adhesion, being more dependent on physical parameters such as gel porosity[120]. Although neutrophils do not seem to move through a collagen gel by proteo-lysis[120], they do have an important role in secreting collagen-degrading enzymes, as do macrophages[13,27,28].

Infiltrating inflammatory cells such as macrophages and neutrophils are intimately involved at sites where collagen is degraded *in vivo*[27,28]. Fibroblasts are also involved at sites of tissue repair and injury, and it is known that these cells produce both specific collagenase and collagen-degrading enzymes[27,33,121]. Collagen degradation by all these cell types can occur either after phagocytosis of partially degraded fibres or by extracellular proteases acting at either neutral or acid pH[13,27,28,43,44,45,121].

When collagen sponge is applied to wounds, cellular ingrowth relates not only to its basic composition (for example, the presence of glycosaminoglycans) but also to its physical attributes, such as porosity[67,69,70,78,80,110,122,123]. It has been suggested therefore[123] that cellular infiltration into the implant can be increased by controlling the pore size and channel content, factors that have already been taken into consideration in the design of artificial skins and collagen sponges for use as burn dressings[69,78–81,122].

Collagen films are different from sponges in that, when prepared from acid- or alkali-swollen suspensions of collagen that have been allowed to dry in air, they tend to have a laminated structure rather than an open porous one[112]. After implantation, cells such as neutrophils, macrophages and fibroblasts, penetrate the film between the layers and cause the layers to come apart, thereby producing a 'ribboning' effect at their ends. Cells can then presumably degrade the implanted film by both phagocytosis and secretion of lytic enzymes. Such films, not being chemically crosslinked, are degraded within 2–7 weeks when implanted into the lumbar muscle of rats[112].

The influence of chemical crosslinking on the degradation of implanted collagen has been investigated in studies[124–127] involving the subcutaneous implantation of collagen purified by incubation with trypsin to remove noncollagenous components. With noncrosslinked metarial, as much as 35% of the original collagen remained at 20 weeks[126]. With crosslinked collagen implants, however, grafts crosslinked with glutaraldehyde maintained their collagen mass over 22 weeks, whereas control, noncrosslinked or formaldehyde-crosslinked implants lost about 30% of their mass over the same period[127]. Some temporary inhibition of breakdown was observed with the formaldehyde-treated material[127], showing the relative instability of this type of crosslink. Crosslinked collagen also has a high resistance to degradation by bacterial collagenase *in vitro*[124,127].

A trypsinized, glutaraldehyde-crosslinked dermal collagen of human origin has been implanted into the dermis/subdermal fat layer of the forearm or abdomen of human subjects[128]. After one month, there was no sign of an inflammatory response or rejection. Even after 12 months there was little reaction or sign of encapsulation. The bulk of the implant was still present but was colonized by fibroblasts, demonstrating the resistance of the crosslinked collagen to biodegradation.

So the fate of implanted collagen depends on its physical form, porosity and degree of crosslinking. Also of importance are its site of implantation and intended use, as well as the individual, animal or human, into which it is introduced. More specific examples of the fate of implanted collagen are given in the sections describing its various surgical applications.

Antigenicity of collagen

At one time it was believed that collagen was nonantigenic. Although it is still considered to be a poor immunogen, animals can produce antibodies to a number of different sites in the collagen molecule[26]. In type I collagen, for example, three classes of antigenic determinants have been described that are recognized to different degrees in different species[129]. These are essentially: (1) helical, conformation-dependent antigenic determinants; (2) terminal, nonhelical antigenic determinants; and (3) central antigenic determinants exposed only after denaturation of the collagen molecule[26,129]. The generally low antigenicity of implanted collagen in the form of films and sponges[60] is increased when it is labelled with molecules such as fluorescein, which is used to visualize implanted collagen histologically[46]. Furthermore, the antigenicity of implanted collagen is altered by chemical cross-linking[130]. The selective removal of the amino- and carboxy-terminal nonhelical regions of the collagen molecule reduces its anti-genicity[131–133]. Pepsin treatment has been particularly used in the manufacture of injectable collagen preparations for correction of soft-tissue defects (see below)[114,117].

Generally, there is little adverse reaction to injectable collagen, about 3% of patients showing a positive skin test before treatment and about 1–5% showing delayed immunogenic effects after treat-ment[117,134–137]. Immune responses to injectable collagen appear clinically as erythema, swelling, pruritus and induration[137], and histo-logically as granulatomous responses[117,137,138]. They tend to be localized reactions that occur within the first two exposures to implanted collagen[117,139]. Circulating antibodies have also been detec-ted in patients showing an adverse reaction to injectable collagen[136,140], as well as in a few individuals who appear to be unaf-fected by the implant and in untreated controls[140].

The immune responses to several collagen-based materials designed for the augmentation of soft tissue have been investigated[114]. The materials compared were Zyderm collagen implant (ZCI), GAX (glutaraldehyde-crosslinked injectable collagen), Koken atelocol-lagen, Gelfoam gelatine powder, Avitene microfibrillar collagen hae-mostat and Collastat collagen haemostat. Guinea-pigs treated with ZCI showed significantly lower antibody titres than those treated with the other materials. Antibody activity against GAX collagen and Gelfoam were not significantly different. Collastat elicited antibodies with the greatest affinity, whereas sera from Avitene-treated animals had the highest antibody titres. Avitene was, therefore, the most immunogenic of the materials tested, ZCI the least. GAX collagen also showed minimal antigenicity as well as the greatest persistence.

Thus collagen is a mildly antigenic fibrous protein, making it eminently suitable for use in new, implantable and injectable biomaterials. Its antigenicity can be further reduced by selective removal of the terminal nonhelical peptides by digestion with pepsin.

Sterilization of collagen biomaterials

Heat sterilization in a moist autoclave is totally unsuitable for collagen because of thermal denaturation[60]. If collagen is thoroughly dried before heating then its stability is increased and sterilizing temperatures may be applied[60]. Under these conditions, however, crosslinking occurs[110,141], as does some thermal denaturation (unpublished work from our own laboratories). So heat cannot be recommended for the sterilization of collagen products.

Irradiation at 2.5 millirad is a convenient method of sterilizing collagen[60], causing minimal damage; indeed, little or no damage has been reported for below 10 millirad. However, there is evidence that at these dosages both chain scission of the polypeptide and crosslinking occur[143], the predominating reaction depending on the amount of moisture in the product[144].

Ethylene oxide can also be used to sterilize collagen reliably[60], although it does react with collagen to some extent, particularly with lysine and hydroxylysine side chains. Consistency of treatment with this method is therefore important.

So all the methods commonly used to sterilize medical products will alter collagen in a way that might affect its rate of biodegradation or mechanical strength. Therefore, before choosing between sterilization by irradiation or ethylene oxide, the relative merits of both techniques should be evaluated.

Chemical modification of collagen by crosslinking

Naturally occurring collagen in tissues such as tendon and skin has a high wet tensile strength and resistance to proteolysis. Reconstituted forms of collagen such as films and sponges, however, do not always have the required tensile strength and resistance, and it is thus often necessary to impart these benefits by crosslinking (or 'tanning') before implantation. Indeed, untanned collagen-based implants may have insufficient wet tensile strength and resistance to proteolysis, although their dry tensile strength may be excellent[60]. Lack of wet strength can give rise to difficulties in handling and implantation, for example by causing sponges to collapse[60] and vascular implants to become disrupted[60], or by rendering film soft and tearable on suturing[86,87].

The need to control biodegradation is also important[60]. Implants made with a biodegradable material, such as collagen, must be present long enough for healing to take place, during which time they will gradually lose strength and be absorbed. The rate of biodegradation required will therefore depend on how the implant is used. For example, when used as a haemostat, the collagen has fulfilled its function once the blood clot has formed. However, when the implant is slowly replaced by host collagen, such as in the augmentation of tissue defects or in burn dressings, a much slower rate of absorption is required. Hence either prolonged absorption or premature loss of strength can cause complications[60].

The tensile strength and rate of biodegradation of collagen-based materials may be altered by the formation of ionic bonds, covalent bonds, hydrogen bonds[60], and composites of collagen with other materials[67,78,110]. In the crosslinking of collagen, the functional groups most commonly involved are the carboxyl groups of aspartic and glutamic residues or the ε-amino groups of lysine and hydroxylysine residues. The amino side-chain groups of asparagine and glutamine may also be involved, but reactions involving the hydroxyl groups of serine, threonine and hydroxyproline are less common[60].

Ionic bonds. Trivalent metals such as chromium are used in the tanning of leather[60,145], and chromium has been used as a crosslinking agent in absorbable medical products, chromic catgut surgical sutures being the most obvious example. While chromic sulphate or chloride *per se* are not effective in tanning, the basic salts formed by the addition of alkali cause polymerization of the chromium coordination sphere so that oligomers of chromic complexes joined by sulphate- and oxygen-bridges are formed. These are large enough to bridge the distance between carboxyl groups in the protein and are effective crosslinking agents[60,145]. A typical chromium crosslink is shown in Fig. 8. Aluminium and other polyvalent cations also form ionic bonds with collagen[60], but they provide less stability than do chromic complexes and are therefore less effective as collagen modifiers for biomedical products.

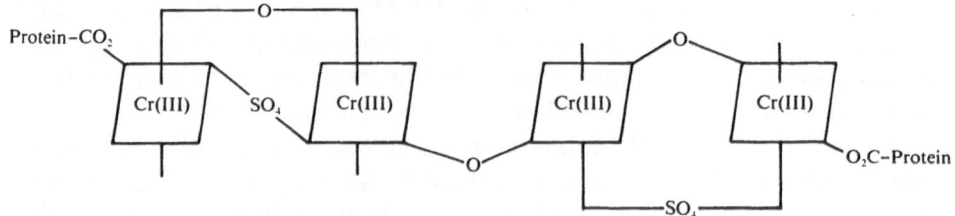

Fig. 8 Schematic representation of chromium complex crosslinking[60,145]

Covalent bonds. Covalent bonds and crosslinks can be readily formed in collagen in a number of ways.

Aldehydes. Formaldehyde, a monoaldehyde, reacts reversibly with the ε-amino group of lysine to form an alkanolamine which further reacts with an amino group of an asparagine or glutamine side chain, giving the crosslink shown in Fig. 9. Tyrosine residues can also add to the adduct via a Mannich reaction (Fig. 10). The reaction of collagen with formaldehyde has been described in more detail by Chvapil *et al.*[60] However, formaldehyde crosslinking is no longer favoured for medical devices. Among other reasons, this is because of the brittleness of the tanned product[60], its reduced stability relative to collagen crosslinked with glutaraldehyde[127,146] and its potential toxicity arising from its ability to leach from implants[146].

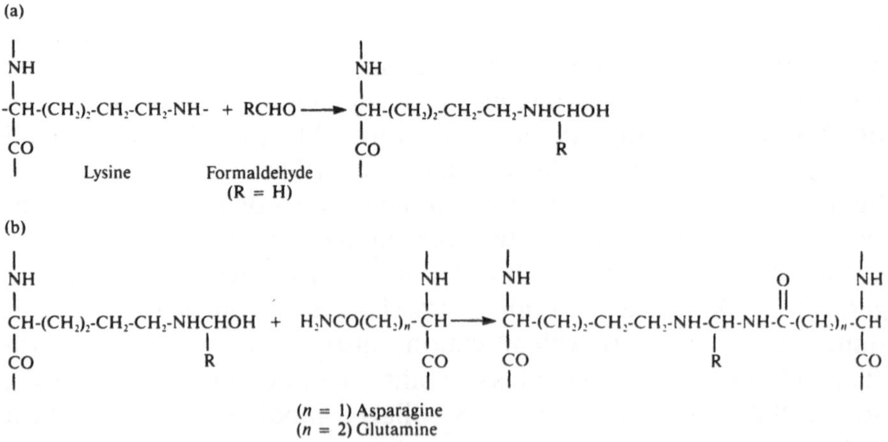

(a)

$$-\overset{\mid}{\underset{\mid}{\text{CH}}}\text{-(CH}_2)_2\text{-CH}_2\text{-CH}_2\text{-NH-} \;+\; \text{RCHO} \longrightarrow \text{CH-(CH}_2)_2\text{-CH}_2\text{-CH}_2\text{-NHCHOH}$$

Lysine   Formaldehyde
(R = H)

(b)

$$\text{CH-(CH}_2)_2\text{-CH}_2\text{-CH}_2\text{-NHCHOH} \;+\; \text{H}_2\text{NCO(CH}_2)_n\text{-CH} \longrightarrow \text{CH-(CH}_2)_2\text{-CH}_2\text{-CH}_2\text{-NH-CH-NH-C-(CH}_2)_n\text{-CH}$$

(n = 1) Asparagine
(n = 2) Glutamine

Fig. 9 Crosslinking of peptide chains with monoaldehydes.

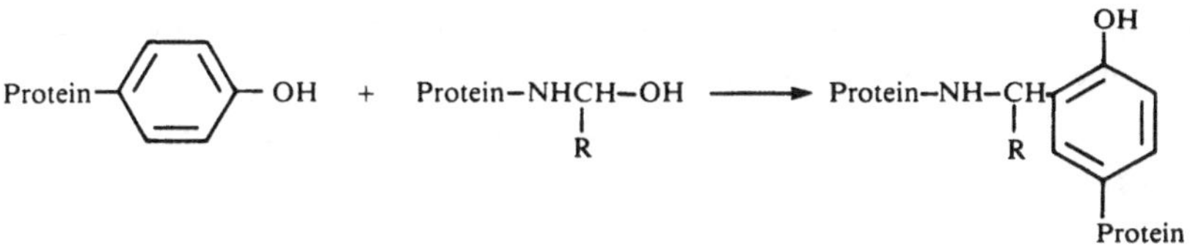

OH

Protein—⬡—OH + Protein—NHCH—OH ⟶ Protein—NH—CH—⬡
                          | |
                          R   R
                                                              Protein

Fig. 10 Reaction of tyrosine residues with an alkanolamine intermediate via a Mannich reaction[60].

Other aldehydes such as glyceraldehyde, glyoxal, acetaldehyde, acrolein, glutaraldehyde and dialdehyde starch can also react with collagen. However, glutaraldehyde is one of the most commonly used fixatives and crosslinking agents in the preparation of collagen-based prostheses[146-155]. Other aldehydes are much less efficient than glutaraldehyde at forming crosslinks, since only a five-carbon dialdehyde can generate ones that are chemically, biologically and thermally stable[146].

Glutaraldehyde crosslinking of collagen has been the subject of several investigations. The complex reaction is dependent on the pH of the reaction and the concentration and purity of the glutaraldehyde[146-160], and is shown in a simplified form in Fig. 11. High concentrations of glutaraldehyde are less efficient than low concentrations, as long polymers of glutaraldehyde can grow on the surface of the collagen fibres and prevent efficient penetration of the molecule

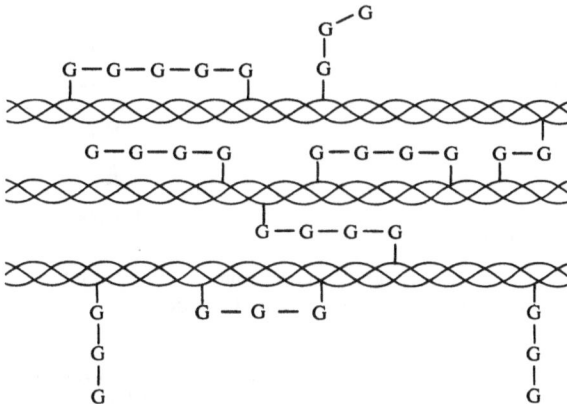

Fig. 11 Crosslinking of collagen with glutaraldehyde. Intra- and inter-molecular crosslinks are shown. G = glutaraldehyde monomer.

into the matrix of the fibres and fibrils themselves[146,154]. The cross-linking reaction is optimal between pH 7 and 8, and requires a free, nonprotonated ε-amino group of lysine for the first addition which forms an imine (Schiff's base).

Carbodiimides. Carbodiimides can be used in the formation of amide linkages between carboxyl and primary amino side groups[146,161,162]. The first stage is the coupling of the carbodiimide to a carboxyl side chain to form an intermediate that is then subjected to nucleophilic attack by a primary amino group to form an amide crosslink (Fig. 12). R and R^1 on the carbodiimide molecule can be chosen so that after the reaction a water-soluble isourea is left which can be easily washed away.

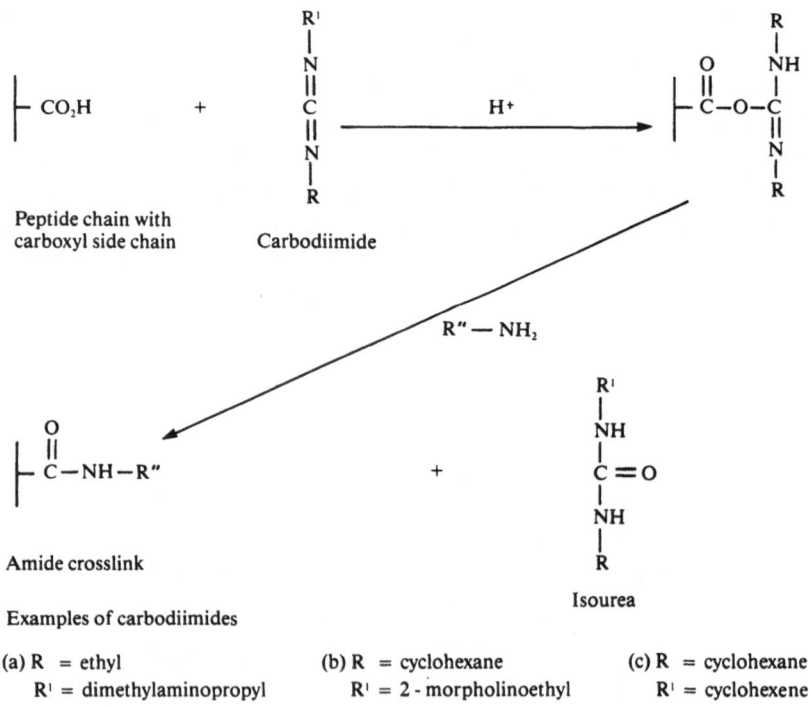

Peptide chain with carboxyl side chain

Carbodiimide

Amide crosslink

Isourea

Examples of carbodiimides

(a) R = ethyl
    R^1 = dimethylaminopropyl

(b) R = cyclohexane
    R^1 = 2 - morpholinoethyl

(c) R = cyclohexane
    R^1 = cyclohexene

Fig. 12 Mechanism of carbodiimide crosslinking. Carbodiimides (a) and (b) give a water-soluble isourea and are therefore suited to aqueous solvents.

Acyl azides. The acyl azide method has recently been used to crosslink collagen-rich tissues such as pericardium[163]. Carboxyl side groups of collagen are transformed into an acyl azide which subsequently reacts with the primary amino group of a lysine or hydroxylysine residue to give an amide-type crosslink (Fig. 13). This method does not have the potential toxic effects associated with glutaraldehyde crosslinking. In addition, acyl azide-crosslinked collagens have similar resistance to digestion by collagenase and cyanogen bromide to those treated with glutaraldehyde[163].

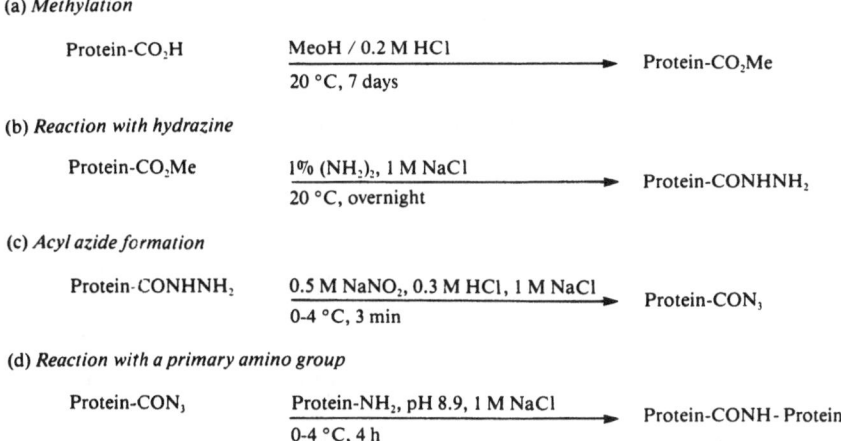

(a) *Methylation*

Protein-CO$_2$H $\xrightarrow[\text{20 °C, 7 days}]{\text{MeOH / 0.2 M HCl}}$ Protein-CO$_2$Me

(b) *Reaction with hydrazine*

Protein-CO$_2$Me $\xrightarrow[\text{20 °C, overnight}]{\text{1\% (NH}_2\text{)}_2\text{, 1 M NaCl}}$ Protein-CONHNH$_2$

(c) *Acyl azide formation*

Protein-CONHNH$_2$ $\xrightarrow[\text{0-4 °C, 3 min}]{\text{0.5 M NaNO}_2\text{, 0.3 M HCl, 1 M NaCl}}$ Protein-CON$_3$

(d) *Reaction with a primary amino group*

Protein-CON$_3$ $\xrightarrow[\text{0-4 °C, 4 h}]{\text{Protein-NH}_2\text{, pH 8.9, 1 M NaCl}}$ Protein-CONH- Protein

Fig. 13 Formation of amide-type crosslinks via an acyl azide intermediate[163].

Isocyanates. Isocyanates can be used to crosslink collagen to produce strong resistant sponges[164,165]. Hexamethylene diisocyanate (1,6-diisocyanatohexane) is the reagent normally used. The first stage is the formation of a urea-type bond after nucleophilic attack by a primary ε-amino group of a lysine residue. The crosslink itself is then formed by the reaction of the second, free, isocyanate group with a further primary amine residue (Fig. 14).

Borohydrides. In skin collagen from young animals, the main intermolecular covalent crosslink is the acid- and heat-labile dehydrohydroxylysinonorleucine[12], which can be stabilized by mild reduction with sodium or patassium borohydride (Fig. 15). In addition, a tetravalent crosslink may be produced, further stabilizing the collagen. The formation of the tetravalent crosslink can be inhibited by buffering at pH 5.5 or by using phosphate (A. J. Bailey, personal communication).

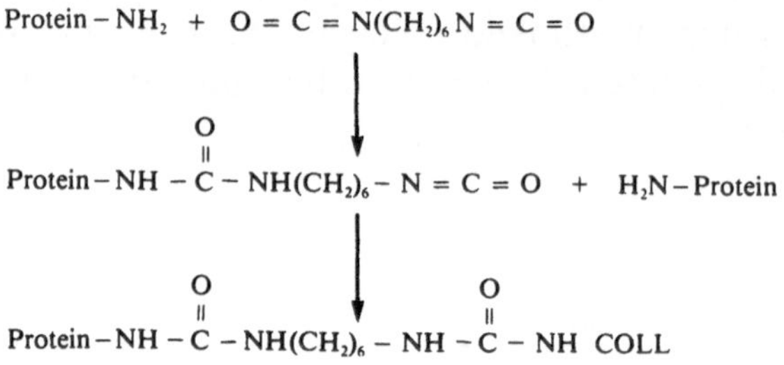

Protein $-$ NH$_2$ $+$ O $=$ C $=$ N(CH$_2$)$_6$ N $=$ C $=$ O

$$\downarrow$$

Protein $-$ NH $-$ $\overset{\overset{\text{O}}{\|}}{\text{C}}$ $-$ NH(CH$_2$)$_6$ $-$ N $=$ C $=$ O $+$ H$_2$N $-$ Protein

$$\downarrow$$

Protein $-$ NH $-$ $\overset{\overset{\text{O}}{\|}}{\text{C}}$ $-$ NH(CH$_2$)$_6$ $-$ NH $-$ $\overset{\overset{\text{O}}{\|}}{\text{C}}$ $-$ NH COLL

Fig. 14 Formation of crosslinks using hexamethylene diisocyanate[164,165].

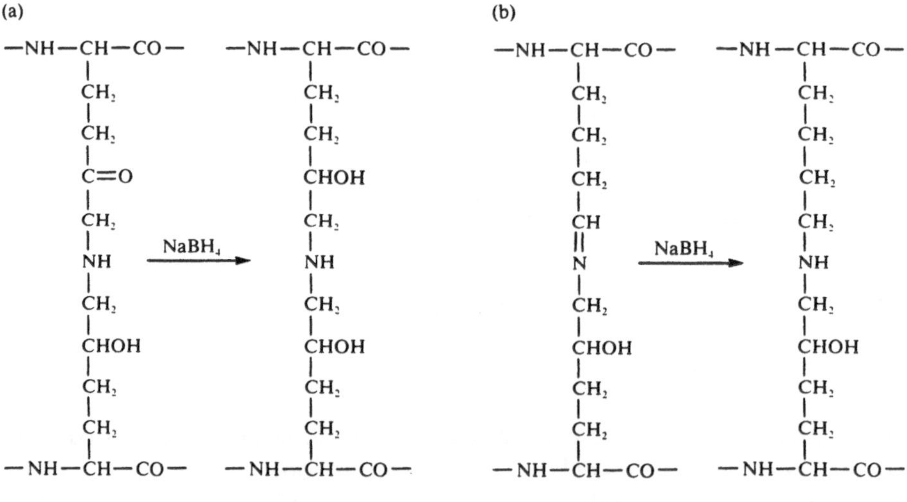

Fig. 15 Reduction of reducible collagen crosslinks with sodium borohydride.

Dehydrothermal crosslinking. When collagen is heated under vacuum, crosslinking can occur[78,110,141,166]. The proposed mechanism of this dehydrothermal crosslinking is amide formation via carboxyl and primary amino side groups[141,166]. It was initially found[166] that heat decreased the solubility of gelatin. This effect was prevented by either methylation of carboxyl side groups (aspartic and glutamic acid) or by acetylation of primary amino groups of lysine residues. It was thus deduced that these side groups are essential for crosslinking via an amide linkage. Esterification via the hydroxyl groups of serine or threonine was considered to be negligible.

This method of crosslinking has since been used in the manufacture of collagen-based dressings[78,110,122] to confer additional stability on the implant and to assist in the binding of glycosaminoglycans. It is important in this type of crosslinking to dry the collagen as thoroughly as possible before heating, as low concentrations of water can cause significant disruption of the helical structure or even hydrolysis at elevated temperature. Dehydration of collagen by gradually increasing the temperature is effective in protecting it against severe degradation[60]. Application of a vacuum assists this dehydration[110], and in doing so helps to drive the formation of amide crosslinks[141]. But even under carefully controlled conditions of dehydration and heating under vacuum, especially at temperatures in excess of 100°C, there is some denaturation, as well as crosslinking (unpublished observations).

Examples of the clinical uses of collagen

COLLAGEN AS A BURN DRESSING

To be effective in the treatment of third-degree burns, any dressing must be able to deal with massive infection and severe fluid loss. Additionally, there may be insufficient healthy skin for grafting over large burns. Patients who overcome these early problems must then face the long-term effects of disfiguring scars and crippling contractures.

The advantages of using collagen as a dressing were pointed out by Chvapil[69,80], who stated that the principal objective in the treatment of burns is to cover large wounds with a substance that has good adherence to a moist wound bed, protects against mechanical and bacterial insults, retains fluid and diminishes pain. Such a substance is collagen sponge, which adheres to the wound bed and, because of its porosity, is actively permeated by inflammatory cells. Furthermore, it is simple to handle, readily sterilized, and provides adequate stimulation of cell reactivity with the formation of a highly vascularized granulation bed.

A considerable step in the development of collagen-based burn dressings was made by Yannas and co-workers[78,79,110,122], who developed a burn dressing that functioned as an artificial skin. The dressing was bilaminar, the outer layer consisting of a silastic membrane and the inner layer of a porous collagen/chondroitin 6-sulphate composite that had been crosslinked dehydrothermally (see above) and then treated with glutaraldehyde. The outer layer acted as a barrier to bacterial invasion and fluid loss, had a similar water vapour transmittance to normal skin, and provided the elasticity and mechanical strength of the dressing. The inner layer, which was placed directly onto the wound surface, had an optimum porosity for invasion by host fibroblasts, that is, it contained pores of 5–150 μm in diameter, with many larger than 40 μm. These fibroblasts colonized the dressing and degraded the collagen/chondroitin 6-sulphate matrix while laying down new host collagen to form a neodermis. After 3–5 weeks, the outer silastic layer could be removed mechanically, and the patient required only a split-skin epidermal graft.

A study[79] of ten patients gave extremely encouraging results with this dressing. Dressings were applied to burn wounds that covered up to 60% of the body surface area. There was no infection in the grafted area, inflammatory or foreign-body response, or rejection of the graft. Additionally, there was no hypertrophic scar formation and only slight wound contracture. The advantages of this skin dressing have been reviewed[81], as have been the advantages of using collagen-based dressings both of the correct porosity and laminated with a semiporous plastic film to prevent moisture loss and bacterial contamination[80].

Autologous epithelial cells have since been incorporated into the collagenous matrix to provide a complete skin substitute[167]. The idea of growing sheets of epithelium *in vitro* from autologous cells harvested from the injured patient to cover burn wounds has been reviewed[81,168], and encouraging results using a sheet of epithelium grown from epithelial cells from cadaver skin have been reported[169]. Living skin-equivalent grafts have been made in rats by casting fibroblasts in collagen lattices and then seeding these with epidermal cells[170]. The material was grafted onto wounds made in the animals from which the cells were taken. The grafts became vascularized, did not evoke a homograft reaction, and inhibited wound contraction. The graft areas, however, remained hairless and smooth.

More recently, keratinocytes have been cultured on a collagen sheet to form a 'quasi' skin used to cover full-thickness defects in rats[171]. The wound was biopsied at 45 and 60 days when complete re-epithelialization had taken place. Barr bodies at the wound site confirmed that quasi-skin keratinocytes were present for up to 60 days, whereas by then the original collagen sheet had completely disappeared. Grafts have been developed[172] by placing confluent, stratified layers of cultured autologous keratinocytes on the surface of a modified collagen/glycosaminoglycan membrane (similar to that developed by Yannas *et al.*[78,110,122] using autologous fibroblasts). These grafts were transferred onto the excised full-thickness wounds of four patients with extensive burn injuries. The composite graft had excellent 'take', with formation of a basement membrane at 9 days together with anchoring fibrils and hemidesmosomes, and showed some epidermal/dermal interdigitation, indicating rete ridge formation. The resulting skin had very good structural integrity 2–3 weeks after grafting. It was suggested that the prior attachment of human keratinocytes to the collagen matrix during culture facilitates the early formation of dermal/epidermal attachment.

The use of keratinocyte-type grafts has been recently reviewed[173]. It was pointed out that their most obvious use is in treating burns that damage over half the body surface area, when too few donor sites are available to cover the wounded area. It is important to consider the condition of the wound bed, a freshly excised wound being preferable to a chronic granulating one, and to ensure that there is no infection. Also, there must be enough time for the growth of sufficient cells to graft onto the patient, especially as autologous cells are needed if the graft is to survive transplantation. Keratinocytes have been grown on a number of collagenous substrates ranging from simple collagen gels to more complex crosslinked matrices.

Whole dermis, both after cryopreservation and as a live cadaveric allograft, has been placed on wound surfaces before application of keratinocyte grafts. Indeed, burn patients have been treated with whole cadaveric allografts after early removal of burn ischar[174,175]. The highly antigenic epidermis of the allograft was removed after 24–30 days before any signs of rejection and when the keratinocyte autografts were ready. The dermabraded bed proved to be ideal for the placement of grafts, giving good take and excellent cosmetic results. DNA analysis more than 5 weeks after keratinocyte grafting suggested that progressive cellular replacement takes place in the dermis[176]. Replacing the dermis by live or cryopreserved human dermis overcomes the need for a substrate of suitable composition and structure for keratinocyte grafts, but introduces the risk of disease transmission, for example via the human immunodeficiency virus[177]. But with adequate screening of donors, this risk should not discourage the use of allograft dermis.

There is, however, some controversy about whether a substrate for keratinocyte grafts is needed at all: a long-term follow-up study[178] of patients showed the eventual regeneration of a normal dermis beneath keratinocyte grafts that had been placed directly onto muscle fascia.

An important aspect of the use of cultured keratinocytes as grafts is that they can synthesize and secrete products such as transforming growth factor-α(TGF-α)[179,180]. These compounds can alter the growth, movement and contraction of wound fibroblasts *in vitro*[181]. The ability to secrete these autologous and parocrine factors survives deep freezing, so the use of cryopreseved keratinocyte allografts taken from storage banks may eventually become widespread[173].

The fate of an artificial skin seeded with the autologous cells has been described in detail[182]. Crosslinked freeze-dried collagen/chondroitin 6-sulphate membranes of the type used as artificial skin dressings by Yannas and co-workers[78,110,122] were prepared and seeded with autologous dermal and epidermal cells. These grafts were placed onto full-thickness defects made on the skin of guinea-pigs, and healing and graft replacement was followed by histology and electron microscopy. After 10 days, more than 50% of the grafts were degraded and there was extensive neovascularization. At 14 days, dermal fibroblasts in the graft site were randomly aligned and only a few had features of myofibroblasts. But in control wounds where no graft was applied, more than 50% of dermal fibroblasts had characteristics of myofibroblasts. At 3 weeks, the graft had been completely resorbed. At 14–17 days, in grafted wound beds only, dermal blood vessels formed a discrete subepidermal plexus orientated parallel to the epidermal plane. Collagen deposition became progressively more randomly oriented at graft sites during the first year, whereas in ungrafted wounds, collagen fibres were aligned in a horizontal plane atypical of a forming scar. After 1 year, the graft sites resembled normal dermis, having well-defined dermal papillae, normal anastomosing superficial vasculature, nerve fibre and random collagen morphology, whereas ungrafted sites resembled a mature scar, having a flattened dermal/epidermal interface, rare and disorganized blood vessels and nerves, and collagen fibres oriented parallel to the epidermis. Thus a highly specific extracellular matrix is important for the induction of true dermal morphogenesis.

Another collagenous structure that is used in the treatment of burns is amniotic membrane[183]. It has been used to treat donor sites, clean shallow burns, freshly excised burn wounds and contaminated open wounds[183]. In donor sites, the rate of epithelialization under amniotic membranes is the same as that with standard dressings of 5% scarlet red ointment and gauze. In burn patients, pain was considerably reduced and the rate of healing was equivalent to that when topical agents or allografts are used as wound dressings. Again, preservation of a healthy excised wound bed and low bacterial count are important. However, practical difficulties, such as the fragile structure of the amnion, may be a drawback to its routine use in the treatment of burns.

COLLAGEN AS A WOUND DRESSING

Collagen sponges, invaluable in the treatment of severe burns, have found use as dressings for many other types of wounds, such as pressure sores, donor sites, leg ulcers and decubitus ulcers[69,164]. They can be crosslinked with glutaraldehyde or hexamethylene diisocyanate[164] to form highly resilient materials that can bind large quantities of fluid. Their porosity can also be controlled to encourage cellular ingrowth[69,164], with pore diameters varying from 8 to 140 μm (average 40 μm).

The advantages of using collagen sponges are that they encourage the formation of new granulation tissue and epithelium on the wound, as well as minimizing contracture and antigenic response. They also show good adherence and conformation to the wound surface, have good handling characteristics and haemostatic properties, and can readily absorb tissue exudate.

Chronic leg ulcers present a surgical problem and are present in pathological conditions such as diabetes and sickle-cell anaemia. In a study of sickle-cell leg ulcers, Reindorf et al.[184] applied the collagen haemostatic dressing Collistat (Hellitrex Inc.) directly to the ulcers of two patients. Dressings were reviewed every 2 weeks and changed every 4 weeks. The collagen matrix caused the chronic ulcers to heal completely in under 3 months, although in one patient a recurrence was seen. It was suggested that even better results with the collagen might be achieved by application of epidermal cells grown *in vitro* to form an epidermal autograft. In an earlier study[185], collagen sponges were applied to six patients with large stasis ulcers of the leg. The formation of granulation tissue appeared to be enhanced, but the formation of epithelium was slow. It was suggested that in this type of application the collagen sponge could be used to carry antibiotics to the wound site.

Armstrong *et al.*[186] compared collagen sponge with Monsel's solution (ferric subsulphate, used to promote haemostasis) in punch biopsies taken from 20 subjects. The collagen sponge had several advantages: there was less inflammation, a lower incidence of wound infection, more rapid formation of epithelium, and a cosmetically better repair.

The effectiveness of collagen sponge on skin wounds has been studied in rats[164]. Collagen sponge, collagen sponge laminated with polyurethane film and conventional gauze dressings were compared in the treatment of (1) superficial wounds, (2) full-thickness wounds, and (3) full-thickness wounds infected with bacteria. The collagen sponge was crosslinked with hexamethylene diisocyanate before use. In all cases, the gauze gave the most problems, adhering to the wound surface. Also, epithelialization of the wounds was slower with this type of dressing. Some adhesion was found with the nonlaminated sponge. In both full-thickness and superficial defects, however, the most encouraging results were obtained with the laminated collagen, both with respect to adhesions and to epithelium formation, which was complete within 5 to 6 days on superficial wounds. This faster epithelialization may have been due to the ability of the laminated sponge to retain fluid, a factor known to favour epithelialization[164], and to the presence of the collagen matrix in contact with the granulation tissue, which probably provides a highly effective cell-supporting medium. In the same study, wound contraction in deep wounds was observed from day 6, irrespective of the dressing, although the nonlaminated collagen produced greater contractures on days 5 and 8. In the deep wounds treated with collagen dressings, thicker granulation tissue and a higher density of microvessels were observed, especially with the laminated material. With bacterial infection of the full-thickness wounds, the most effective reduction in bacteria was obtained with the plain collagen sponge (50,000 times), followed by the laminated material (1,000 times), which gave a significantly greater reduction than did gauze treatment or control treatment in which no dressing was applied. Thus collagen sponge, especially when laminated, would seem to be an effective wound dressing.

Van Gulik *et al.*[165] studied the effects in 40 patients of a sheep split-skin graft crosslinked with either hexamethylene diisocyanate or glutaraldehyde and subsequently sterilized by γ-irradiation. Haemostasis was invariably achieved and there was excellent adherence and reduction in postoperative pain. The graft was simple to use and there was no sign of local intolerance or systemic response. In 35 cases, complete epithelialization was achieved and the mode of crosslinking did not affect the overall performance of the graft. However, the glutaraldehyde-tanned grafts were more flexible and comfortable.

Using a crosslinked collagen sponge containing carbon-fibre electrodes, Dunn *et al.*[187] showed that wound healing can be effected by electrical stimulation. Fibroblast ingrowth and collagen-fibre alignment were increased in the sponge matrix to which direct currents of between 20 and 100 μA were applied. Maximum fibroblast ingrowth into the collagen sponge was observed near the cathode at 100 μA. The anode attracted inflammatory cells and therefore had to be removed from the wound site to prevent tissue destruction. It was concluded that cathodic stimulation using inert carbon electrodes in a collagen sponge could be a promising method of treating chronic dermal wounds.

COLLAGEN COMPOSITE MATERIALS AS DRESSINGS AND IMPLANTS

When designing collagen-based materials either as topical dressings or as implants, the incorporation of other components of the extracellular matrix may be advantageous. Indeed, this has already been successfully done in the porous collagen/chondroitin 6-sulphate composite dressing used as an artificial skin in the treatment of burns[78,110,122]. As well as assisting in delaying contracture in animal wounds[78], coordinating the deposition of organized dermal repair tissue[188] and acting as a support for cells in culture[67,68], collagen matrices containing hyaluronic acid and fibronectin can increase fibroblast proliferation and improve the deposition of organized repair tissue[66,67].

Collagen substrates can influence the growth characteristics of cells[63-65] and also modulate various aspects of cell behaviour, including adhesion and spreading[67,189]. In addition to the cellular response to collagen alone, the various glycosaminoglycans that constitute an essential part of the extracellular matrix influence cell adhesion[68,190,191], migration and mobility *in vitro*[64], as well as the assembly of the collagen molecules themselves into fribrils and fibres[192-196]. Chemical modification of collagen also influences cell attachment and growth [190,191,197], and the presence of hyaluronic acid and/or fibronectin in collagen sponges can enhance the repair of dermal wounds[66].

The effect of glycosaminoglycans on cell adhesion and spreading on collagen film and sponge matrices has been studied in some detail[190,191]. When 2.5% hyaluronic acid (based on the weight of collagen) was incorporated into a collagen film, attachment and growth of mouse L929 fibroblasts was greater than with a simple collagen film, whereas concentrations of hyaluronic acid in excess of 5% inhibited cellular attachment and growth. Conversely, chondroitin sulphate (both 4-sulphate and 6-sulphate isomers) enhanced cellular attachment and growth in almost a linear fashion up to 20% concentration. The reason for this is not clear, but it does indicate that the incorporation of chondroitin sulphate into collagen-based biomaterials might be beneficial. In all cases, cellular attachment and growth was enhanced by the presence of fibronectin. Interestingly, enhanced cellular attachment and growth could be obtained by acetylating the lysine side chains of the collagen molecule.

In a subsequent *in vivo* study, collagen-based films and sponges were implanted into the lumbar muscles of rats[113,190]. The films and sponges were made of native collagen, acetylated collagen, collagen with 10% chondroitin sulphate or collagen with either 2.5% or 20% hyaluronic acid. After 7 days, all materials elicited an acute inflammatory cell response characterized by the presence of numerous neutrophils and histiocytes. Films and sponges followed a similar pattern. The cell population after 14 days was mainly mononuclear, that is, consisted mostly of neutrophils, macrophages, lymphocytes and fibroblasts. Native collagen elicited a subacute inflammatory response after 7 days. However, at 14 days a marked infiltration by neutrophils was apparent and existing collagen material was subsequently degraded. Acetylated collagen film evoked a much greater inflammatory cell response than native collagen. Both collagen/hyaluronic acid composites elicited a similar response. Collagen/10% chondroitin sulphate composites elicited the least inflammatory cell response at 7 days, with infiltration by host fibroblasts clearly seen after 14 days.

The influence of heparin on the wound-healing response to collagen implants has been studied by McPherson *et al.*[198]. With an injectable form of collagen (see below) applied to a rat subcutaneous wound, addition of heparin caused a dose-dependent increase in fibroblast invasion. Furthermore, collagen and collagen/heparin implants were cleared from the subcutis at identical rates. When applied to dermal wounds of guinea-pigs, the collagen implant alone was pushed upwards by developing granulation tissue at the base of the wound, and served as a support for epidermal cell migration, proliferation and differentiation as the wound began to close. The implant was slowly invaded and turned over as granulation tissue developed from the base and margins of the wound bed. The inclusion of heparin in these implants resulted in a significantly different pattern of wound healing. Histologically, the collagen/heparin implants appeared to be more broken-up or porous following implantation, probably owing to the greater penetration of developing granulation tissue into the implant. Radiolabelling showed that again the clearance rates of the collagen implants from wound sites were the same with or without heparin. Laser doppler flowometry suggested that the heparin-containing implants were more vascular than either control wound sites or sites treated with collagen alone.

A collagen/vicryl composite membrane has recently been developed that is strong, leakproof, biodegradable, weakly antigenic, haemostatic, and can be replaced by native host scar collagen[87,199,200]. It is ideal for the repair of internal organs, and has proved to be particularly useful in urology and duramater repair (see below). In addition, this material has found considerable use in other areas of surgery, such as in the protection of cervical anastomoses and those of the small bowel[201,202], as a synthetic tube between a colic transplant and cervical oesophagus[203], in the prevention of air leakage in thoracic surgery[106], and in the repair of abdominal wall defects (Meddings *et al.*, Department of Urology, Royal Infirmary, Glasgow, UK; unpublished results). In all cases, encouraging results have been obtained, and the material may well point to a way forward for the development of other collagen-based bioprostheses.

COLLAGEN AS A HAEMOSTAT

In the initiation of a blood clot, platelets (thrombocytes) adhere to the collagenous surface of an exposed blood vessel or wounded tissue. Following adhesion, the platelets swell, change from being disc- to sphere-shaped, send out long thin pseudopodia, release factors, such as Ca^{2+}, ADP, 5-hydroxytryptamine, prostaglandins, serotonin and thromboxane A2, and initiate the blood-clotting cascade via the extrinsic pathway by activating Hagemann factor (XII). These substances cause other platelets immediately to bind to the initial layer of adhered thrombocytes and to each other, resulting in the formation of a platelet aggregate. After aggregation, platelets secrete part of the contents of their granules and together with fibrin form the tightly fused mass known as the haemostatic plug[13,60,204]. The triple-helix structure of collagen appears to be essential for platelet aggregation[204]. Aggregate activity of the collagen is lost if more than 90% of the free amino groups are blocked by chemical means, whereas free carboxyl groups appear to be of little importance in this respect, although they are critical in the activation of Hagemann factor[60].

Hence reconstituted collagen is a particularly effective haemostat, finding considerable use in biomedicine, usually in the form of a sponge or fine powder[60,69]. Collagen haemostats have also been successfully used in the treatment of injuries to highly vascularized soft tissue such as liver and spleen[205–209].

Collagen sponge and microcrystalline collagen haemostat (MCH) have been successfully used to control bleeding from the palatal donor site in mucogingival surgery[73,210], and microcrystalline collagen has been applied to control bleeding from vascular anastomoses[211], the bleeding time from the suture line being considerably reduced. Microcrystalline collagen has also been used to induce haemostasis following partial hepatectomy[209] and to treat injuries of brain, pancreas, liver and kidney[212–214].

In experiments in which coagulopathy has been induced in dogs by various agents such as warfarin and heparin, MCH was compared with absorbable gelatin powder in its ability to produce haemostasis[215]. The haemostatic potency of MCH was superior to gelatin in both control animals and in those in which coagulopathy was induced. However, a 50% loss of the haemostatic ability of MCH was found in dogs treated with heparin or warfarin compared with control animals.

A fibrin sealant soaked into a porcine collagen sponge has been used to produce haemostasis in experimental rat liver injuries[207], and splenic injuries in humans have been treated using a fibrin adhesive combined with collagen fleece[206]. Encouraging results have also been reported[208] with the use of both fibrin sealants and collagen haemostatic agents in liver transplantation studies in pigs.

In a study of collagen fleece and oxidized cellulose on brain tissue of rabbits[74], collagen proved to be the better haemostat and was resorbed more rapidly than cellulose. Collagen was found to be superior to oxidized cellulose as a haemostat after dental extractions[216]. Both appeared to be effective haemostats initially, but the rate of secondary bleeding 10 minutes after the first arrest was significantly lower with collagen.

Elroy *et al.*[217] recently evaluated *in vitro* haemostatic activity of various topical agents by recording the coagulation time and the generation of fibrinopeptide A. Five commercially available haemostatic agents were compared, of which four were based on collagen, and one on oxidized cellulose. The collagen-based materials were the more effective haemostats.

Clearly, the potent haemostatic ability of collagen means that it must not be allowed to pass into the circulation or lumen of blood vessels: if it is, then embolization could either directly or indirectly cause organ damage[218]. Although the addition of low-porosity filters to devices such as blood-saving systems and pump oxygenators can prevent the passage of over 93% of MCH particles, the filtrate still has platelet-aggregating ability[218]. Hence, once blood has been contaminated with MCH, it should not be returned to a patient's circulation.

Niebauer *et al.*[219] emphasized the potential problems caused by particles of collagen haemostat in the bloodstream when intraoperative autotransfusion and collagen haemostats are used simultaneously to minimize blood loss. For example, during cardiac and organ transplantation surgery, filtering devices are used to collect and return autologous blood to the venous circulation. It was shown that 2% of microfibrillar collagen haemostat particles can pass through the $20\,\mu m$ filter contained in the autotransfusion device. With canine kidney perfusion, multi-focal perivascular inflammatory reactions occurred within the renal parenchyma 5 days after transfusion of filtered autologous blood containing minute amounts of collagen haemostat. The findings demonstrate that a strong inflammatory response to heterologous collagen can result despite the use of a $20\,\mu m$ filter, and that the simultaneous use of intra-operative autotransfusion and collagen haemostats is potentially risky, since adverse reactions can be elicited, especially in the microvasculature of tissues containing end-arterial circulation.

Collagen is, however, an extremely effective haemostat, providing ready haemostasis following injury to highly vascularized soft tissues such as liver, kidney and spleen. In addition to the preparations described in this section, an absorbant and strong crosslinked collagen sponge called Instat (Johnson & Johnson Ltd) should be an excellent haemostat for surgical use.

COLLAGEN AS A VAGINAL CONTRACEPTIVE BARRIER

Because of the high water-binding capacity of collagen sponge, its use as a tampon, and as a vaginal contraceptive barrier in which the sponge can be used to contain a spermicide, has been suggested[60,69,98]. Although initial results were encouraging, clinical studies[98] indicated that as a contraceptive barrier it was inconvenient to both partners, insufficiently effective to compete with other methods of vaginal contraception, and possibly unsafe because of its capacity to support the growth of bacteria. Further testing of collagen sponge for this particular application has been discontinued.

INJECTABLE COLLAGEN FOR SOFT TISSUE AUGMENTATION

Injectable collagen is manufactured essentially in two forms. The Zyderm collagen implant, ZCI, is a neutral sterile suspension of bovine collagen that has been predigested with pepsin to remove the nonhelical terminal regions of the molecule[117]. The physical and rheological properties of this material have been discussed elsewhere[117] and will not be described further. The second injectable collagen, GAX, is similar to ZCI, but is lightly crosslinked with glutaraldehyde to give increased stability and persistence *in vivo*[114–116].

ZCI has been used for several years in the treatment of facial soft tissue deformities, such as small scars and wrinkles[114,117,135,140,220–223], where it is injected under the site intended for augmentation and becomes invaded by host fibroblasts, which in turn lay down new collagen, thus filling the defect.

Multiple treatments are, however, often required and permanent correction is not always achieved[135,221,224]. Histologically, it appears that the injected collagen stimulates a host response resulting in implant degradation and replacement by newly generated host collagen[76]. The fate of the implant has been studied over 16 weeks in three male volunteers. At 5 weeks, degradation of the implant was observed and by 16 weeks bovine collagen was no longer detectable in the dermis by either H&E staining or by immunofluorescence. However, immunofluorescent localization of human type III collagen in the implant site together with metabolically active fibroblasts strongly indicated that host granulation tissue was replacing the bovine collagen with new human collagen[76].

Although factors such as the type of skin, its thickness and the patient's age can affect the outcome of ZCI augmentation[225], the two most important factors are the type of defect and the technique of injection[135,221,224–226]. Injection should be given superficially in the dermis[221,225,226], because when injected subcutaneously and in the fat, ZCI is quickly absorbed[226]. In the case of a scar, it is better to inject the collagen within it than beneath it. An adequate volume must also be injected and 'overcorrection' is always necessary[135,221,225,227], as only 25–30% of the injected volume will remain after the dispersion has condensed. Multiple injections are often necessary to maintain correction of the defect, especially for deeper clefts, and, even then, permanent correction is not always achieved[135,221,224].

Stegman and Tromovitch[228] first reported the use of Zyderm injections and obtained a clinical improvement of 50–80% in scars resulting from acne, varicella, trauma, and in some wrinkles and creases on the face. The need to overcorrect at each injection was stressed. Kamer and Churukian[135] presented a 3-year study of 300 patients treated with ZCI for cutaneous defects arising from acne, trauma or previous surgery. A generally good result was obtained when patients were examined for treatment efficiency and duration. ZCI was found to be easy to use, safe and effective, although multiple injections were often necessary to achieve a lasting effect, especially in acne cases.

In a separate clinical study, 87 patients were treated over 2.5 years with ZCI[224]. Intradermal augmentations were found to be particularly valuable in frown grooves, grooves between cheek, lips and chin, the groove beneath the malar prominence, and the groove just lateral to the labial commissure, as well as for enhancement of malar pads or prominences. Fine lines, postrhinoplastic depressions and punched out depressed scars were less successfully treated. ZCI has also been used to correct depressed scars on the nose following Moh's surgery[220].

Zyderm is not indicated in the treatment of hypertrophic scars, keloids, facial hemiatrophy and depressions in which there is extensive scarring such as ice-pick acne scars[226]. Static scars and lesions without acute inflammatory ongoing disease will show a longer-lasting effect than lesions arising from ongoing stress[226]. Hence sites that are exposed to regular imposed stress such as frowning, smiling or a contracting surgical scar, will probably show a greater loss of correction than sites not under stress[225]. Indeed, Stegman and Tromovitch[228] estimated that corrections of creases and wrinkles lasts for only 6–9 months, whereas much longer correction has been reported for more static scars[135,224,226].

Injectable collagen (ZCI) has also been used in laryngeal rehabilitation. Following a study in dogs[229], ZCI was applied for vocal chord augmentation in patients with glottic insufficiency[230,231]. In a 1-year follow-up, collagen augmentation did not result in a normal voice, but dramatic improvement was achieved in those patients who did not present with appreciable vocal-fold scarring[231].

Kligman and Armstrong[232] compared ZCI with the mildly glutaraldehyde-crosslinked implant Zyplast (ZI) in humans. Biopsies were taken at regular intervals from sites in the dermis of the lower back (both ZCI and ZI) or the dermis of the forearm (ZI only). The fate of the two materials differed in that ZCI was dispersed among host collagen fibres, gradually lost the appearance of a cohesive implant and was absorbed by 90 days, presumably without becoming colonized by host fibroblasts. These observations are somewhat at variance with those reported by Burke *et al.*[76]. By contrast, the cross-linked implant remained as distinct foci for 3–6 months, accompanied by massing of fibroblasts around the periphery. These fibroblasts subsequently invaded the implant and provoked a vigorous inflammatory reaction, which in turn caused the replacement of the implant by new host collagen. These findings indicate that the gutaraldehyde-treated material might provide longer-lasting correction of soft-tissue defects.

The greater persistence of GAX has also been demonstrated in guinea-pigs[115]. Radioactive techniques showed that the half-life for ZCI was 4 days, whereas that for GAX was 21 days.

Stegman *et al.*[223] used both ZCI and ZI in a clinical study. Four patients periodically received ZCI or ZI in pre-auricular and infra-auricular regions of facial skin for between 1 and 9 months. Both materials were identified microscopically in the mid to deep dermis throughout the study, and in 60–70% of the injected sites some material was also present subdermally. A slow, gradual colonization of ZCI by fibroblasts was noted, compared with the delayed intense interaction of these cells with ZI. Also, there was some new collagen deposition associated with remodelling of the ZI, while there was no synthetic activity with ZCI. The results suggested that ZCI and ZI 'migrate' deeper and eventually move into the subcutaneous plane. This movement could explain the loss of correction at 6–9 months that is noted when this implant is used for age-related changes.

CARDIOVASCULAR SURGERY

Collagen-based vascular prostheses have generally consisted of a porous mesh fabric such as Dacron, impregnated with gelatin or collagen, or lined with a continuous collagen tubing[60]. Studies *in vitro* suggest that collagen tubes alone may be compliant but lack suture-holding capacity[100]. Although there have been many successes in this area, one difficulty owing to the thrombogenic nature of collagen has been in the development of collagen-containing substitutes for small vessels[60,233]. Small-diameter vascular prostheses that contain collagen should therefore have an antithrombogenic surface to increase their patency[234]. One method of creating a nonthrombogenic surface is to bind heparin to the collagen[234-236]. There has been concern that the ionic linkage between collagen and heparin might be insufficient to give a lasting effect, and that leaching of this glycosaminoglycan into the bloodstream might occur[234]. However, heparin can be bound covalently to collagen using the carbodiimide reaction, leading to a more stable and durable product[234,235].

In large vessels, the potential thrombogenicity of the prosthetic surface is less important, as the high velocity of blood flow prevents accumulation and growth of surface thrombi; indeed, Dacron prostheses are preclotted to prevent blood loss before application[233]. Collagen can replace the need for preclotting when it is used to seal the pores of knitted grafts, and offers considerable advantages by providing an excellent seal during implantation and acting as a temporary scaffold, promoting cell growth and graft healing[223].

In a comparison of collagen-impregnated Dacron and non-impregnated Microvel grafts implanted into the aortas of dogs[233], it was shown histologically that approximately 40% of the collagen was resorbed after 2 weeks, and that at 4–6 weeks it had been replaced by host tissue between the Dacron fibres. Complete resorption of the collagen was found at 3 months, accompanied by full tissue ingrowth with a smooth neo-intima formation.

In dogs with Microvel grafts, the results were similar to those found with the collagen/Dacron prostheses. However, further experimental work including ultrastructural and morphological analyses of the explants[233] demonstrated that the collagen/Dacron grafts had not only a more uniform coverage of endothelial cells, but also, more significantly, several times the extent of microvessel formation than did the controls.

A calf-skin collagen patch has been applied to the aorta of rats, showing a good long-term patency (85%)[237]. The thrombogenic nature of the collagen is clearly overcome by the haemodynamic conditions in the aorta, which result in little platelet adhesion or aggregation. At the collagen/blood interface, a tissue composed of active myofibroblasts, collagen bundles and elastic fibres progressively developed. After 4 months, nests of fibroendothelial cells were present and, between 6 and 14 months, surface cell differentiation was still incomplete on the bovine collagen patch (although complete on the adjacent aorta), amorphous fibres and fibroepithelial cells coexisting. Heterologous patch debris was found at 14 months, when some calcification was observed. Mineralization of vascular grafts, although not fully under-stood, is known to take place in bioprosthetic heart valves. It can, however, be reduced by crosslinking the collagen[238,239].

The use of monoclonal antibodies to collagen for assessing host–implant interactions has recently been examined[240]. The data, obtained from collagen-based vascular prostheses in dogs, were highly encouraging and allowed the host invasion of a prostheses to be assessed. Host-derived collagens were shown to be present within the prosthetic material. This technique clearly has applications in other areas of surgery.

In extracardial conduits, the porosity of the graft is important. Tightly woven Dacron tends to form a poorly adherent obstructing layer of pseudointima, whereas knitted high-porosity conduits allow fibrous and vascular ingrowth and more secure anchoring of the pseudointima. However, at the time of insertion, the porosity must be sufficiently low to prevent bleeding in the heparinized patient. Jonas *et al.*[241] compared knitted Dacron grafts with and without collagen and fibrin-glue pretreatments. The collagen was either 'weakly' crosslinked with formaldehyde or 'strongly' crosslinked with glutaraldehyde. *In vitro* studies suggest that both weakly and strongly crosslinked collagen, as well as fibrin glue, give a sufficient reduction in the porosity of the graft to allow it to be applied to heparinized patients. When the materials were implanted subcutaneously in rats, there was calcification with the fibrin-impregnated prostheses, which also allowed good capillary ingrowth. The tissue reaction to the collagen impregnation was variable and, as one would expect, the intensity of the inflammatory response and tissue adhesion, as well as the rate of absorption of the collagen, appeared to be related to its degree of crosslinking. However, in none of the collagen materials was any calcification observed. In a further *in vivo* study of both sheep and dogs[242], knitted Dacron conduits impregnated with glutaraldehyde and formaldehyde, as well as fibrin glue, were again evaluated. The results showed that delayed sealant absorption was a disadvantage and that use of the crosslinked collagens resulted in lack of adhesion between inner capsule and conduit, thereby causing focal haemorrhage dissection in the circulatory implants.

In a recent pilot study, a collagen-coated Dacron graft gave encouraging clinical results with 17 patients[236]. The collagen coating resulted in a sufficiently low-porosity graft which retained the handling of knitted prostheses and yet was virtually impervious to heparinized blood.

Senatore *et al.*[243] immobilized the fibrinolytic enzyme urokinase onto Dacron-reinforced autogenous fribrocollagenous tubes derived from dogs. The urokinase grafts were more patent than similarly derived control grafts and their surface was continually free of thrombus. These results should be encouraging, particularly for those developing small-diameter vascular prostheses.

Further attempts to improve the patency of collagen-impregnated Dacron vascular grafts have been made by seeding vascular endothelial cells into the prostheses[101,244]. Weinberg and Bell[101] constructed an *in vitro* artificial blood vessel having a multilayered structure similar to an artery. The middle layer corresponding to the media of an artery was prepared by casting culture medium, collagen and smooth-muscle cells in an annular mould. Following contraction of the collagen gel, a tubular lattice was produced around a central mandrel reinforced by a Dacron mesh. The outer layer corresponding to the adventitia was cast around the first lattice with adventitial fibroblasts rather than smooth-muscle cells. After 2 weeks, the resulting tube was removed from the mandrel and the lumen was seeded with endothelial cells. Electron microscopy showed that both the endothelial cells lining the lumen and the smooth-muscle cells in the wall were healthy and well differentiated. *In vitro* studies showed that the lining of endothelial cells functioned as a permeability barrier and could produce von Willebrand factor and prostacyclin. It was suggested that such a model artery could heal at an anastomosis to become truly integrated with the host's vasculature. Shindo *et al.*[244] showed in dogs that the patency of collagen-impregnated grafts could be improved by *in vitro* seeding with autogenous endothelial cells. Seeded grafts showed a considerably higher patency rate (86%) than nonseeded grafts (14%). The results were sufficiently encouraging to indicate that this method may be used to improve the patency of vascular grafts and to reduce thrombogenicity of small-calibre arterial substitutes.

Collagen, either in the form of bovine pericardium or as porcine aortic valves, has been much used in the construction of bioprosthetic heart valves. Indeed it is the most important component being used to manufacture the cusps of the valves, which are submitted to continuous mechanical wear and tear. Maximum crosslinking of the collagen component with glutaraldehyde reduces thrombogenicity, calcification and biodegradation[146,238,239,245,246]. The proteolysis of collagen when crosslinked with glutaraldehyde is greatly decreased, giving the valve an extremely long residence time *in vivo*. However, the continuous exposure to proteolytic enzymes from blood and inflammatory cells over a period of years may result in some degradation. To what extent these changes are significant in causing tears and perforations is not known, but it seems clear that morphologically detectable breakdown of the collagen takes place when stresses are high.

During their lifetime *in vivo*, collagen components of prosthetic heart valves are subjected to continuous processes that place limits on their durability. The most serious of these is the constant bending (compression tension) that occurs with each cycle of valvular opening and closure and that causes the mechanical disruption of collagen fibres[246]. Calcification of bioprosthetic heart valves can also occur over a number of years. The calcific deposits can cause stenosis by stiffening the cusps and hence reducing their mobility; they can also cause regurgitation, because they are often associated with cuspal tears and perforations.

OPHTHALMOLOGY

An early reference to collagen in ophthalmology[247] described the testing of collagen sponge for tissue reaction and absorption rate in rabbits' eyes. The collagen was almost completely absorbed over 5 weeks, and it was suggested that collagen may well find a use as a medium to administer drugs to the eye, in scleral resection, or in the production of experimental glaucoma. Later, L'Esperance[248] suggested that a reconstituted collagen tape could be used as a scleral buckling band in retinal detachment surgery. Tanner *et al.*[249] used a crosslinked collagen film as a graft for corneal replacement in dogs and cats. All the grafts were accepted with no evidence of rejection and became completely clear (there was some initial corneal clouding in dogs). The grafts resulted in a normal corneal structure, except for a slight thickening of the stratified squamous epithelium at the junction of the collagen graft and surrounding cornea. Collagen has also been used as a suture material in entropion repair, as lamellar keratectomy tissue, as a vitreous substitute and in glaucoma filter tubes[60].

Geggel *et al.*[93] proposed that collagen gels could be a suitable substrate in biofabricated occular surfaces. Collagen gels supported the growth of corneal epithelial cells *in vitro* and were well-tolerated in rabbits' eyes for up to 6 weeks. Wasserman[97] proposed the use of collagen pads as an adjunctive treatment for blepharitis, and Weber and Baker[250] indicated how cyanoacrylate adhesion could be used with collagen to deal with filtering blebs following glaucoma surgery.

However, perhaps the most widely studied and accepted use for collagen in ophthalmology is in the form of collagen shields. For example, Weissman and Lee[96] described the use of these shields as bandage contact lenses. Physical characteristics such as oxygen permeability and transmissibility, water content, and thickness were investigated. The shields behaved as 63% water content hydrogel contact lenses. Oxygen transmission levels (to allow corneal metabolism) were consistent with the values required for a good open-eye contact lens. There was, however, some doubt as to the suitability of the shields for use under extended-wear conditions. In the case of keratotomy, Marmer[251] showed more rapid and complete epithelial healing, more complete epithelial bonding and reduced inflammatory response with collagen shields. There was less stromal oedema at the wound sites in corneas treated with collagen, as opposed to non-collagen, shields. The collagen shield seemed to protect keratocytes adjacent to the wound, with diminished keratocyte reaction.

Collagen shields have more recently found considerable use as an occular drug-delivery system. For example, collagen shields immersed in tobramycin can enhance epithelial healing after surgery as well as provide antibiotic prophylaxis against infection[252]. O'Brien *et al.*[90] showed that tobramycin penetration into the anterior chamber of the eye was significantly greater with a collagen shield than that with either hydrophilic soft contact lenses or controls where no shield or lens was applied. Unterman *et al.*[253] demonstrated that collagen shields were effective in delivering sustained therapeutic concentrations of tobramycin to the cornea and aqueous humour of rabbits. Sawuch *et al.*[154] showed that a topical application of tobramycin in the presence of a collagen shield was significantly more effective than application of the drug alone in the treatment of bacterial keratitis in animals. And Hobden *et al.*[94] showed that collagen shields containing tobramycin were as effective in treating experimental keratitis in animals as regular drop treatment. With other antibiotics such as gentamycin and vancomycin, Phinney *et al.*[95] demonstrated that the levels of these drugs in cornea, tears and aqueous humour when using a corneal shield as a delivery system were higher or comparable to those obtained by frequent drop therapy.

In an *in vitro* model, Vasantha *et al.*[255] examined the use of collagen, crosslinked collagen and collagen hydrazide for the controlled release of pilocarpine. They were able to regulate the release of the drug over a period of 5–15 days depending on the carrier. Although encouraging, *in vivo* data is not yet available.

UROLOGY

The application of artificial materials in the repair of the urinary tract is in many ways more demanding than in other areas. In addition to the other properties required of a biomaterial, the prosthesis must prevent not only the leakage of urine but also the crystallization of urinary salts on its surface. The search for a suitable biodegradable material able to be successfully applied to the repair of the injured urinary tract has therefore been long and frustrating.

Early attempts at bladder reconstruction using either preserved bladder grafts in dogs[256] or absorbable gelatin sponge[257] showed some success[60]. With the bladder graft, a new epithelialized connective-tissue membrane appeared and there was some sign of smooth-muscle regeneration. However, there was leakage of urine, and the graft failed to incorporate with new bladder tissue, separating at 3–4 weeks. With the gelatin sponge, however, there was replacement with new connective tissue after 4 weeks, smooth-muscle regeneration at 6 weeks, and a new bladder wall with a complete muscle layer after 10 weeks.

Tanner *et al.*[84] used a collagen film to cover heminephrectomy in dogs. After 5 months, the film had disappeared and was replaced by a somewhat thickened renal capsule. No collections of blood or urine were found beneath the film during the experiment and little renal atrophy adjacent to the film was observed. The use of this collagen film on a human subject gave satisfactory results. Other early work has been reviewed by Chvapil *et al.*[60].

More recently, Tachibana *et al.*[85] used collagen sponge in a tubular form as ureteral grafts in dogs. The results were highly dependent on whether a urinary stent was applied. Without a stent, persistent urinary leakage and extensive stricture at the anastomotic site led to severe hydronephrosis. With a stent, hydronephrosis was avoided although there was some dilation above the anastomotic site; stone formation was not a problem. However, when the same collagen sponge was implanted into the urinary bladder of rats, there was some calcification of the implant, as well as urinary infection in 20% of the animals[85].

A simple noncrosslinked collagen film was tested for its suitability for the repair of injured urinary tract[86,258]. Following an *in vitro* trial of the film, in which were demonstrated its leakproof properties and compatibility with human urine[258], the film was used to seal experimental ureterotomies in rabbits[86]. Although it prevented leakage of urine and thus aided healing, biodegraded without trace and did not encourage crystallization of urinary salts, it was soft and difficult to suture. This problem was overcome by using a water-soluble polyester mesh (polyglactin 910, vicryl) to reinforce the film, hence producing a novel biodegradable composite membrane that was extremely strong and pliable, biodegradable and easy to suture.

Before using the membrane *in vivo*, an extensive *in vitro* investigation was carried out to test the resistance of the material to breakdown in human urine over a range of pH values[199], as well as the tendency of urinary salts to crystallize on its surface[259] and its predisposition to colonization by urinary tract pathogens[260]. In subsequent experiments, the material was successfully used to repair full-thickness defects of the urinary bladder[87,261] and partial nephrectomy[87,262]. In the case of the urinary bladder, the membrane proved to be leakproof and was replaced by new host scar tissue with a minimum of shrinkage over the course of the experiment (26 weeks). The repair area was lined with a normal-looking transitional epithelium, and there was smooth-muscle regeneration from 6 weeks onwards. Any initial problems with calcification were due to nonabsorbable sutures – the problems were immediately overcome when the use of such sutures was abandoned. The regeneration of smooth-muscle was further investigated by Gorham *et al.*[263], their results strongly suggesting that it arose as a result of cellular differentiation from within the repair site rather than from ingrowing detrusor muscle.

In the partial nephrectomy experiments, similar results to those reported by Tanner *et al.*[84] were obtained. Rapid haemostasis was achieved at operation, and there was no leakage or collection of blood or urine underneath the membrane, which itself was replaced by a thickened renal capsule. When both partial nephrectomy and bladder experiments were repeated using the composite membrane sterilized by α-irradiation (as opposed to 70% ethanol[87]), excellent results were obtained. There was no leakage or calculus formation in any of the bladder defects, and an excellent repair to the kidney was again found. Finally, results of a limited named patient trial have been most encouraging (R. Scott, personal communication).

PERIODONTOLOGY

A frequent problem in periodontal disease is the apical migration of the gingival epithelium into the defect between the tooth and the gingiva. This in turn prevents the regeneration of periodontal tissues and causes tooth detachment.

Collagen membranes are effective in preventing this downward migration of epithelium in dogs, and in assisting the formation of connective tissue to the root surface[102,104,105]. Similar findings were reported by Yaffe *et al.*[264] for an enriched collagen solution. However, Tanner *et al.*[265] found that a microfibrillar collagen in the form of a nonwoven web neither prevented epithelial migration nor allowed the regeneration of connective tissue in the periodontal pocket in humans. The physical form of the collagen could therefore be an important factor in this application.

Increased periodontal attachment and regeneration of alveolar defects were reported by Ciancio *et al.*[266], who applied a solution of collagen 'stabilizer' to periodontal defects in dogs. In addition, there was more bone formation than in 'control' animals, suggesting that such a material may reduce the need for surgical treatment of osseous defects and enhance surgically induced bone-regeneration procedures. Arrest of epithelial migration and reattachment of periodontal ligament was shown in dogs by Yaffe and Shoshan[267], who also reported formation of bone and cementum as well as proliferation of fibrous connective tissue. In these experiments an acid-soluble collagen was used to treat experimental induced alveolar bone dehiscences.

Recently, Minabe *et al.*[91,92] demonstrated the benefits of using collagen to immobilize tetracycline in periodontal treatment. Antimicrobial activity was observed up to 10 days following implantation. Indeed, the clinical indices of periodontal disease in patients were significantly decreased at 4 and 7 weeks, with a corresponding decrease in the number of bacteria in the periodontal pockets. In addition to its antimicrobial activity, tetracycline inhibits the destructive effect of collagenase enzymes[268].

NERVE REGENERATION AND REPAIR

The use of collagen film as a resorbable wrapper in nerve repair was first described in 1964[269,270]. Braun[269] reported the advantage of a tubulation technique using collagen film in the repair of the sciatic nerve, and Kline and Hayes[270] showed that by wrapping several peroneal nerves in adult chimpanzees with collagen film, repair was better than that with a simple suture technique. The methodology of these early experiments, however, involved severance rather than excision of nerve tissue.

Since these early beginnings, there have been mixed reports about this particular use of collagen. Colin and Donoff[75] used 2-mm-diameter collagen tubes in a study of the repair of rat tibial nerve. A silicone rod-induced fibrovascular sheath was used to evaluate the regeneration of the nerve through either a 2-mm collagen tube or a contralateral nerve autograft. First, 5 mm of right tibial nerve was resected from experimental rats and either replaced with a 5 × 2-mm silicone rod or the nerve ends sutured to intermuscular fascia. After 4 weeks, animals were repaired by replacing the silicone rod with either a collagen tube or a contralateral nerve autograft. The proportion of nerve fibres transversing the surgical site was not influenced by the method of repair or by the absence of sheathing. Tubulized repairs most closely resembled unoperated nerves; autografted repairs had a large diameter, but much fibrosis; and control repairs were immature and disorganized. It was concluded that there was no difference between repairs formed with or without a vascular pseudosheath. However, collagen tubes supported regeneration better than contralateral autograft repair and there was evidence for the adequacy of collagen conduits in bridging nerve defects.

Parker *et al.*[271] compared a simple suturing technique for repair of severed facial nerve in rabbits with the use of microfibrillar collagen. In 12 animals, six facial nerves were repaired with five nylon sutures in the epineurium, and five were repaired with one suture in the epineurium, the remaining epineurium being coated with microfibrillar collagen. One nerve was repaired by approximation of the epineurium using only microfibrillar collagen. In a 4-month study, the success rate of the two methods appeared to be functionally equivalent, but histological results showed that the suture repair was superior.

Satou *et al.*[272] studied the effect of collagen gel on the regeneration of severed rat sciatic nerve in silicone tubes. The sciatic nerves were resected, and distal and proximal stumps were inserted into a silicone tube to leave a 5-mm gap between them. The gaps were injected with a collagen solution that subsequently formed a gel; those in the control tubes were left empty. Examination up to 35 days later revealed considerably more rapid growth of sprouting axons towards the distal stump in the gel tubes than in the control tubes. Furthermore, there were fewer cellular components such as fibroblasts and Schwann cells in the gel tubes than in the control tubes. A more-orderly direction of growth was observed in the gel tubes. The results suggest that in the initial stages of regeneration, collagen-gel matrix has an effect in maintaining growth factors, eliminating disorderly proliferative fibroblasts and directing proliferating Schwann cells. Thus it would seem that sprouting of axons can be enhanced by collagen gel, that regeneration of peripheral nerves in silicone tubes is aided by the addition of collagen-gel matrix.

But Valentini *et al.*[273] reported that collagen gel impedes peripheral nerve regeneration through semipermeable nerve-guidance channels. These authors again used the sciatic nerve, but this time in mice. Semipermeable guidance channels (M_r 50,000 cut-off polyvinyl-chloride acrylic polymer tubes) were filled with saline or with collagen- or laminin-containing gels, and used to repair a 4-mm gap. After 12 weeks, nerve cables that had regenerated in gel-filled channels displayed fewer myelinated axons than those regenerating in saline-filled channels. Remnants of the exogenous substrates were still in evidence, in amounts relating to the original gel concentration. The impairment of nerve regeneration by collagen- or laminin-containing gels suggests that the regenerative environment in the semipermeable channels is not improved by the addition of growth substrates in a gel form.

DURAMATER REPLACEMENT IN NEUROSURGERY

Following neurosurgery, it is often necessary to apply prosthetic material to replace excised duramater. A dural replacement should prevent both the leakage of cerebrospinal fluid and the formation of pia-arachnoid adhesions which could otherwise lead to epilepsy. Lyophilized human duramater may be used as a dural substitute, but the potential complications of using human tissue, such as the possible transmittance of the causative agents of AIDS and Kreuzfeld-Jacob syndrome, make this far from ideal. Autologous fascia may also be used, but this results in additional trauma and discomfort for the patient.

Janetta and Whayne[107] evaluated the use of a formaldehyde-crosslinked film, as well as a laminate in which two layers of the film overlie a fabric of woven collagen yarn, as a dural replacement in adult mongrel dogs. In two animals, the film was placed in a 1.5 × 3.0-cm defect in duramater over intact pia-arachnoid. In four dogs, bilateral 2.0 × 4.0-cm areas of duramater were excised and the underlying pia-arachnoid traumatized. The side (one per animal) to be covered with film was selected at random, and the defect was covered by a 3.0 × 5.0-cm sheet of material. In a further group of nine dogs, similar bilateral dural defects were created, except that more-severe trauma using coagulation and cortical resection was used. One side was covered by the film laminate, and the opposite side was covered by either a dural autograft or by temporalis muscle and pericranium. Animals were examined at intervals between 30 and 270 days. Both film and film laminate were inert, leakproof, smooth, nonirritating and prevented adhesions. Absorption was gradual with a mild foreign-body reaction as new duramater formed. There was less adhesion than with the pericranium and temporates muscle. The film laminate was preferable to the film, as it was strong, easily sutured, resembled human duramater in thickness and appeared to fulfil all the requirements of a dural substitute.

Similar experiments were performed in dogs by Kline *et al.*[108]. The control animals had autologous fascia as a dural replacement. The collagen film proved to be satisfactory as a dural replacement, being absorbed and replaced by connective tissue in 6–8 months. The film laminate produced a greater reaction and was absorbed more slowly than either the film or fascia.

Studies in dogs show how a composite material consisting of a collagen film reinforced with a vicryl mesh (polyglactin 910, a water soluble polyester) can provide an extremely useful prosthesis to replace duramater. Furthermore, in a series of experiments using rabbits, the composite membrane caused little adhesion and prevented leakage of cerebrospinal fluid during healing (R. N. Meddings *et al.*, unpublished results). Again, these experiments indicate that this composite membrane makes a good dura replacement.

In addition, clinical studies of several hundred patients in France

(S. A. Ethnor, unpublished results) have provided extremely encouraging data on the usefulness of the collagen/vicryl composite in dural-replacement neurosurgery.

COLLAGEN AS A DRUG-DELIVERY SYSTEM

Collagen can interact with several substances to form complexes which vary in stability[60] and which may be degraded slowly when administered subcutaneously, intramuscularly or intraperitoneally. Hence collagen can act as a drug-delivery system and prolong the pharmacological action of drugs such as local anaesthetics, analgesics and antibiotics[60,90–92,95].

It has already been discussed how collagen shields can be used to deliver antibiotics such as tobramycin, gentamycin and vancomycin to the eye to treat conditions such as bacterial keratitis.

In the treatment of periodontal disease (see above), antibiotics such as tetracycline have been incorporated into collagen films for introduction into the periodontal pocket to good effect.

Collagen sponge has also been used as a drug carrier for tetracyclines[274]: after implantation into rabbits, the drug could be detected in the plasma for more than 7 days. Peng *et al.*[275] used a collagen-sponge cervical cap to deliver all-*trans* retinoic acid to patients with cervical dysplasia. A short elimination half-life of the drug was found and there was no systemic absorption following administration. The device was advantageous as it allowed high concentrations of all-*trans* retinoic acid to be delivered locally without undue systemic absorption.

The ability of collagen films to support the release of drugs has been studied in an examination of the mechanical properties and release rates of medroxyprogesterone acetate[88]. Collagen films crosslinked by chrome tanning, formaldehyde or a combination of both, were examined as release media. The highest diffusion rate was obtained with the chrome-tanned film, whereas the slowest rate was produced with formaldehyde-crosslinked films. Formaldehyde-tanning produced films of higher Young's modulus than chrome-treated films, which in turn had a Young's modulus 2–3 times greater than untreated film. Combined chrome and formaldehyde treatments produced films of intermediate properties.

Extracts of demineralized rat bones containing factors stimulating bone induction were reconstituted with highly purified human type I collagen to provide a suitable and easily manipulated delivery system for surgical implantation[276]. When implanted subcutaneously in rats, the implants both governed and delineated the dimensions of the resulting bony tissue. It was proposed that this implant system may have clinical applications in the filling of osseous defects within the scope of orthopaedic, oral and maxillofacial surgery.

A novel sustained drug-delivery system consisting of liposomes sequestered in a collagen gel has been described[89]. Two peptide hormones, insulin and growth hormone, encapsulated in vesicles sequestered in the matrix, were slowly released into the circulation from either an intracellular or subcutaneous site. The maximum period of release for insulin was 3–5 days, whereas that for growth hormone was 14 days. Modification of the liposome surface with fribronectin gave enhanced sequestration within the collagen gel. This liposome/collagen-gel drug-delivery system appears to have several advantages over other liposome formulations or simple gel formulations constructed with free drug. It was suggested that such a system may have topical applications in the treatment of surgical and nonsurgical wounds and burns, as well as uses as a local implant in regions such as the vaginal, rectal or auricular cavities.

TENDON AND LIGAMENT REPAIR

Damage to tendons occurs most commonly during physical activity, and both natural and synthetic materials have been used for their repair[277]. The ideal material for repair of ligament or tendon should have the biomechanical properties of normal tissue and should incorporate repair tissue rapidly without appreciable loss of strength, thus reducing the period of protection that is needed[278]. The repair material should provide a scaffold for aligned deposition of collagen, appear mechanically and histologically normal, and last long enough to protect the maturing neotendon until it can function.

No currently available prosthesis can provide the function of normal tendon or ligament indefinitely[278], and indeed, artificial 'permanent' implants often fail as a result of fatigue, the remnants subsequently needing to be removed surgically. Autologous tendon transfers are commonly used to treat traumatic loss of tendon or ligament, but are limited by the availability of autologous material, loss of normal functioning structures and a slow rate of incorporation of neotendon.

Type I collagen supports and accelerates ingrowth of tissue cutaneously and subcutaneously in animals[60,66,67,69,70,78,80,110,117,122,123], and can be formed into thin fibres that have high tensile strength[279,280].

Indeed, materials based on reconstituted collagen have been recently developed for the repair of ligament and tendon[281].

Goldstein *et al.*[278] developed a reconstituted collagen tendon prosthesis that was implanted into rabbit Achilles tendons. Type I collagen was extruded into fibres and crosslinked with either glutaraldehyde or by dehydrothermal treatment followed by reaction with a carbodiimide. The tendon prosthesis was then assembled by coating a longitudinal array of the fibres with noncrosslinked collagen. In one leg of each rabbit, the Achilles tendon was replaced by the prosthesis, whereas in the contralateral leg, the tendon was excised, devascularized and anastomosed as an autogenous graft. The autogenous grafts were infiltrated centrally by fibroblasts and capillaries after 10 weeks and were partially replaced by repair tissue at 20 weeks. In 3 weeks, all collagen implants were infiltrated with fibrous tissue. At 10 weeks, reorganization of collagenous tissue was observed in and around the prostheses, and the carbodiimide-treated implants were resorbed and replaced by normal-looking neotendon. The glutaraldehyde-treated materials were resorbed more slowly and surrounded by more inflammatory cells, with neotendon formation proceeding less rapidly. The initial mechanical strength of the carbodiimide-treated implants was lower than the fresh autogenous grafts. At 20 weeks, their strength and Young's modulus approached those of fresh tendon whereas the autogenous tendon grafts were significantly weaker and more extensible than the carbodiimide-treated implant. The strength and Young's modulus of the glutaraldehyde-treated implants were initially higher than those of fresh tendon, but at 20 weeks they were the same as those of devascularized autogenous tendon grafts. The study showed that neotendon formation can be stimulated by suitable collagenous tendon prostheses.

Wasserman *et al.*[277] developed a tendon prosthesis again to repair the Achilles tendon in rabbits. The material consisted of collagen tow prepared from insoluble type I collagen crosslinked dehydrothermally and then reacted with cyanamide. Implants were examined at 3 and 10 weeks postoperatively. Extensive cell infiltration, biodegradation and neotendon formation were observed as early as 3 weeks. These processes were associated with fribroblastic activity. At 10 weeks, the prosthesis had been completely resorbed, and highly organized neotendon was apparent which contained collagen-fibre bundles resembling normal tissue.

Glutaraldehyde-treated pericardium has been used as a ligament substitute in the patellar ligament of rat knee joints[282]. Animals were examined at 18 days and 12 weeks postoperatively. At 18 days, ingrowth of cells and vessels was fast, and at 12 weeks, new collagen depositions within the pericardial patch were found together with biodegradation of the implant.

Law *et al.*[283] again made use of reconstituted type I collagen fibres crosslinked with either glutaraldehyde or by dehydration and cyanamide treatment. *In vitro* experiments showed that cyanamide-treated material was better than glutaraldehyde-tanned collagen at promoting fibroblast growth. Intramuscular implantation showed excellent biocompatibility of both implants. Aligned ingrowth into the implant from the medial collateral ligament was observed, and it was suggested that crosslinked collagen fibres represent a strong candidate for scaffold ligaments or tendon prostheses.

Conclusion

Only an overview has been given here of the many uses of collagen in the biomedical field, providing an insight into its tremendous versatility and a review of the vast amount of literature on the subject.

As we have seen, collagen has many properties favouring its use as a biomaterial. It can be reconstituted into a number of forms such as films, sponges and injectables, all having many actual or potential applications. Because collagen is a protein, it is readily degradable in the body. It is haemostatic and only weakly antigenic, and can be modified chemically to alter its physical, chemical and biological properties.

Collagen can also be combined with many other materials to form a range of composites. Hence, collagen-based biomaterials can be tailor-made to suit a diversity of surgical needs. Indeed, the full potential of this versatile material has still to be realized.

Acknowledgements

I thank Dr N. Light for his comments and helpful suggestions in the preparation of this chapter, and Mrs J. Wells for typing the manuscript.

References

1. Huc, A. (1985) Collagen biomaterials characteristics and applications. *J. Am. Leather Chem. Assoc.* **80**: 195–212.
2. Bogue, R.H. (1922) *The Chemistry and Technology of Gelatin and Glue.* McGraw-Hill, New York.
3. Martin, G.R., Timpl, R., Muller, P.K. & Kuhn, K. (1985) The genetically distinct collagens. *Trends in Biochemical Sciences* **10**: 285–287.
4. Dublet, B. & van der Rest, M. (1987) Type XII collagen is expressed in embryonic chick tendons. *J. Biol. Chem.* **262**: 17724–17727.
5. Smith, G.N., Williams, J.M. & Brandt, K.D. (1987) Effect of polyanions on fibrillogenesis by type XI collagen. *Coll. Rel. Res.* **7**: 17–25.
6. Summers, T.A., Irwin, M.H., Mayne, R. & Balian, G. (1988) Monoclonal antibodies to type X collagen. *J. Biol. Chem.* **263**: 581–587.
7. Bailey, A.J. & Light, N.D. (1989) *Connective Tissue in Meat and Meat Products.* Elsevier Applied Science, London.
8. Piez, K.A. (1976) 'Primary Structure'. In *Biochemistry of Collagen* (eds Ramachandran, G.N. & Reddi, A.H.), pp. 1–44. Plenum, London, New York.
9. Fietzek, P.P. & Kuhn, K. (1976) The primary structure of collagen. In *International Review of Connective Tissue Research* (eds Hall, D.A. & Jackson, D.S.), vol. 7, pp. 1–60. Academic Press, London, New York.
10. Miller, E.J. (1976) Biochemical characteristics and biological significance of the genetically distinct collagens. *Mol. Cell. Biochem.* **13**: 165–192.
11. Kuhn, K. & Glanville, R.W. (1980) *Biology of Collagen* (eds Viidik, A. & Vuust, J.), ch. 1 'Molecular Structure and Higher Organizations of Different Collagen Types', pp. 1–14. Academic Press, London, New York.
12. Light, N.D. & Bailey, A.J. (1980) Molecular structure and stabilisation of the collagen fibre. In *Biology of Collagen* (eds Viidik, A. & Vuust, J.), pp. 15–38. Academic Press, London, New York.
13. Duance, V. & Bailey, A.J. (1981) Biosynthesis and degradation of collagen. In *Handbook of Inflammation. Vol. 3. Tissue Regeneration and Repair* (ed. Glynn, L.E.), pp. 51–109. Elsevier, Amsterdam.
14. Last, J.A. & Reiser, K.M. (1984) Collagen biosynthesis. *Environmental Health Perspectives* **55**: 169–177.
15. Fessler, J.H., Doege, K.J., Duncan, K.G. & Fessler, L.I. (1985) Biosynthesis of collagen. *J. Cell. Biochem.* **28**: 31–37.
16. Tanzer, M.L., Church, R.L., Yaeger, J.A., Wampler, D.E. & Park, E.D. (1974) Procollagen: intermediate forms containing several types of peptide chains and non-collagen peptide extensions at the NH_2 and COOH ends. *Proc. Natl. Acad. Sci. USA* **71**: 3009–3013.
17. Byers, P.M., Click, E.M., Harper, E. & Bornstein, P. (1975) Interchain disulphide bonds in procollagen are located in a large non-triple helical carboxy terminal domain. *Proc. Natl. Acad. Sci. USA* **72**: 3009–3013.
18. Fessler, L.I., Morris, N.P. & Fessler, J.H. (1975) Procollagen. Biological scission of amino and carboxyl extension peptides. *Proc. Natl. Acad. Sci. USA* **72**: 4905–4909.
19. Kivirriko, K.I., Ryhanen, L., Anttinen, H., Borstein, P. & Prockop, D.J. (1973) Further hydroxylation of lysyl hydroxylase in-vitro. *Biochemistry* **12**: 4966–4971.
20. Glanville, R.W. & Kuhn, K. (1979) Primary structure of collagen. In *Fibrous Proteins: Scientific Industrial and Medical Aspects* (eds Parry, D.A. & Creamer, L.K.), vol. 1, pp. 133–150. Academic Press, London, New York.

21. Grand, M.E., Schofield, J.D., Kefalides, N.A. & Prockop, D.J. (1973) Biosynthesis of basement membrane collagen in embryonic chick lens. *J. Biol. Chem.* **248**: 7432–7437.

22. Fessler, L.I. & Fessler, J.H. (1974) Protein assembly of procollagen and the effects of hydroxylation. *J. Biol. Chem.* **249**: 7637–7646.

23. Schofield, J.D., Uitto, J. & Prockop, D.J. (1974) Formation of interchain disulphide bonds and helical structures during the biosynthesis of procollagen by tendon cells. *Biochemistry* **13**: 1801–1806.

24. Fessler, J.H. & Fessler, L.I. (1976) Procollagen of different types and their processing in chick tendons and aorta. *Fed. Proc.* **35**: 1355.

25. Trelstad, R.L. & Birk, D.E. (1985) The fibroblast in morphogenesis and fibrosis: cell topography and surface-related functions. In *Fibrosis: Ciba Foundation Symposium 114*, pp. 12–19. Pitman, London.

26. Linsenmayer, T.F. (1981) Collagen. In *Cell Biology of Extracellular Matrix* (ed. Hay, E.), pp. 5–37. Plenum, London, New York.

27. Gross, J. (1981) An essay on biological degradation of collagen. In *Cell Biology of Extracellular Matrix* (ed. Hay, E.D.), pp. 217–258. Plenum, London, New York.

28. Sellers, A. & Murphy, G. (1981) Collagenolytic enzymes and their naturally occurring inhibitors. In *International Review of Connective Tissue Research* (eds Hall, D.A. & Jackson, D.S.), vol. 9, pp. 151–190. Academic Press, London, New York.

29. Harris, E.D. & Cartwright, E.C. (1977) Mammalian collagenases. In *Proteinases in Mammalian Cells and Tissues* (ed. Barrett, A.J.), pp. 249–283. Elsevier, Amsterdam.

30. Gross, J. & Lapiere, C.M. (1962) Collagenolytic activity in amphibian tissues: a tissue culture assay. *Proc. Natl. Acad. Sci. USA* **48**: 1014–1022.

31. Lazarus, G.S., Brown, R.S., Daniels, J.R. & Fulmer, H.M. (1968) Human granulocyte collagenase. *Science* **159**: 1483–1485.

32. Lazarus, G.S., Daniels, J.R., Lian, J. & Burliegh, M.C. (1972) Role of granulocyte collagenases in collagen degradation. *Am. J. Pathol.* **68**: 565–579.

33. Gross, J., Highberger, J.M., Johnson-Wint, B. & Biswas, C. (1980) Collagenase in normal–pathological connective tissue. In *Mode of Action and Regulation of Tissue Collagens* (eds Woolley, D.E. & Evanson, J.M.), pp. 11–35. Wiley, New York.

34. Gross, J. & Nagai, Y. (1965) Specific degradation of the collagen molecule by tadpole collagenolytic enzyme. *Proc. Natl. Acad. Sci. USA* **54**: 1197–1204.

35. Murphy, G., Bretz, U., Baggiolini, M. & Reynolds, J.J. (1980) The latent collagenases and gelatinase of human polymorphonuclear leukocytes. *Biochem. J.* **192**: 517–525.

36. Wang, H.M., Chan, J., Pettigrew, D.W. & Sodek, J. (1978) Cleavage of native type III collagen in the collagenase susceptible region by thermolysin. *Biochem. Biophys. Acta* **533**: 270–277.

37. Cawston, T.E. & Murphy, G. (1981) Mammalian collagenases. In *Methods in Enzymology* (eds Lorand, L., Colowick, S.P. & Kaplan, N.O.), vol. 80, part C, pp. 711–722. Academic Press, London, New York.

38. Hibbs, M.S., Hasty, K.A., Kang, A.H. & Mainardi, C.L. (1984) Secretion of collagenolytic enzymes by human polymorphonuclear leukocytes. *Coll. Rel. Res.* **4**: 467–477.

39. Weiss, J.B., Sedowofia, K. & Jones, C. (1980) Collagen degradation: a defended multi-enzyme system. In *Biology of Collagen* (eds Viidik, A. & Vuust, J.), pp. 113–134. Academic Press, London, New York.

40. Harris, E.D., Welgus, H.G. & Krane, S.M. (1984) Regulations of the mammalian collagenases. *Coll. Rel. Res.* **4**: 493–512.

41. Capodici, C. & Berg, R.A. (1989) Cathepsin G degrades denatured collagen. *Inflammation* **13**: 137–145.

42. Capodici, C., Muthukumaran, G., Amoruso, M.A. & Berg, R.A. (1989) Activation of neutrophil collagenase by cathepsin G. *Inflammation* **13**: 245–258.

43. Etherington, D.J., Maciewicz, R.A., Taylor, M.A.J., Wardale, R.J., Silver, I.A., Murrills, R.A. & Pugh, D. (1986) The role of collagen degrading cysteine proteinases in connective tissue metabolism. In *Cysteine Proteinases and Their Inhibitors* (ed. Turk, V.), pp. 269–282. Walter de Gruyter, Berlin.

44. Maciewicz, R.A., Etherington, D.J., Kos, J. & Turk, V. (1987) Collagenolytic cathepsins of rabbit spleen: a kinetic analysis of collagen degradation and inhibition by chicken cystatin. *Col. Rel. Res.* **7**: 295–304.

45. Maciewicz, R.A. & Etherington, D.J. (1988) A comparison of four cathepsins (B,L,N,S) with collagenolytic activity from rabbit spleen. *Biochem. J.* **256**: 433–440.

46. Etherington, D.J., Silver, I.A. & Restall, D.J. (1979) Resorption of insoluble heterologous fluorescein collagen sponges in sensitised and non-sensitised rats. *Br. J. Exp. Path.* **60**: 549–559.

47. Etherington, D.J., Pugh, D. & Silver, I.A. (1981) Collagen degradation in an experimental inflammatory lesion: studies on the role of the macrophage. *Acta. Biol. Med. Germ.* **40**: 1625–1636.

48. Silver, I.A., Murrills, R.A. & Etherington, D.J. (1988) Microelectrode studies on the acid environment beneath adherent macrophages and osteoclasts. *Exp. Cell. Res.* **175**: 266–276.

49. Pugh, D., Etherington, D.J. & Silver, I.A. (1984) Acid hydrolase distribution in sensitised and naive macrophages in-vivo and in-vitro. *Z. Mikrosk-Anat-Forsch. Leipzig* **98**: 375–384.

50. Bailey, A.J. & Etherington, D.J. (1980) Metabolism of collagen and elastin. In *Comprehensive Biochemistry* (eds Florkin, M. & Stotz, E.), vol. 19B, pp. 299–460. Elsevier, Amsterdam.

51. Everts, V. & Beeisten, W. (1988) The cellular basis of tooth eruption, the role of collagen phogacytosis. In *The Biological Mechanisms of Tooth Eruption and Root Resorption* (ed. Davidovitch, Z.), pp. 237–242. EBSCO Media, Birmingham, Alabama 35233, USA.

52. Etherington, D.J., Taylor, M.A.J. & Henderson, B. (1988) Elevation of cathepsin L levels in the synovial lining of rabbits with antigen induced arthritis. *Br. J. Exp. Path.* **69**: 281–289.

53. Van Noorden, C.J.F., Smith, R.C. & Rusnick, D. (1988) Cysteine proteinase activity in arthritic rat knee joints and the effects of a selective systemic inhibitor, Z-Phe-Ala-CH$_2$F. *J. Rheumatol.* **15**: 1525–1535.

54. Van Noorden, C.J.F., Vogels, I.M.C. & Smith, R.E. (1989) Localisation and cytophotometric analysis of cathepsin B activity in unfixed and undecalcified cryostat sections of whole rat knee joints. *J. Histochem. Cytochem.* **37**: 617–624.

55. Maciewicz, R.A., Wardale, R.J., Etherington, D.J. & Paraskeva, C. (1989) Immunodetection of Cathepsins B and L present in and secreted from human pre-malignant and malignant colorectal tumour cell lines. *Int. J. Cancer* **43**: 478–486.

56. Etherington, D. J. (1973) Collagenolytic-cathepsin and acid proteinase activities in the rat uterus during post-partum involution. *Eur. J. Biochem.* **32**: 126–128.

57. Cawston, T.E., Mercer, E., de Silva, M. & Hazleman, B.L. (1984) Metallo-proteinases and collagenase inhibitors in rheumatoid synovial fluid. *Arthritis Rheum.* **27**: 285–290.

58. Gadher, S.J., Eyre, D.R., Duance, V.C., Wotton, S.F., Heck, L.W., Schmid, T.M. & Woolley, D.J. (1988) Susceptibility of cartilage collagens type II, IX, X and XI to human synovial collagenase and neutrophil elastase. *Euro. J. Biochem.* **175**: 1–7.

59. Katz, A.R. & Turner, R.J. (1970) Evaluation of tensile and absorption properties of polyglycolic acid sutures. *Surg. Gynecol. Obstet.* **134**: 701–716.

60. Chvapil, M., Kronenthal, R.L. & van Winkle, W. Jr (1973) Medical and surgical applications of collagen. In *International Review of Connective Tissue Research* (eds Hall, D.A. & Jackson, D.S.), vol. 6, ch. 1, pp. 1–61. Academic Press, London, New York.

61. Stenzel, K., Miyata, T. & Rubin, A. (1974) Collagen as a biomaterial. *Ann. Rev. Biophys. Bioeng.* **3**: 231–253.

62. Elsdale, T. & Bard, J. (1972) Collagen substrata for studies on cell behaviour. *J. Cell. Biol.* **54**: 626–637.

63. Schor, S.L. (1980) Cell proliferation and migration on collagen substrate invitro. *J. Cell. Sci.* **41**: 159–175.

64. Kleinman, H., Klebe, R.J. & Martin, G.R. (1981) Role of collagenous matrices on the adhesion and growth of cells. *J. Cell. Biol.* **88**: 473–485.

65. Yang, J. & Nandi, S. (1983) Growth of cultured cells using collagen as a substrate. *Int. Rev. Cytol.* **81**: 249–286.

66. Doillon, C.J. & Silver, F.H. (1986) Collagen based wound dressings, effects of hyaluronic acid and fibronectin on wound healing. *Biomaterials* **7**: 3–8.

67. Doillon, C.J., Silver, F.H. & Berg, R.A. (1987) Fibroblast growth on a porous collagen sponge containing hyaluronic acid and fibronectin. *Biomaterials* **8**: 195–200.

68. Doillon, C.J., Wasserman, A.J., Berg, R.A. & Silver, F.G. (1988) Behaviour of fibroblasts and epidermal cells cultivated on analogues of extracellular matrix. *Biomaterials* **9**: 91–96.

69. Chvapil, M. (1977) Collagen sponge: theory and practice of medical applications. *J. Biomed. Mater. Res.* **11**: 721–741.

70. Chvapil, M. (1980) Reconstituted collagen. In *Biology of Collagen* (eds Viidik, A. & Vuust, J.), pp. 313–324. Academic Press, London, New York.

71. Pharriss, B.B. (1980) Collagen as a biomaterial. *J. Am. Leather Chem. Assoc.* **75**: 474–485.

72. Klopper, P.J. (1986) Collagen in surgical research. *Eur. Surg. Res.* **18**: 218–223.

73. Stein, M.D., Salkin, L.M., Freedman, A.L. & Glushko, V. (1985) Collagen sponge as a topical haemostatic agent in mucogingival surgery. *J. Periodontol.* **56**: 35–38.

74. Voormolen, J.H.C., Ringers, J., Bots, G.T.A.M., van der Heide, A. & Hermans, J. (1987) Haemostatic agents: brain tissue reaction and effectiveness. *Neurosurgery* **20**: 702–709.

75. Colin, W. & Donoff, R.B. (1984) Nerve regeneration through collagen tubes. *J. Dent. Res.* **63**: 987–993.

76. Burke, K.E., Naughton, G., Waldo, E. & Cassai, N. (1983) Bovine collagen implant: histologic chronology in pig dermis. *J. Dermatol. Surg. Oncol.* **9**: 889–895.

77. DeLustro, F., Smith, S.T., Sundsmo, J., Salem, G., Kincaid, S. & Ellingsworth, L. (1987) Reaction to injectable collagen: results in animal models and clinical use. *Plast. Reconstr. Surg.* **79**: 581–592.

78. Yannas, I.V. & Burke, J.F. (1980) Design of an artificial skin. I. Basic design principles. *J. Biomed. Mater. Res.* **14**: 65–81.

79. Burke, J.F., Yannas, I.V., Quinby, W.C., Bondoc, C.C. & Yung, W.K. (1981) Successful use of a physiologically acceptable artificial skin in the treatment of extensive burn injury. *Ann. Surg.* **194**: 413–428.

80. Chvapil, M. (1982) Considerations on manufacturing principles of a synthetic burn dressing: a review. *J. Biomed. Mater. Res.* **16**: 245–263.

81. Pruitt, B.A. & Levine, N.S. (1984) Characteristics and uses of biologic dressings and skin substitutes. *Arch. Surg.* **119**: 312–322.

Biomaterials

82. Holl-Allen, R.T.J. (1984) Porcine dermal implants in man. *J. Royal Coll. Surg. Edinb.* **29**: 151–153.

83. Holl-Allen, R.T.J. (1984) Porcine dermal repair of inguinal hernias. *J. Royal Coll. Surg. Edinb.* **29**: 154–157.

84. Tanner, J.C., Marcucci, M.A., Bradley, W.H. & Morgan, J.W. (1968) Partial nephrectomy and use of collagen grafts for renal wound closure. *J. Urol.* **109**: 710–712.

85. Tachibana, M., Nagamatsu, G.R. & Addonizio, J.C. (1985) Ureteral replacements using collagen sponge tube grafts. *J. Urol.* **133**: 866–869.

86. Scott, R., Baraza, R., Gorham, S.D., McGregor, I. & French, D.A. (1986) Assessment of collagen film for use in urinary tract surgery. *Br. J. Urol.* **58**: 203–207.

87. Scott, R., Mohammed, R., Gorham, S.D., French, D.A., Monsour, M.J., Shivas, A. & Hyland, T. (1988) The evolution of a biodegradable membrane for use in urological surgery. A summary of 109 in-vivo experiments. *Br. J. Urol.* **62**: 26–31.

88. Bradley, W.G. & Wilkes, G.L. (1977) Some mechanical property considerations of reconstituted collagen for drug release supports. *Biomat. Med. Dev. Art. Org.* **5**: 159–175.

89. Weiner, A.L., Carpenter-Green, S.S., Soehgen, E.C., Lenk, R.P. & Popescu, M.C. (1985) Liposome–collagen gel matrix: a novel sustained drug delivery system. *J. Pharm. Sci.* **74**: 922–925.

90. O'Brien, T.P., Sawusch, M.R., Dick, J.J., Hamburg, T.R. & Gottsch, J.D. (1988) Use of collagen corneal shields versus soft contact lenses to enhance penetration of topical tobramycin. *J. Cataract Refract. Surg.* **14**: 505–507.

91. Minabe, M., Takeuchi, K., Tamura, T., Hori, T. & Umemoto, T. (1989) Subgingival administration of tetracycline on a collagen film. *J. Periodontol.* **60**: 552–556.

92. Minabe, M., Vematsu, A., Nishijima, K., Tomomatsu, E., Tamura, T., Hori, T., Umemoto, T. & Hino, T. (1989) Application of a local drug delivery system to periodontal therapy. *J. Periodont.* **60**: 113–117.

93. Geggel, H.S., Friend, J. & Thoft, R.A. (1985) Collagen gel for ocular surface. *Invest. Ophthalmol. Vis. Sci.* **26**: 901–905.

94. Hobden, J.A., Reidy, J.J., O'Callaghan, R.J., Hill, J.M., Inster, M.S. & Rootman, D.S. (1988) Treatment of experimental pseudomonas keratitis using collagen shields containing tobramycin. *Arch. Ophthalmol.* **106**: 1605–1607.

95. Phinney, R.B., Schwarz, S.D., Lee, D.A., Mondino, B.J. (1988) Collagen-shield delivery of gentamicin and vancomycin. *Arch. Ophthalmol.* **106**: 1599–1604.

96. Weissman, B.A. & Lee, D.A. (1988) Oxygen transmissibility, thickness, and water content of three types of collagen shields. *Arch. Ophthalmol.* **106**: 1706–1708.

97. Wasserman, E.L. (1989) Blepharitis and the collagen eye patch. *Ann. Ophthalmol.* **21**: 124–131.

98. Chvapil, M., Droegenmueller, W., Heine, M.W., McGregor, J.C. & Dotters, D. (1985) Collagen sponge as vaginal contraceptive barrier. Critical summary of seven years of research. *Am. J. Obst. Gynecol.* **151**: 325–329.

99. Ishihara, T., Ferans, V.J., Jones, M., Boyce, S.W. & Roberts, W.C. (1981) Structure of bovine parietal pericardium and of unimplanted Ionescu–Shiley pericardial valvular bioprostheses. *J. Thorac. Cardiovasc. Surg.* **81**: 747–757.

100. Menasche, P., Flaud, P., Huc, A. & Piwnica, A. (1984) Collagen vascular grafts: a step towards improved compliance in small-calibre bypass surgery; preliminary report. *Life Support Systems* **2**: 233–237.

101. Weinberg, C.P. & Bell, E. (1986) A blood vessel model constructed from collagen and cultured vascular cells. *Science* **231**: 397–400.

102. Blumenthal, N.M. (1988) The use of collagen membranes to guide regeneration of new connective tissue attachment in dogs. *J. Periodont.* **59**: 830–836.

103. Minabe, M., Takeuchi, K., Tomomatsu, E., Hori, T. & Umemoto, T. (1989) Clinical effects of local application of collagen film – immobilized tetrocycline. *J. Clin. Periodont.* **16**: 291–294.

104. Pitaru, S., Tal, H., Soldinger, M., Grosskopf, A. & Noff, M. (1988) Partial regeneration of periodontal tissues using collagen barriers. *J. Periodont.* **59**: 380–386.

105. Pitaru, S., Tal, H., Soldinger, M. & Noff, M. (1989) Collagen membranes prevent apical migration of epithelium and support new connective tissue attachment during periodontal wound healing in dogs. *J. Periodont. Res.* **24**: 247–253.

106. Dahan, M., Berjaud, J., Renella-Coll, J. & Gaillard, J. (1988) Utilisation du filet de vicryl enduit de collagene en chirurgie d'exerese pulmonaire. *Ann. Chir.* **42**: 297–299.

107. Janetta, P.J. & Whayne, T.F. (1965) Formaldehyde-treated regenerated collagen film and film-laminate as a substitute for dura-mater. *Surgical Forum* **16**: 435–437.

108. Kline, D.G. (1965) Dural replacement with resorbable collagen. *Arch. Surg.* **91**: 924–929.

109. San-Galli, F. (1989) Etude experimentale d'une prosthese de dura-mere en treillis VICRYL collagene implantée chez le chien. *DEA Biologie Sante Option 'Biomateriaux'.* University of Bordeaux, INSERM U.306.

110. Yannas, I.V., Burke, J.F., Gordon, P.L., Haung, C. & Rubenstein, R.W. (1980) Design of an artificial skin. II. Control of chemical composition. *J. Biomed. Mater. Res.* **14**: 107–131.

111. Weadock, K.S., Wolff, D. & Silver, F.H. (1987) Diffusivity of I-labelled collagen: mechanism of diffusion and effect of absorption. *Biomaterials* **8**: 105–112.

112. Gorham, S.D., Hyland, T.P., French, D.A. & Willins, M.J. (1990) Cellular invasion and breakdown of three different collagen films in the lumbar muscle of the rat. *Biomaterials* **11**: 113–118.

113. Srivastava, S., Gorham, S.D., French, D.A., Shivas, A.A. & Courtney, J.M. (1990) In-vivo evaluation and comparison of collagen acetylated, collagen and collagen/glycosaminoglycan composite films and sponges as candidate biomaterials. *Biomaterials* **11**: 155–161.

114. DeLustro, F., Condell, R.A., Nguyen, M.A. & McPherson, J.M. (1986) A comparative study of the biologic and immunologic response to medical devices derived from dermal collagen. *J. Biomed. Mater. Res.* **20**: 109–120.

115. McPherson, J.M., Sawamura, S. & Armstrong, R. (1986) An examination of the biologic response to injectable, glutaraldehyde crosslinked collagen implants. *J. Biomed. Mater. Res.* **20**: 93–107.

116. McPherson, J.M., Sawamura, S.J. & Conti, A. (1986) Preparation of [^{3}H] collagen for studies of the biologic fate of xenogenic collagen implants in-vivo. *J. Invest. Dermatol.* **86**: 673–677.

117. Wallace, D.G., McPherson, J.M., Ellingsworth, L., Cooperman, L., Armstrong, R. & Piez, K.A. (1988) Injectable collagen for tissue augmentation. In *Collagen. Vol. III. Biotechnology* (ed. Nimni, M.), pp. 117–144. CRC Press, Florida.

118. Forrester, J.V., Docherty, R., Kerr, C. & Lackie, J. (1986) Cellular proliferation in the vitreous: the use of vitreous explants as a model system. *Invest. Ophthalmol. Vis. Sci.* **27**: 1085–1094.

119. Guidry, G. & Grinnell, F. (1986) Contraction of hydrated collagen gels by fibroblasts: evidence for two mechanisms by which collagen fibrils are stabilized. *Coll. Rel. Res.* **6**: 515–529.

120. Brown, A.F. (1982) Neutrophil granulocytes, adhesion and locomotion on collagen substrata and in collagen matrices. *J. Cell. Sci.* **58**: 455–467.

121. Welgus, H.G., Jeffrey, J.J., Stricklin, G.P. & Eisen, A.Z. (1982) The

gelatinolytic activity of human skin fibroblast collagenase. *J. Biol. Chem.* **257**: 11534–11539.

122. Dagalakis, N., Flink, J., Stasikelis, P., Burke, J.F. & Yannas, I.V. (1980) Design of an artificial skin. III. Control of pore structure. *J. Biomed. Mater. Res.* **14**: 511–528.

123. Doillon, C.J., Whyne, C.F., Brandwein, S. & Silver, F.H. (1986) Collagen-based wound dressings: control of the pore structure and morphology. *J. Biomed. Mater. Res.* **20**: 1219–1228.

124. Oliver, R.F., Grant, R.A., Cox, R.W., Hulme, M.J. & Mudie, A. (1976) Histological studies of subcutaneous and intra-peritoneal implants of trypsin prepared dermal collagen in the rat. *J. Clin. Orthopaed. Rel. Res.* **115**: 291–302.

125. Oliver, R.F., Barker, H., Cooke, A. & Grant, R.A. (1982) Hydroxyproline turnover in dermal collagen grafts in reconstructed skin wounds in the rat. *Connect. Tis. Res.* **9**: 59–62.

126. Oliver, R.F., Barker, H., Cooke, A. & Grant, R.A. (1982) Dermal collagen implants. *Biomaterials* **3**: 38–40.

127. Oliver, R.F., Barker, H., Cooke, A. & Stephen, L. (1982) ^{3}H-Collagen turnover in non-cross-linked and aldehyde cross-linked dermal collagen grafts. *Br. J. Exp. Path.* **63**: 13–17.

128. Shakespeare, P.G. & Griffiths, R.W. (1980) Dermal collagen implants in man. *Lancet* i: 795–796.

129. Von der Mark, K. (1981) Localization of collagen types. In *International Review of Connective Tissue Research* (eds Hall, D.A. & Jackson, D.S.), vol. 9, pp. 265–324. Academic Press, London, New York.

130. Meade, K.R. & Silver, F.H. (1990) Immunogenicity of collagenous implants. *Biomaterials* **11**: 176–180.

131. Schmitt, F.O., Levine, L., Drake, M.P., Rubin, A.L., Pfahl, O. & Davison, P.F. (1964) The antigenicity of tropocollagen. *Proc. Natl. Acad. Sci. USA* **51**: 493–497.

132. Davison, P.F., Levine, L., Drake, M.P., Rubin, A.L. & Bump, S. (1967) The serological specificity of tropocollagen telopeptides. *J. Exp. Med.* **126**: 331–345.

133. Knapp, T.R., Luck, E. & Daniels, J.R. (1977) Behaviour of solubilised collagen as a bioimplant. *J. Surg. Res.* **23**: 96–105.

134. Cooperman, L. & Michaeli, D. (1984) The immunogenicity of injectable collagen. I. A 1-year prospective study. *J. Am. Acad. Dermatol.* **10**: 638–646.

135. Kramer, F.M. & Churukian, M.M. (1984) Clinical use of injectable collagen. *Arch. Otolaryngol.* **110**: 93–98.

136. McCoy, J.P. Jr, Schade, W.J., Siegle, R.J., Waldinger, T.P., Vanderveen, E.E. & Swanson, N.A. (1985) Characterisation of the humoral immune response to bovine collagen implants. *Arch. Dermatol.* **121**: 990–994.

137. Elson, M.L. (1989) The role of skin testing in the use of collagen injectable materials. *J. Dermatol. Surg. Oncol.* **15**: 301–303.

138. Barr, R.J. & Stegman, S.J. (1984) Delayed skin test reaction to injectable collagen implant (Zyderm). *J. Am. Acad. Dermatol.* **10**: 652–658.

139. Cooperman, L. & Michaeli, D. (1984) The immunogenicity of injectable collagen. II. A retrospective review of seventy-two tested and treated patients. *J. Am. Acad. Dermatol.* **10**: 647–651.

140. Siegle, R.J., McCoy, J.P, Schade, W. & Swanson, N.A. (1984) Intradermal implantation of bovine collagen. *Arch. Dermatol.* **120**: 183–187.

141. Yannas, I.V. & Tobolsky, A.V. (1967) Cross-linking of gelatine by dehydration. *Nature* **215**: 509–510.

142. Bailey, A.J. & Tromans, W.J. (1964) Effects of ionizing radiation on the

ultrastructure of collagen fibrils. *Radiation Res.* **23**: 145–155.

143. Cheung, D.T., Perelman, N., Tony, D. & Nimni, M. (1990) The effect of a γ-irradiation on collagen molecules, isolates α -chains, and crosslinked native fibres. *J. Biomed. Mater. Res.* **24**: 581–589.

144. Bailey, A.J. (1988) Effect of ionizing radiation on connective tissue components. In *International Review of Connective Tissue Res.* (eds Hall, D.A. & Jackson, D.S.), vol. 4, pp. 233–279. Academic Press, London, New York.

145. Heidemann, E. (1988) The chemistry of tanning. In *Collagen. Vol. III. Biotechnology* (ed. Nimni, M.), pp. 39–62. CRC Press, Florida.

146. Nimni, M.E., Cheung, D.T., Strates, B., Kodama, M. & Sheikh, K. (1988) Bioprosthesis derived from cross-linked and chemically modified collagenous tissues. In *Collagen. Vol. III. Biotechnology* (ed. Nimni, M.), pp. 1–38. CRC Press, Florida.

147. Cater, C.W. (1963) The evaluation of aldehydes and other difunctional compounds as cross-linking agents for collagen. *J. Soc. Leather Trades Chem.* **47**: 259–272.

148. Chambers, R.W., Bowling, M.C. & Grimley, P.M. (1968) Glutaraldehyde fixation in routine histopathology. *Arch. Path.* **85**: 18–30.

149. Hardy, P.M., Nicholls, A.C. & Rydon, H.N. (1976) The nature of the crosslinking of proteins by glutaraldehyde. I. Interaction of glutaraldehyde with amino groups of 6-amino hexanoic acid and of N-acetyl-lysine. *J. Chem. Soc. Perkin* **1**: 958–962.

150. Woodroof, E.A. (1979) The chemistry and biology of aldehyde treated tissue heart valve xenografts. In *Tissue Heart Valves* (ed. Ionescu, M.I.), pp. 349–362. London, Butterworths.

151. Cheung, D.T. & Nimni, M. (1982) Mechanism of crosslinking of proteins by glutaraldehyde. I. Reaction with model compounds. *Conn. Tiss. Res.* **10**: 187–199.

152. Cheung, D.T. & Nimni, M. (1982) Mechanism of crosslinking of proteins by glutaraldehyde. II. Reaction with monomeric and polymeric collagen. *Conn. Tiss. Res.* **10**: 201–216.

153. Gavilanes, J.G., Gonzales de Buitrago, G., Lizarbe, M.A., Municio, A.M. & Olmo, N. (1984) Stabilization of pericardial tissue by glutaraldehyde. *Conn. Tiss. Res.* **13**: 37–44.

154. Cheung, D.T., Perelman, N., Ko, E.C. & Nimni, M. (1985) Mechanism of crosslinking of proteins by glutaraldehyde. III. Reaction with collagen in tissues. *Conn. Tiss. Res.* **13**: 109–115.

155. Salgaller, M.L. & Bakpai, P.K. (1985) Immunogenicity of glutaraldehyde-treated bovine pericardial tissue xenografts in rabbits. *J. Biomed. Mater. Res.* **19**: 1–12.

156. Bowes, J.H. & Cater, C.W. (1966) The reaction of glutaraldehyde with proteins and other biological materials. *J. Royal Microscopical Soc.* **85**: 193–200.

157. Richards, F.M. & Knowles, J.R. (1968) Glutaraldehyde as a protein crosslinking reagent. *J. Mol. Biol.* **37**: 231–233.

158. Gillett, R. & Gull, K. (1972) Glutaraldehyde – its purity and stability. *Histochemie* **30**: 162–167.

159. Monson, P., Puzo, G. & Mazarguil, H. (1975) Etude de mecanisme d'establissement des liaisons glutaraldehyde-proteines. *Biochemie* **57**: 1281–1292.

160. Swenson, M.K., Meir, E., Yanai, P., Zvilichovsky, B. & Blauer, G. (1975) The interaction of glutaraldehyde with poly (αL-lysine), n-butylamine, and collagen. II. Hydrodynamic, electron microscopic, and optical investigations on the reaction products. *Biopolymers* **14**: 2599–2612.

161. Goodfriend, T.L., Leving, L. & Fasman, G.D. (1984) Antibodies to

bradykinin and angiotensin. A use of carbodiimides in immunology. *Science* **144**: 1344–1346.

162. Nimni, M.E., Cheung, D., Strates, B., Kodama, M. & Sheikh, K. (1987) Chemically modified collagen: a natural biomaterial for tissue replacement. *J. Biomed. Mater. Res.* **21**: 741–771.

163. Petite, H., Rault, I., Huc, A., Menasche, P. & Herbage, D. (1990) Use of the acylazide method for crosslinking collagen rich tissues such as pericardium. *J. Biomed. Mater. Res.* **24**: 179–187.

164. Chvapil, M., Chvapil, T.A. & Owen, J.A. (1986) Reaction of various skin wounds in the rat to collagen sponge dressing. *J. Surg. Res.* **41**: 410–418.

165. Van Gulik, T.M., Christiano, K.A., Broekhuizen, A.H., Raaymakers, E.L.F.B. & Klopper, P.J. (1989) A tanned sheep dermal collagen graft as a dressing for split-skin graft donor sites. *Neth. J. Surg.* **41**: 41–43.

166. Bello, J. & Riese-Bello, H. (1958) Liaisons transversales de la gelatine sous l'influence de la chaleur. *Sci. Indust. Photograph* **29**: 361–364.

167. Yannas, I.V., Burke, J.F., Orgill, D.P. & Skrabutt, E.M. (1982) Wound tissue can utilize a polymeric template to synthesize a functional extension of skin. *Science* **215**: 174–176.

168. Laing, J.E. & Shakespeare, P.G. (1983) Skin replacement after burns. *Br. Med. J.* **286**: 245–246.

169. Hefton, J.M., Madden, M.R., Finkelstein, J.L. & Shires, G.T. (1983) Grafting of burn patients with allografts of cultured epidermal cells. *Lancet* **ii**: 428–430.

170. Bell, E., Ehrlich, H.P., Buttle, D.J. & Nakatsu, T. (1981) Living tissue formed in-vitro and accepted as skin-equivalent tissue of full-thickness. *Science* **221**: 1052–1054.

171. Morykwas, M.J., Stevenson, T.R., Marcelo, C.L., Thornton, J.W. & Smith, D.J. (1989) The in-vitro and in-vivo testing of a collagen sheet of support keratinocyte growth for use as a burn wound covering. *J. Trauma* **29**: 1163–1167.

172. Hansborough, J.F., Boyce, S.T., Cooper, M.L. & Foreman, T.J. (1989) Burn wound closure with cultured autologous keratinocytes and fibroblasts attached to a collagen–glycosaminoglycan substrate. *J. Am. Med. Ass.* **212**: 2125–2130.

173. Hancock, K. & Leigh, I.M. (1989) Cultured keratinocytes and keratinocyte grafts. *Br. Med. J.* **299**: 1179–1180.

174. Cuono, C., Langdon, R. & McGuire, J. (1986) Use of cultured epidermal autografts and dermal allografts as skin replacement after burn injury. *Lancet* **i**: 1123–1124.

175. Cuono, C.B., Langdon, R., Birchall, N., Barttelboot, S. & McGuire, J. (1987) Composite autologous-allogenic skin replacement: development and clinical application. *Plast. Reconstr. Surg.* **80**: 626–635.

176. Young, D., Langdon, R., Kahn, R. *et al.* (1989) Analysis of the fate of allografted dermis using a DNA fingerprinting technique. *Proc. Am. Burn Assoc.* **12**: 71–75.

177. Clarke, J. (1987) HIV transmission and skin grafts. *Lancet* **i**: 983.

178. Compton, C.C., Gill, J.M. & Bradford, D.A. (1989) Skin regenerated from cultured epithelial autografts from 6 days to 5 years after grafting. A light, electron microscopic and ultrastructural study. *Lab. Invest.* **60**: 600–612.

179. Coffey, R.J. Jr, Derynck, R., Wilcox, J.N., Bringman, T.S., Goustin, A.S., Moses, H.L. & Pitelkow, M.R. (1987) Production and auto-induction of transforming growth factor-α in human keratinocytes. *Nature* **328**: 817–820.

180. Schulz, G.S., White, M., Mitchell, R., Brown, D., Lynch, J., Twarzdick, D.R. & Todaro, G.J. (1987) Epithelial wound healing enhanced by transforming growth factor-α and vaccinia growth factor. *Science* **235**: 350–352.

181. Eisenger, M., Sadan, S., Silver, I.A. & Flick, R.B. (1988) Growth regulation of skin cells by epidermal cell-derived factors: implications for wound healing. *Proc. Natl. Acad. Sci. USA* **85**: 1937–1941.

182. Murphy, C.F., Orgill, D.P. & Yannas, I.V. (1990) Partial dermal regeneration is induced by biodegradable collagen-glycosaminoglycan grafts. *Lab. Invest.* **63**: 305–313.

183. Quinby, W.C., Hoover, H.C., Scheflan, M., Walters, P.T., Slavin, S.A. & Bondoc, C.C. (1982) Clinical trials of amniotic membranes in burn wound care. *Plast. Reconstr. Surg.* **70**: 711–716.

184. Reindorf, C.A., Walker-Jones, D., Adelike, A.D., Lawal, O. & Oluwole, S.F. (1989) Rapid healing of sickle-cell leg ulcers treated with collagen dressing. *J. Natl. Med. Assoc.* **81**: 866–868.

185. Burton, J.L., Etherington, D.J. & Peachey, R.D. (1978) Collagen sponge for leg ulcers. *Br. J. Dermatol.* **99**: 681–685.

186. Armstrong, R.B., Nichols, J. & Pachance, J. (1986) Punch biopsy wounds treated with Monsel's solution of a collagen matrix. *Arch. Dermatol.* **122**: 546–549.

187. Dunn, M.G., Doillon, C.J., Berg, R.A., Olson, R.M. & Silver, F.H. (1988) Wound-healing using a collagen matrix: effect of DC electrical stimulation. *J. Biomed. Mater. Res.* **22**: 191–206.

188. Doillon, C.J., Whyne, C.F., Berg, R.A., Olsen, R.M. & Silver, F.H. (1984) Fibroblast–collagen sponge interactions and the spatial deposition of newly formed collagen fibres in-vitro and in-vivo. *Scanning Electron Microsc.* **3**: 1313–1320.

189. Ruoslati, E., Hayman, E.G. & Pierschbacher, M.D. (1985) Extracellular matrices and cell adhesion. *Arteriosclerosis* **5**: 581–594.

190. Srivastava, S. (1988) Implantable collagen-based biomaterials: influence of polymer modifications on cellular interactions. PhD Thesis, University of Strathclyde, UK.

191. Srivastava, S., Gorham, S.D. & Courtney, J.M. (1990) The attachment and growth of an established cell line on collagen, chemically modified collagen, and collagen composite surfaces. *Biomaterials* **11**: 162–168.

192. Scott, J.E., Orford, C.R. & Hughes, E.W. (1981) Proteoglycan–collagen arrangements in developing rat-tail tendon. *Biochem. J.* **195**: 573–581.

193. Olberg, A. & Ruoslahti, B. (1982) Interactions between chondroitin sulphate proteoglycan, fibronectin and collagen. *J. Biol. Chem.* **257**: 4859–4863.

194. Lilja, S. & Barrach, H.J. (1983) Normally sulphated and highly sulphated glycosaminoglycans CGAC/affecting fibrillogenesis of type I and type II collagen in-vitro. *Exp. Pathol.* **23**: 173–181.

195. Chandrasehkar, S., Kleinman, H., Hassell, J.R., Martin, G.R., Termine, J.D. & Trelstad, R.L. (1984) Regulation of type I collagen fibril assembly by link protein and proteoglycans. *Coll. Rel. Res.* **4**: 323–338.

196. Katz, E.P., Wachtel, E.J. & Maroudas, A. (1986) Extrafibrillar proteoglycans osmotically regulate the molecular packing of collagen in cartilage. *Biochim. Biophys. Acta.* **882**: 136–139.

197. Kasai, S., Miyatu, T., Kunimoto, T. & Nitta, K. (1983) Enhanced effect of chemically modified collagen substratum on attachment and growth of fibroblasts in serum free media. *Biomed. Res.* **4**: 147–154.

198. McPherson, J.M., Ledger, P.W., Ksander, G., Sawamura, S.J., Conti, A., Kincaid, S., Michaeli, D. & Clark, A.F. (1988) The influence of heparin on the wound healing response to collagen implants in-vivo. *Coll. Rel. Res.* **1**: 83–100.

199. Gorham, S.D., Monsour, M.J. & Scott, R. (1987) The in-vitro assessment of a collagen/vicryl (polyglactin) composite film together with candidate suture material for use in urinary tract surgery. I. Physical testing. *Urol. Res.* **15**: 53–59.

200. Carachi, R., Azmy, A., Gorham, S., Reid, J. & French, D.A. (1989) Use of a bioprosthesis to relieve tension in oesophageal anastomosis: an experimental study. *Br. J. Surg.* **76**: 496–498.

201. Gossot, D. & Lefebvre, J.-F. (1988) Ischaemic atrophy of the cervical portion of a substernal colic transplant: successful reconstruction using a synthetic resorbable tube. *Br. J. Surg.* **75**: 801–802.

202. Siksik, J.M., Cugnenc, P.H., Frileux, P., Gallix, P., Calise, D. & Arsca, M. (1989) Small bowel anastomoses in a septic environment protected by a polyglactin 910–collagen mesh. Experimental study in the rat. *Ann. Chirurg.* **43**: 236–240.

203. Gossot, D., Celerier, M. & Sarfati, E. (1988) Protection des anastomoses cervicales après coloplastie retrosternal par un treillis resorbable impregne de collagene. *La Presse Médicale* **17**: 30–32.

204. Shoshan, S. (1981) Wound healing. In *International Review of Connective Tissue Research* (eds Hall, D.A. & Jackson, D.S.), vol. 9, pp. 1–26. Academic Press, London/New York.

205. Peacock, E.E., Seigler, H.F. & Biggers, P.W. (1965) Use of tanned collagen sponges in the treatment of liver injuries. *Ann. Surg.* **161**: 238–247.

206. Scheele, J., Gentsch, H.H. & Matteson, E. (1982) Splenic repair by fibrin tissue adhesive and collagen fleece. *Surgery* **95**: 6–13.

207. Jakob, H., Campbell, C.D., Stemberger, A., Wriedt-Lubbe, I., Blumel, G. & Replogle, R.L. (1984) Combined application of heterologous collagen and fibrin sealant for liver injuries. *J. Surg. Res.* **36**: 571–577.

208. Margarit, C., Martinez-Ibanez, V., Lloret, J., Ventura, H., Broto, J., Isnard, R., Blanco, J. & Allende, H. (1987) Segmental liver transplantation in pigs: use of a fibrin sealant and collagen as hemostatic agents. *Transplantn. Proc.* **19**: 3835–3837.

209. Sakon, M., Monden, M., Gotoh, M., Kobayoshii, K., Kambayashi, J., Mori, T. & Okamura, J. (1989) Use of microcrystalline collagen powder and fibrinogen tissue adhesive for haemostasis and prevention of rebleeding in patients with hepatocellular carcinoma associated with cirrhosis of the liver. *Surg. Gynecol. Obstet.* **168**: 453–454.

210. Saroff, S.A., Chasens, A.T., Eisen, S.F. & Levey, S.H. (1982) Free soft tissue autografts. Haemostasis and protection of the palatal donor site with a microfibrillar collagen preparation. *J. Periodontol.* **53**: 425–428.

211. Asenii, P., Vertemati, M., Beati, C., Brenna, S. & Belli, L. (1985) Control of bleeding from vascular anastomoses using a microcrystalline collagen haemostat. *IRCS Med. Sci.* **13**: 1185.

212. Hait, M.R., Robb, C.A., Baxter, C.R., Borgmann, A.R. and Tippett, L.O. (1973) Comparative evaluation of avitene microcrystalline collagen haemostat in experimental animal wounds. *Am. J. Surg.* **125**: 284–287.

213. Mason, R.G. & Read, M.S. (1974) Some effects of a microcrystalline collagen preparation on blood. *Haemostasis* **3**: 31–45.

214. Pertuiset, B. & Sichez, J.P. (1982) Use of a new antihaemorrhagic agent in neurosurgery. Trial in 35 cases. *Neurochurgie* **28**: 41–43.

215. Hatsuoka, M., Seiki, M., Sasaki, K. & Kashii, A. (1986) Haemostatic effects of microfibrillar collagen haemostat (MCH) in experimental coagulopathy model and its mechanism of haemostasis. *Thromb. Res.* **42**: 407–412.

216. Jasmin, J. & Fontaine, M. (1987) Effectiveness of a haemostatic collagen dressing compared with regenerated oxidised cellulose in oral surgery. *Curr. Therap. Res.* **42**: 172–181.

217. Eloy, R., Baguet, J., Christie, G., Rissoan, M.C., Paul, J. & Belleville, J. (1988) An in-vitro evaluation of the haemostatic activity of topical agents. *J. Biomed. Mater. Res.* **22**: 149–157.

218. Robicsek, F., Duncan, G.D., Born, G.V.R., Wilkinson, H.A., Masters, T.N. & McClure, M. (1986) Inherent dangers of simultaneous applications of microfibrillar collagen haemostat and blood-saving devices. *J. Thorac. Cardiovasc. Surg.* **92**: 766–770.

219. Niebauer, G.W., Mehmet, C., Goldschmidt, M. & Lemole, G. (1989) Simultaneous use of microfibrillar collagen haemostat and blood-saving devices in a canine kidney perfusion model. *Ann. Thorac. Surg.* **48**: 523–527.

220. Bailin, P.L. & Bailin, M.D. (1982) Correction of depressed scars following Moh's surgery: the role of collagen implantation. *J. Dermatol. Surg. Oncol.* **8**: 845–849.

221. Tromovitch, T.A., Stegman, S.J. & Glogau, R.G. (1984) Zyderm collagen implant techniques. *J. Am. Acad. Dermatol.* L **10**: 273–278.

222. Klein, A.W. & Rish, D.C. (1985) Substances for soft tissue augmentation: collagen and silicone. *J. Dermatol. Surg. Oncol.* **11**: 337–339.

223. Stegman, S.J., Chu, S., Bensch, K. & Armstrong, R. (1987) A light and electron microscopic evaluation of Zyderm collagen and Zyplast implants in aging human facial skin. *Arch Dermatol.* **123**: 1644–1649.

224. Webster, R.C., Kattner, M.D. & Smith, R.C. (1984) Injectable collagen for augmentation of facial areas. *Arch. Otolaryngol.* **110**: 652–656.

225. Tolleth, H. (1985) Long-term efficacy of collagen. *Aesth. Plast. Surg.* **9**: 155–158.

226. Matton, G., Anseeuw, A. & De Keyser, F. (1985) The history of injectable biomaterials and the biology of collagen. *Aesth. Plast. Surg.* **9**: 133–140.

227. Stegman, S.J. & Tromovitch, T.A. (1980) Implantation of collagen for depressed scars. *J. Dermatol. Surg. Oncol.* **6**: 450–453.

228. Stegman, S.J. & Tromovitch, T.A. (1983) *Injectable Collagen in Cosmetic Dermatologic Surgery* (eds Stegman, S.J. & Tromovitch, T.A.). Year Book Medical Publishers, Chicago.

229. Ford, C.N., Martin, D.W. & Warner, T.F. (1984) Injectable collagen in laryngeal rehabilitation. *Laryngoscope* **94**: 513–518.

230. Ford, C.N. & Bless, D.M. (1986) A preliminary study of injectable collagen in human vocal fold augmentation. *Otolaryngol. Head. Neck. Surg.* **94**: 104–112.

231. Ford, C.N. & Bless, D.M. (1986) Clinical experience with injectable collagen for vocal fold augmentation. *Laryngoscope* **96**: 863–869.

232. Kligman, A.M. & Armstrong, R.C. (1986) Histologic reponse to intradermal Zydem and Zyplast (glutaraldehyde cross-linked) collagen in humans. *J. Dermatol. Surg. Oncol.* **12**: 351–357.

233. Li, Shu-Tung (1988) Collagen and vascular prosthesis. In *Collagen. Vol. III. Biotechnology* (ed. Nimni, M.), pp. 253–272. CRC Press, Florida.

234. Rao, K.P. & Joseph, K.T. (1988) Collagen graft polymers and their biomedical applications. In *Collagen. Vol. III. Biotechnology* (ed. Nimni, M.), pp. 63–86. CRC Press, Florida.

235. Venkataramani, E.S., Senatore, F.F., Lokapur, A.K., Shankar, H. & Feola, M. (1986) Studies on heparin immobilization to collagen. *Res. Comm. Chem. Path. Pharmacol.* **54**: 421–424.

236. Lee, M.E., Chaux, A., Kass, R.M. & Matloff, J.M. (1988) Elimination of graft haemorrhage in heparinized patients with a collagen coated Dacron prosthesis. *Mount Sinai J. Med.* **55**: 144–146.

237. Chignier, E., Eloy, R., Huc, A., Gimeno, R. & Gleizal, C. (1985) Long-term behaviour of bovine collagen membrane used as vascular substitute. Experimental study in rats. *J. Biomed. Mater. Res.* **19**: 115–131.

238. Carpentier, A., Nashref, A., Carpentier, S., Goussef, N., Relland, J., Levy,

R.J., Fishbein, M.C., El Asmar, B., Benomar, M., Benomar, M., El Sayed, S. & Donzeau-Gauge, P.G. (1982) Prevention of tissue valve calcification by chemical techniques. In *Cardiac Bioprostheses* (eds Cohn, A.H. & Galluchi, V.), pp. 320–327. Yorke Medical Books, New York.

239. Lentz, D.J., Pollock, E.M., Olsen, D.B., Andrews, E.J., Murashita, J. & Hastings, W.L. (1982) Inhibition of mineralization of glutaraldehyde-fixed Hancock bioprosthetic heart valves. In *Cardiac Bioprostheses* (eds Cohn, A.H. & Galluchi, V.), pp. 306–319. Yorke Medical Books, New York.

240. Werkmeister, J.A., Peters, D.E. & Ramshaw, J.A.M. (1989) Development of monoclonal antibodies to collagens for assessing host–implant interactions. *J. Biomed. Mater. Res.* **23**: 273–283.

241. Jonas, R.A., Schoen, F.J., Levy, R.J. & Castaneda, A.R. (1986) Biological sealants and knitted Dacron: porosity and histological. *Ann. Thorac. Surg.* **41**: 657–663.

242. Jonas, R.A., Schoen, F.J., Ziemar, G., Britton, L. & Castaneda, A.R. (1987) Biological sealants and knitted Dacron conduits: comparison of collagen and fibrin glye pretreatments in circulatory models. *Ann. Thorac. Surg.* **44**: 283–290.

243. Senatore, F., Bernath, F. & Meisner, K. (1986) Clinical study of urokinase-bound fibrocollagenous tubes. *J. Biomed. Mater. Res.* **20**: 177–188.

244. Shindo, S., Takagi, A. & Whittemore, A.D. (1987) Improved patency of collagen impregnated grafts after in-vitro autogenous endothelial cell seeding. *J. Vasc. Surg.* **6**: 325–332.

245. Dewanjee, M.K., Singh, S.K., Wooley, P.H., Mackey, S.T., Solis, E. & Kaye, M.P. (1986) Identification of new collagen formation with ^{125}I-labelled antibody in bovine pericardial tissue valves implanted in calves. *Nucl. Med. Biol.* **13**: 413–422.

246. Ferrans, V.J., Hilbert, S.L., Tomita, Y., Jones, M. & Roberts, W.C. (1988) Morphology of collagen in bioprosthetic heart valves. In *Collagen. Vol. III. Biotechnology* (ed. Nimni, M.), pp. 145–190. CRC Press, Florida.

247. Mulberger, R.D. & Carmichael, P.L. (1962) Experimental implants of collagen sponge material in rabbit eyes. *Am. J. Ophthalmol.* **54**: 19–20.

248. L'Esperance (1965) Reconstituted collagen tape in retinal detachment surgery. *Arch. Ophthalmol.* **73**: 472–475.

249. Tanner, J.C., Smith, J.P., Bradley, W.H. & Rife, C.C. (1968) Lamellar keratoplasty: use of collagen for corneal replacement. *Eye Ear Nose Throat Monthly* **47**: 27–30.

250. Weber, P.A. & Baker, N.D. (1989) The use of cyanoacrylate adhesive with a collagen shield in breaking filtering blebs. *Ophthalmic Surg.* **20**: 284–285.

251. Marmer, R.H. (1988) Therapeutic and protective properties of the corneal collagen shield. *J. Cataract Refract. Surg.* **14**: 496–499.

252. Poland, D.E. & Kaufman, H.E. (1988) Clinical uses of collagen shields. *J. Cataract Refract. Surg.* **14**: 489–491.

253. Unterman, S.R., Rootman, D.S., Hill, J.M., Parelman, J.J., Thompson, H.W. & Kaufman, H.E. (1988) Collagen shield drug delivery: therapeutic concentrations of tobramycin in the rabbit cornea and aqueous humour. *J. Cataract Refract. Surg.* **14**: 500–504.

254. Sawusch, M.R., O'Brien, T.P., Dick, J.D. & Gottsch, J.D. (1988) Use of collagen corneal shields in the treatment of bacterial keratitis. *Am. J. Ophthalmol.* **106**: 279–281.

255. Vasantha, R., Sehgal, P.K. & Panduranga Rao, K. (1988) Collagen ophthalmix inserts for pilocarpine drug delivery system. *Int. J. Pharmaceutics* **47**: 95–102.

256. Tsuji, I., Kuroda, K., Fujieda, J., Shiraishi, Y., Kaissai, T. & Ishida, H.

(1963) A clinical and experimental study of cystoplasty not using the intestine. *J. Urol.* **89**: 214–225.

257. Tsuji, I., Shiraishi, Y., Kaissai, T., Kunishima, K., Orikasa, S. & Abe, N. (1967) Further experimental investigation on bladder reconstruction without using the intestine. *J. Urol.* **97**: 1021–1028.

258. Gorham, S.D., McCafferty, I., Baraza, R. & Scott, R. (1984) Preliminary development of a collagen membrane for use in urological surgery. *Urol. Res.* **12**: 295–299.

259. Gorham, S.D., Anderson, J.D., Monsour, M.J. & Scott, R. (1988) The in-vitro assessment of a collagen/vicryl (polyglactin) composite film together with candidate suture materials for use in urinary tract surgery. II. Surface deposition of urinary salts. *Urol. Res.* **16**: 111–117.

260. Gemmell, C.G., Gorham, S.D., Monsour, M.J., McMillan, F. & Scott, R. (1988) The in-vitro assessment of a collagen/vicryl (polyglactin) composite film together with candidate suture materials for potential use in urinary tract surgery. III. Adherence of bacteria to the material surface. *Urol. Res.* **16**: 381–384.

261. Monsour, M.J., Mohammed, R., Gorham, S.D., French, D.A. & Scott, R. (1987) An assessment of a collagen/vicryl composite membrane to repair defects of the urinary bladder in rabbits. *Urol. Res.* **15**: 235–238.

262. Monsour, M.J., Mohammed, R., Gorham, S.D., French, D.A. & Scott, R. (1987) An assessment of a collagen/vicryl composite membrane to repair defects of the urinary bladder in rabbits. *Urol. Res.* **15**: 235–238.

263. Gorham, S.D., French, D.A., Shivas, A.A. & Scott, R. (1989) Some observations on the regeneration of smooth muscle in the repaired urinary bladder of the rabbit. *Eur. Urol.* **16**: 440–443.

264. Yaffe, A., Ehrlich, J. & Shoshan, S. (1984) Restoration of periodontal attachment employing enriched collagen solution in the dog. *J. Periodont.* **55**: 623–628.

265. Tanner, M.G., Solt, C.W. & Vuddhakanok, S. (1988) An evaluation of new attachment formation using a microfibrillar collagen barrier. *J. Periodont.* **59**: 524–530.

266. Ciancio, S.G., Golub, L.M., Matter, M.L. & Bunnell, H. (1985) The application of a collagen stabiliser to the gingiva of the Beagle dog. *J. Periodont.* **56**: 148–153.

267. Yaffe, A. & Shoshan, S. (1987) Re-attachment of periodontal ligament by collagen in experimentally induced alveolar bone dehiscence in dogs. *Arch. Oral. Biol.* **32**: 69–75.

268. Greenwald, R.A., Golub, L.M., Lavietes, B., Ramamurthy, N.S., Gruber, B., Laskin, R.S. & McNamara, T.F. (1987) Tetracyclines inhibit human synovial collagenase in-vivo and in-vitro. *J. Rheumatol.* **14**: 28–32.

269. Braun, R.M. (1964) Experimental peripheral nerve repair tubulation. *Surgical Forum* **15**: 452–454.

270. Kline, D.G. & Hayes, G.J. (1963) The use of a resorbable wrapper for peripheral-nerve repair. *Neurosurgery* **21**: 737–740.

271. Parker, G., White, T. & Jenkins, R. (1984) Surgical repair of extratemporal facial nerve: a comparison of suture repair and microfibrillar collagen repair. *Laryngoscope* **94**: 950–953.

272. Satou, T., Nishida, S., Hiruma, S., Tanji, K., Takahashi, M., Fujita, S., Mizuhara, Y., Akai, F. & Hashimoto, S. (1986) A morphological study on the effects of collagen gel matrix on regeneration of severed rat sciatic nerve in silicone tubes. *Acta. Pathol. Japan* **36**: 199–208.

273. Valentini, R.F., Aebischer, P., Winn, S.R. & Galletti, P.M. (1987) Collagen

and laminin containing gels impede peripheral nerve regeneration through semipermeable nerve guidance channels. *Exp. Neurol.* **98**: 350–356.

274. Kinel, F.A., Ciaccio, L.A. & Henderson, S.B. (1984) Collagen as a drug carrier. *Arch. Pharm.* **317**: 657–661.

275. Peng, Y.M., Alberts, D.S., Graham, V., Surwit, E.A., Weiner, S. & Meyskens, F.L. (1986) Cervical tissue uptake of all-*trans*-retinoic acid delivered via a collagen sponge–cervical cap delivery device in patients with cervical dysplasia. *Invest. New Drugs* **4**: 245–249.

276. Deatherage, J.R. & Miller, E.J. (1987) Packaging and delivery of bone induction factors in a collagenous implant. *Coll. Rel. Res.* **7**: 225–231.

277. Wasserman, A.J., Kato, Y.P., Ganim, D., Dunn, M.G. & Silver, F.H. (1989) Neotendon formation using a collagen fibre tendon prosthesis. *Proc. 47th A. Meet. Electron Microscopy Soc. Am.* (ed. Bailey, G.W.), p. 880. San Francisco Press, San Francisco.

278. Goldstein, J.D., Tria, A.J., Zawadsky, J.P., Kato, Y.P., Christiansen, D. & Silver, F.H. (1989) Development of a reconstituted collagen tendon prosthesis. *J. Bone Joint Surg.* **A17**: 1183–1191.

279. Gelman, R.A., Poppke, D.C. & Piez, K.A. (1979) Collagen fibril formation in-vitro. The role of the nonhelical terminus regions. *J. Biol. Chem.* **254**: 11741–11745.

280. Kato, Y.P., Christiansen, D.L., Hahn, R.A., Shieh, S.H., Goldstein, J.D. & Silver, F.H. (1989) Mechanical properties of collagen fibres: a comparison of reconstituted and rat-tail tendon fibres. *Biomaterials* **10**: 38–42.

281. Hughes, K.E., Fink, D.J., Hutson, T.B. & Veis, A. (1984) Oriental fibrillar collagen and its application to biomedical devices. *J. Am. Leather Chem. Assoc.* **79**: 146–158.

282. Chvapil, M., Gibeault, D. & Wang, T.-F. (1987) Use of chemically purified and cross-linked bovine pericardium as a ligament substitute. *J. Biomed. Mater. Res.* **21**: 1383–1393.

283. Law, T.K., Parsons, J.T., Silver, F.H. & Weiss, A.B. (1989) An evaluation of purified reconstituted type I collagen fibres. *J. Biomed. Mater. Res.* **23**: 961–977.

3 Polyhydroxyalkanoic acids

Alexander Steinbüchel

Institut für Mikrobiologie, Georg-August-Universität Göttingen,
Grisebachstraße 8, D-3400 Göttingen, Germany

5 Polyhydroxyalkanoic acids

Alexander Steinbüchel

Institut für Mikrobiologie, Georg-August-Universität Göttingen, Grisebachstraße 8, D-3400 Göttingen, Germany

Introduction

Polyhydroxyalkanoic acids (PHA) are water-insoluble polyesters of alkanoic acids containing a hydroxyl group as at least one functional group in addition to the carboxy group, and possess the general formula shown in Fig. 1. Although some of these polymers are also available from chemical synthesis, many of them are synthesized by bacteria and are deposited in abundant amounts in the cytoplasm of the cells. The variability of the position of the hydroxyl group and of the type of the R-pendant group of the constituents (see below), a large variety of different constituent monomers in copolyesters, as well as the varying degree of polymerization, allow the biosynthesis of many different polymers exhibiting different physical properties.

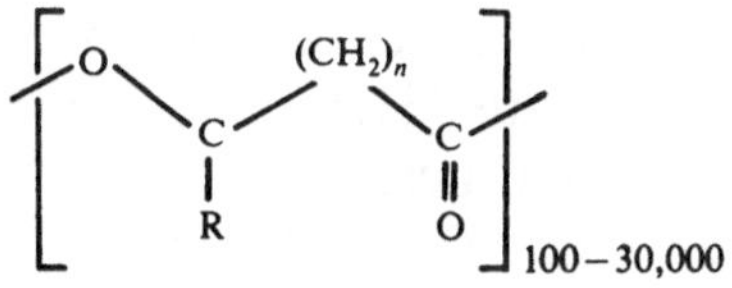

Fig. 1 General structural formula of polyhydroxyalkanoic acids.

The results emerging from research on PHA have been summarized in numerous reviews and have been discussed from a general and physiological viewpoint[1–17]. A very comprehensive review was published recently[18]. Milestones in the research of these bacterial polyesters were recently reviewed by Marchessault[19] and Schlegel[20] as were the results of studies on the molecular analysis of the *Alcaligenes eutrophus* PHA-biosynthetic genes by the author[21–23]. Other reviews focused on the development of biotechnological processes for the production of PHA and on various technical applications of these polyesters[9,24–35].

At special sessions on the occasions of larger meetings, mostly organized by Chemical Societies (for example, June 1988 in Toronto, Canada), or at meetings which focused on biotechnological aspects (for example, June 1988 in Barcelona, Spain), results of research on

PHA have been presented. In May 1990, at the NATO Advanced Research Workshop on 'New Biosynthetic Biodegradable Polymers of Industrial Interest from Microorganisms' in Sitges, Spain, and in October 1990 at the International Symposium on Biodegradable Polymers in Tokyo, Japan, PHA were the most important polymers on the agenda. The proceedings of the NATO ARW were published recently[8]. The increasing interest in PHA reflects the achievements of basic research in this field, the anticipation of biotechnological production and the increasing numbers of applications of these polyesters (for example, see ref. 36).

Hydroxyalkanoate Units in Bacterial Copolyesters

Poly(3-hydroxybutyric acid), or PHB, whose structural formula is shown in Fig. 2, was detected 70 years ago as a constituent of bacterial cells[37]. For almost 60 years it was the only known member of this type of polyester. Apart from the polyester isolated from butane-grown cells of *Nocadia* sp.[38], which was tentatively identified in 1964 as a copolyester of 3-hydroxybutyrate (3HB) and 3-hydroxy-2-butenoate, constituents other than 3HB synthesized by defined bacterial cultures were only described after 1982. This was surprising, because Wallen and Rohwedder isolated a polymer consisting of 3HB and 3-hydroxyvalerate (3HV) from flocks of activated sludge as early as 1972. Traces of 3-hydroxyhexanoate (3HHx) and 3-hydroxyheptanoate (3HHp) were also present in this polymer preparation[39,40]. The authors also noted that the polymer melted at 97–100°C, whereas PHB homopolyester melted at 160–170°C. Other physical properties of the polymer isolated by Wallen were investigated by Marchessault[41,42].

Application of infrared spectroscopy (for example, see refs 43, 44) and of advanced gas chromatographic (for example, see refs 45–50) or NMR spectroscopic (for example, refs 51, 52) methods to the analysis of PHA revealed the presence of many other hydroxyalkanoic acids in bacterial storage polyesters. In 1982, Imperial Chemical Industries

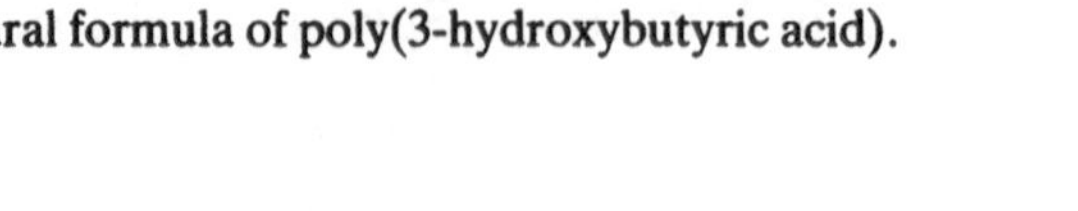

Fig. 2 Structural formula of poly(3-hydroxybutyric acid).

patented a process for the production of a copolyester consisting of 3HB and 3HV, poly(3HB-*co*-3HV), from cells of *A. eutrophus*[53]. In 1983, 3-hydroxyoctanoate (3HO) was detected as a main constituent of a polyester accumulated by *Pseudomonas oleovorans* during cultivation on octane in the laboratory of B. Witholt[54]. At the same time, Findlay and White detected 3HV, 3HHp and three unidentified 3-hydroxyalkanoic acids, in addition to 3HB, as constituents of a polyester extracted from cells of *Bacillus megaterium*. One of the latter was most likely an iso-branched 3-hydroxyalkanoic acid[55].

Extension of the search for new building blocks in bacterial PHA has been reported from the laboratories of Y. Doi (Tokyo Institute of Technology, Tokyo, Japan), B. Witholt (University of Groningen, Groningen, The Netherlands), R. C. Fuller and R. W. Lenz (both University of Massachusetts, Amherst, USA), and has revealed the presence of almost 35 different constituents (Table 1). Straight-chain saturated 3-hydroxyalkanoic acids with a chain length ranging from 3 to 12 carbon atoms, as well as unsaturated 3-hydroxyalkenoic acids with a chain length ranging from 5 to 14 carbon atoms and one or two double bonds, were detected in polymer. Hydroxyalkanoic acids with an aromatic side chain[56] were also isolated from PHA produced by bacteria. In addition, various branched and brominated 3-hydroxyalkanoic acids were incorporated. Not only 3-hydroxyalkanoic acids but also 4-hydroxybutyric acid and 5-hydroxyvaleric acid were identified as constituents of PHA (Table 1).

A related homopolyester of malic acid has been detected in fungi. *Penicillium cyclopium*[57] and *Physarum polycephalum*[58] synthesize poly(β-malic acid). This polyester is the only water-soluble biosynthetically produced PHA. Its function is not known, but it is assumed that it binds to DNA[58].

The bacterial PHA are related to poly(2-hydroxyalkanoic acids) such as poly(lactic acid) or poly(glycolic acid), a hydrolytically unstable class of compounds used as drug-delivery systems or in surgery primarily for their degradative properties (for reviews, see refs 59, 60). However, 2-hydroxyalkanoic acids, 6-hydroxyalkanoic acids or alkanoic acids with the hydroxyl group at an even more remote position relative to the carboxy group have never been detected as constituents of bacterial polyesters. They are only available from chemical synthesis.

Table 1 Constituents detected in microbial PHAs

Constituent	Reference
3-hydroxypropionic acid	194
3-hydroxybutyric acid	37
3-hydroxyvaleric acid	53, 162, 206
3-hydroxyhexanoic acid	43, 46
3-hydroxyheptenoic acid	43, 46
3-hydroxyoctanoic acid	43, 46, 54
3-hydroxynonanoic acid	43, 46
3-hydroxydecanoic acid	43, 46
3-hydroxyundecanoic acid	43, 46
3-hydroxydodecanoic acid	43
Unidentified straight-chain 3-hydroxyalkanoic acid	55
4-hydroxybutyric acid	192, 193
5-hydroxyvaleric acid	195
3-hydroxy-2-butenoic acid	38
3-hydroxy-4-pentenoic acid	56
3-hydroxy-4-hexenoic acid	197
3-hydroxy-5-hexenoic acid	43, 197, 239
3-hydroxy-6-octenoic acid	197
3-hydroxy-7-octenoic acid	43, 197, 239
3-hydroxy-8-nonenoic acid	43
3-hydroxy-9-decenoic acid	43, 239
3-hydroxy-6-dodecenoic acid	198
3-hydroxy-5-tetradecenoic acid	198
3-hydroxy-5,8-tetradecenoic acid	198
3-hydroxy-6-bromohexanoic acid	56
3-hydroxy-8-bromooctanoic acid	56
3-hydroxy-11-bromoundecanoic acid	56
3-hydroxy-4-methylhexanoic acid	52
3-hydroxy-5-methylhexanoic acid	52
3-hydroxy-5-methyloctanoic acid	52
3-hydroxy-6-methyloctanoic acid	52
3-hydroxy-7-methyloctanoic acid	52
Unidentified iso-branched 3-hydroxyalkanoic acid	52
Malic acid	57, 58, 240

Location of PHA in Cells

Inclusion bodies

PHA occur in two different forms in the bacterial cell. In most micro-organisms PHA occur as inclusion bodies and are deposited in granules in the cytoplasm[61]. The granules can be as large as 0.5 μm, and usually several granules are present in the cell (Fig. 3). These granules cause light-scattering and can contribute significantly to the opacity of colonies. *A. eutrophus* for instance, accumulates PHB in quantities of up to 80% of the biomass[62]; there has been one report[63] of a value of 96%.

The native structure of the granules and the physical state of PHA in these granules is not fully understood. Beside PHA, the granules contain about 2% (w/w) protein and 0.5% (w/w) lipid[64,65]. It was assumed that these lipids are located on the surface of the core granule forming a membrane, and that proteins involved in the metabolism of

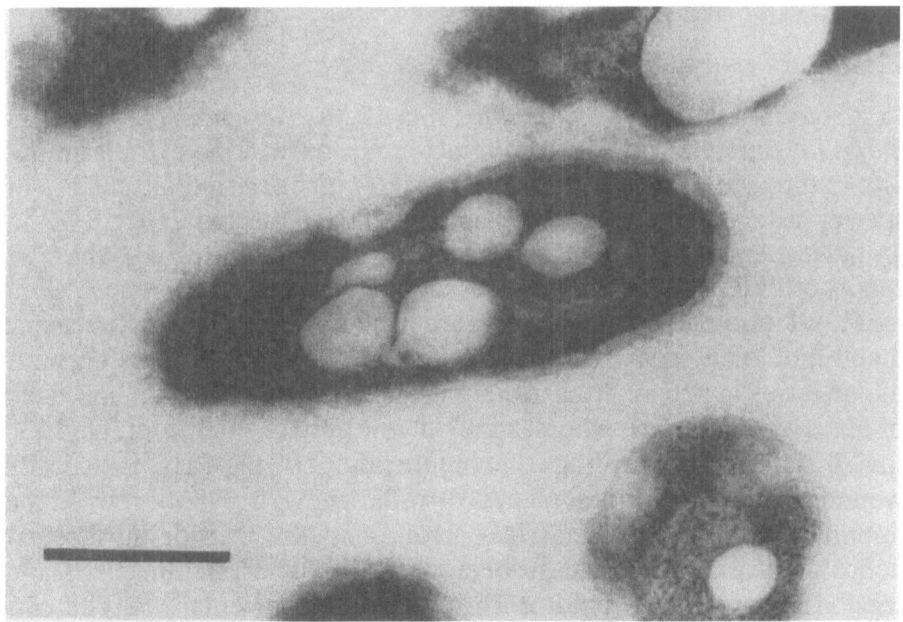

Fig. 3 Electron micrograph of a thin section of *Alcaligenes eutrophus* cells. Scale bar 0.5 μm (Courtesy of S. Wiese, Göttingen).

129

PHA, such as PHA synthase and PHA depolymerase, are most probably associated with the granules in the membrane. It has been demonstrated in *A. eutrophus* that PHA synthase is associated with the granules in the stationary phase of growth during the accumulation of PHB[66]. However, the primary structure of PHA synthase as deduced from the nucleotide sequence did not distinguish this enzyme as a characteristic hydrophobic membrane protein[67,68]. Recently, evidence was obtained by ^{13}C-NMR spectroscopy[69,70] as well as by X-ray diffraction studies[255] that at least 70% of the polyester in the native granules occurs in a very mobile amorphous state. Freezing and storage of the granules at low temperatures, treatment with, for example, acetone or sodium hypochlorite solution, or even application of strong centrifugational forces to the cells, triggered the irreversible conversion of the polyester to the crystalline state that had been observed in earlier studies by X-ray diffraction[71,72]. The granules contain approximately 40% (w/w) water[65] and NMR spectroscopy suggested that this water is an integral component of the granules. It acts as a plasticizer for PHA, and it was proposed that PHA synthase and PHA depolymerase are operative in the mobile hydrated polymer[70].

PHB:Ca^{2+}:polyphosphate complexes

In *Azotobacter vinelandii*, *Bacillus subtilis*, *Escherichia coli* and *Haemophilus influenzae*, PHB was detected not only in the cytoplasm but also in the cytoplasmic membrane[73–75]. *H. influenzae* and *E. coli* are of special interest because they are unable to accumulate PHB in granules. Apart from that of Gilbert and Brown[76], this was the first report of PHB in *E. coli*. Quasi-crystalline complexes of PHB, calcium ions and polyphosphate (PP$_i$) were isolated from the cytoplasmic membrane of *E. coli*[77]. In these complexes, the constituents occur at molar ratios approximating 1:0.5:1 (Fig. 4). The PHB chain consisted of 120 to 200 subunits, whereas the polyphosphate chain contained 130 to 170 subunits. According to computer models, the PHB is helically wound, with 14 monomers per turn, forming a cylinder with the lipophilic methyl and methylene groups on the outside surface. A helical conformation has also been reported for crystalline PHB and for PHB dissolved in chloroform or in other solvents[78–80]. The carbonyl oxygen atoms of the ester linkage are located on the inner surface of the cylinder and are complexed via Ca^{2+} with the PP$_i$ strand

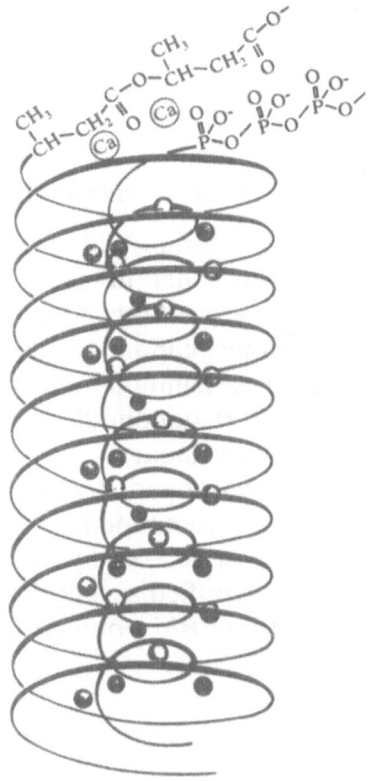

Fig. 4 Structure of PHB:Ca^{2+}:polyphosphate complexes spanning the cytoplasmic membrane as proposed by Reusch. Reproduced from ref. 81.

which is also helically wound[77]. The outer diameter of these trans-membrane channels is 24Å; the length was 45Å and approaches the dimensions of the unit membrane. Because the highest concentrations of these PHB:Ca^{2+}:PP$_i$ complexes were found in genetically competent cells, and the inner diameter of the cylinder provides enough space for single-stranded DNA, it was suggested that the complexes are involved in the uptake of DNA. Transport of calcium ions and/or phosphate were other suggested functions.

Interestingly, PHB:Ca^{2+}:PP$_i$ complexes were also detected and isolated from various plant and animal tissues[81]. This is the only report of the occurrence of PHB in eukaryotic cells. It is completely unknown how PHB is synthesized in these organisms and in *E. coli*. Enzymatic analysis and hybridization studies done during the cloning of the *A.*

131

eutrophus PHB-biosynthetic genes have not revealed the presence of PHB synthase activity or of DNA molecules in *E. coli* exhibiting homologies with *A. eutrophus* PHB synthase (A. Steinbüchel, unpublished data).

Occurrence of 3-hydroxyalkanoic acids in other cell constituents

Hydroxyalkanoic acids occur together with the other components as monomeric or dimeric constituents in some other biomolecules. Rhamnolipids, for instance, are extracellular glycolipids consisting of rhamnose and 3-hydroxyalkanoic acids. Rhamnolipid R-1 (2-*O*-α-L-rhamnopyranosyl-α-L-rhamnopyranosyl-3-hydroxydecanoyl-3-hydroxydecanoate) and rhamnolipid R-2 (α-L-rhamnosyl-β-hydroxydecanoyl-β-hydroxydecanoate) are synthesized by, for example, *Pseudomonas aeruginosa* and *P. fluorescens*, and are excreted into the medium[82,83]. These components are ionic surfactants and are probably involved in the uptake of hydrocarbons. Ether-linked 3-hydroxybutyrate has been identified as a constituent of extracellular polysaccharides of some strains of *Rhizobium trifolii*[84,85].

Distribution of PHA in Organisms

Table 2 Studies on the metabolism of polyhydroxyalkanoic acids in microorganisms

Species	A	B	C	D	E	F
Achromobacter sp.	241					
Achromobacter xylosoxidans	242					
Acidovorax delafieldii	252	263	154, 293			
Acidovax facilis	252, 260, 426	162, 223		154		
Acinetobacter sp.	243, 244, 245		244			
Acinetobacter calcoaceticus	244	246	244			
Acinetobacter lwoffii	244					
Actinomyces sp.	247					
Alcaligenes sp.	248					

Species	References or relevant studies					
	A	B	C	D	E	F
Alcaligenes aestus	249					
Alcaligenes cupidus	249					
Alcaligenes denitrificans	242, 248, 250			251		
Alcaligenes eutrophus	47, 65, 252, 258, 263	51, 56, 99, 160, 192, 193, 194, 195, 206, 222, 233, 253, 254, 259, 264, 266, 267, 268, 275	70, 255, 260	14, 62, 63, 72, 99, 115, 116, 118, 120, 144, 150, 151, 156, 157, 204, 206, **216, 217,** 221, **222,** 223, 233, 251, 256, 257, 261, 265, 270, 271, 272, 273, 274, 275, 276, 280, 281, 282, 283, 284, 285, 286, 287, 288, 289, 290, 291, 292, 293	66, 117, 119, 175, 182, 183, 186, 262, **269**	67, 152, 158, 166, 167, 169, 171, 172, 173, 174, 213, 214, 223, 277, 278, 279
Alcaligenes faecalis	242, 250	160		251	228*, 229*, 230*, 231*, 233*, 294*	232
Alcaligenes hydrogenophilus		99		251		
Alcaligenes latus		99, 281	295, 296	139, 211, 251, 297, 298		
Alcaligenes oderans	250					
Alcaligenes pacificus	249					
Alcaligenes paradoxus	252, 260					
Alcaligenes rhulandii	260, 299					
Alcaligenes venestus	249					
Alteromonas macleodii	249					
Amoebobacter roseus		99				

Biomaterials

Species	References or relevant studies					
	A	B	C	D	E	F
Amoebobacter pendens		99				
Aphanocapsa sp.	107					
Aphanothece sp.		106				
Aquaspirillum autotrophicum	260	99				
Azorhizobium caulinodans				300		
Azospirillum sp.						
Azospirillum brasilense	301		302	303, 304, 307, 308, 309, 310	**305**, 306	
Azospirillum lipoferum			302	**311**, 312		
Azotobacter sp.	313	99		314		
Azotobacter agilis	315			316		
Azotobacter beijerinckii	315, 317			5, 318, 321, 322, 323, 324, 325, 326, 327	319, 320, 323	
Azotobacter chroococcum	62, 90, 315, 328, 329		330	72, 331		
Azotobacter insigne	315					
Azotobacter macrocytogenes	315					
Azotobacter vinelandii	315, 332, 333, 334, 335, 336, 337	338	339, 340, 341, 342	73, 75, 179, 314, 339, 341, 342, 343, 344, **345**, 347, 348, 349, 350	**346**	
Bacillus sp.	351					
Bacillus anthracis	352					
Bacillus cereus	352, 353, 358, 359, 363	160, 281, 354	64, 71, 299, 355, 360, 364, 365	71, **350**, 356, **358**, 361, **362**, 366, 367	357	
Bacillus circulans	368					
Bacillus megaterium	37, 72, 352, 353, 354, 358, 369, 370, 371, 372, 373, 374,	55, 160, 383	64, 153, 218, 330, 384, 385, 386, 387	153, 221, 358, 384, 388, 389, 390, 391, 392, 393, 394, 395,	189, 220, **335**	

Species	References or relevant studies					
	A	B	C	D	E	F
Bacillus megaterium continued	375, 376, 377, 378, 379, 380, 381, 382			396, 397, 398		
Bacillus polymyxa	352					
Bacillus subtilis	335	160		73		
Bacillus thuringiensis	399		400	401		
Beggiatoa sp.	402		403			
Beggiatoa alba				404, 405		
Beijerinckia sp.	313, 406					
Beijerinckia indicus	315	160				
Beijerinckia fluminensis	315					
Beijerinckia mobilis	315					
Beijerinckia lacticogenes	315					
Beneckea sp.	407, 408					
Beneckea natrigens	407, 408					
Beneckea nereida	407, 408					
Beneckea nigra-pulchrituda	408					
Beneckea pelagia	407					
Bordetella pertussis	409					
Bradyrhizobium japonicum				410	241, 411	
Caryophanon latum			412	412		
Caulobacter sp.	413					
Caulobacter bacteroides	414		414			
Caulobacter crescentus	414, 415, 416, 417		414			
Chloroflexus aurantiacus	100		101	102		
Chlorogloea sp.						
Chlorogloea fritschii	103, 107		418			
Chromatium sp.				419		
Chromatium minutissimum		99				
Chromatium okenii		99	420, 421	72, 421		
Chromatium purpuratum		99				
Chromatium tepidum	422		422			
Chromatium vinosum	423, 424	99				
Chromatium warmingii		99				
Chromobacterium sp.	332					

Species	References or relevant studies					
	A	B	C	D	E	F
Chromobacterium violaceum	332					
Clostridium botulinum						
Clostridium perfringens	90					
Clostrium sphenoides	90					
Comamonas sp.	425					
Comamonas acidovorans	426, 427	162, 223		162		
Comamonas testosteroni	426, 427	160, 161, 162, 176, 223				
Corynebacterium sp.	378					
Corynebacterium autotrophicum	260					
Corynebacterium equi		163				
Corynebacterium hydrocarboxydans		163, 200				
Derxia sp.	313					
Derxia gummosa	315		160			
Desulfococcus multivorans	92					
Desulfonema limicola	92					
Desulfonema magnum	92					
Desulfosarcina variabilis	92					
Desulfovibrio sapovorans	91, 92		91			
Ectothiorhodospira halochloris				428		
Ectothiorhodospira mobilis		99				
Ectothiorhodospira shaposhnikovii	428	99				
Ectothiorhodospira vacuolata		99				
Escherichia coli			77	74, 75, 76, 77		
Ferrobacillus ferrooxidans	72, 429		429			
Flavobacterium sp.	241, 430					

Species	References or relevant studies					
	A	B	C	D	E	F
Haemophilus influenzae				73		
Halobacterium sp.		113				
Halobacterium gibbonsii	55					
Halobacterium hispanicum	55					
Halobacterium volcanii	55					
Haloferax mediterranei		201		55, 112, 431		
Hydroclathratus clathratus	432					
Hydrogenomonas facilis			433			
Hydrogenophaga flava	252, 260	162, 223		154		
Hydrogenophaga palleronii	252, 260	162, 223		154		
Hydrogenophaga pseudoflava		162, 223, 281, 503	260	503		
Hydrogenophaga taeniospiralis		162, 223				
Hyphomicrobium sp.	434, 435, 436			437		
Hyphomicrobium vulgare	434					
Ilyobacter delafieldii				234*		
Labrys monachus	438					
Lamprocystis reseopersicina		99				
Lampropedia hyalina	72		439	439		
Legionella sp.	440					
Leptothrix discophorus	441		442	442, **443**, **444**		
Methylobacterium sp.		160				
Methylobacterium AM1			69, 70	**445**		
Methylobacterium extorquens		160, 262	70, 446	411, 446, 447, 448		
Methylobacterium organophilum				449, **450**		
Methylococcus thermophilus	451					

Biomaterials

| Species | References or relevant studies | | | | | |
	A	B	C	D	E	F
Methylocystis parvus	452, 453, 454					
Methylomonas methanica	455		456			
Methylosinus sporium	452					
Methylosinus trichosporium	452, 457		458	**459**		
Methylovibrio soehngenii	460					
Micrococcus denitrificans				72, **461**		
Micrococcus halodenitrificans	72, 462, 463	281		464, 465		
Mycobacterium album	378					
Mycobacterium vaccae	378					
Mycoplana rubra	466			287		
Nitrobacter sp.	467					
Nitrobacter agilis	468		469	418		
Nitrobacter winogradskyi			470, 471	472		
Nocardia sp.	378	38				
Nocardia alba		163				
Nocardia asteroides				473		
Nocardia brasiliensis	473					
Nocardia lucida		163, 200				
Nocardia otitidis-caviarum	473					
Nocardia petroleophila		163				
Nocardia rubra	378					
Paracoccus dentrificans	252, 266	99		154		
Oscillatoria limosa		474	474			
Penicillium cyclopium		57				
Photobacterium mandapamensis	408, 475					
Photobacterium phosphoreum	408, 476		408			
Physarum polycephalum	240	58		58		

| Species | References or relevant studies | | | | | |
	A	B	C	D	E	F
Prosthecomicrobium hirschii	438					
Prosthecomicrobium pneumatis	438					
Pseudomonas sp.	163, 164, 241, 249, 252, 332, 378, 427, 477, 478	160, 162, 163, 223, 479, 480		71, 163, 164, 446, 447, 481, 482, 483	62*, 484*	
Pseudomonas aeruginosa	484	160, 161, 162, 163, 164, 170, 223, 480	162, 223	**162**, 223, 480		173
Pseudomonas alcaligenes		162, 223				
Pseudomonas antimycetica	332, 484					
Pseudomonas asplenii		162, 223				
Pseudomonas aureofaciens		162, 163, 223				
Pseudomonas butanovora		162, 223		485, 486		
Pseudomonas caryophylli	487	162, 223				
Pseudomonas cepacia	488	162, 223, 281, 489				
Pseudomonas cichorii		162, 223				
Psuedomonas citronellolis		162, 163, 223	162			
Pseudomonas chlororaphis		162, 223				
Pseudomonas coronafaciens		162, 223				
Pseudomonas dacunhae		162, 223				
Pseudomonas denitrificans		162, 163, 223	162			
Pseudomonas diminuta	498	162, 223				
Pseudomonas doudoroffii	249					
Pseudomonas echinoides		162, 223				
Pseudomonas fluorescens		161, 162, 163, 164, 170, 223				173

Species	A	B	C	D	E	F
Pseudomonas indigofera		162, 223				
Pseudomonas gladioli		162, 223				
Pseudomonas glathei		162, 223				
Pseudomonas hydrogenovora	161	162				
Pseudomonas indigofera		162, 223				
Pseudomonas lemoignei	426, 427			426*, 491*	189*, 225*, 226*, 492*, 493*	
Pseudomonas lemonieri		161, 170			442a*	
Pseudomonas mallei	427					
Pseudomonas marginalis		162, 163, 223				
Pseudomonas marginata	487					
Pseudomonas marina	249					
Pseudomonas mendocina		162, 163, 223	162			
Pseudomonas mesoacidophila	494					
Pseudomonas methanolica				392		
Pseudomonas mixta		162, 223	495			
Pseudomonas multivorans	427					
Pseudomonas oleovorans	35	35, 43, 46, 52, 54, 56, 99, 155, 160, 161, 162, 170, 196, 197, 223, 239, 480, 496, 497, 498, 499, 500	43, 54	**155**, 170, 174, 480, 500, 501		155, 168, 169, 173
Pseudomonas oxalaticus		162, 223				
Pseudomonas pantotropha	426					
Pseudomonas pseudoalcaligenes	427	162, 223				

Species	References or relevant studies					
	A	B	C	D	E	F
Pseudomonas pseudomallei	426, 427, 504					
Pseudomonas putida		155, 160, 161, 162, 163, 164, 170, 198, 223, 480	161, 366	155		155, 168, 173
Pseudomonas rubrilineas	505					
Pseudomonas rubrisubalbicans	505					
Pseudomonas ruhlandii	252					
Pseudomonas saccharophila	72, 252		506	214, 477		
Pseudomonas solanacearum	332, 484, 487	162, 223				
Pseudomonas stutzeri		162, 223				173
Pseudomonas syringae		162, 223				173
Pseudomonas thermophilus	507					
Pseudomonas viburni	487					
Pseudomonas viridiflava		162, 223				
Rhizobium sp.	71, 72, 332, 484, 508, 509	510	508	511, 512, 513, 514, 515, 516	**517**	
Rhizobium hedysarum	509			518		
Rhizobium japonicum	484, 519, 520, 521			522, 523	517	
Rhizobium leguminosarum	90, 509, 524, 525, 526, 527			528	**517**	
Rhizobium lupini	525			529, 530, 531	**517**, 532	
Rhizobium meliloti	149, 348, 533, 534, 535	509		528, 536, 537, 538	**517**	
Rhizobium phaseoli	533, 534		539	528	**517**	
Rhizobium trifoli	540			528	**517**	
Rhodobacillus		97		97, 98		
Rhodobacter capsulatus		99		541		
Rhodobacter sphaeroides	56, 99, 180, 181			180		

Species	References or relevant studies					
	A	B	C	D	E	F
Rhodococcus sp.		163, 200		163, 200		
Rhodococcus rhodochrous		163				
Rhodocyclus gelatinosus		99				
Rhodocyclus tenuis		99				
Rhodomicrobium vannielii	542	99		541	543	
Rhodopseudomonas sp.	544				545	
Rhodopseudomonas acidophila		99				
Rhodopseudomonas blastica		99				
Rhodopseudomonas capsulata	546			547, 548		
Rhodopseudomonas palustris		99	549	550		
Rhodopseudomonas spheroides	551		552	553, 554, 556	543, 545, 555, 557, 558, 559	
Rhodopseudomonas viridis		99	560			
Rhodospirillum fulvum		99				
Rhodospirillum molischianum		99				
Rhodospirillum rubrum	72, 285	56, 99, 180	64, 561, 562, 563	180, 375, 477, 563, 564	159, **189**, **218**, **219**, **221**, **566**, 567	
Sphaerotilus discophorus				**443**, **444**		
Sphaerotilus natans	568, 569, 570, 571, 572		569, 571, 573			
Sphingomonas paucimobilis	162, 223, 502					
Spirillum sp.			574	574		
Spirillum itersonii	72, 575					
Spirillum lipoferum				576	576	
Spirillum normaal				72		
Spirillum serpens	72, 484, 575					
Spirulina jenneri	109					
Spirulina laxissima	109					
Spirulina maxima	109					

Species	References or relevant studies					
	A	B	C	D	E	F
Spirulina platensis	108, 109					
Spirulina subsalsa		474				
Staphylococcus aureus	111					
Staphylococcus xylosus	111					
Staphyloccus epidermidis	111					
Stella humosa	438					
Stella vacuolata	438, 579					
Streptomyces sp.	247					
Streptomyces antibioticus	247, 578					
Streptomyces coelicolor	579				579	
Syntrophomonas wolfei	93, 95	238				
Thiobacillus A2		99				
Thiobacillus acidophilus	580					
Thiobacillus ferrooxidans	72, 429, 581		429			
Thiobacillus versutus		582		582		
Thiocapsa sp.		583				
Thiocapsa pfennigii		99				
Thiocystis violacea		99				
Vibrio parahaemolyticus	584, 585					
Xanthobacter autotrophicus	586			251		
Xanthomonas maltophilia		162, 163				
Zoogloea sp.	425					
Zoogloea ramigera	241, 587, 588, 589, 590, 591	233		15, 590, 592, 593, 594, 595	177, 178, 186, 187, 593, 596, 597, 598, 599, 600, 601, 602, 603, 604	165, 176

143

The references have been organized in six categories according to the type of studies: A, only accumulation of PHA was mentioned; B, composition of PHA was analysed; C, morphological and/or structural studies on the appearance of PHA in the cells have been performed or inclusion of a photograph of cells with PHA granules is shown; D, extended physiological studies on the metabolism of PHA were done; E, enzymatic studies on proteins involved in the metabolism of PHA were done; F, genetic studies and molecular analysis of genes involved in the metabolism of PHA were done. References for studies, which included or even focused on the intracellular mobilization of PHA or on the extracellular degradation of PHA, appear in **bold letters** or are marked by an asterisk, respectively.

As shown in Table 2, synthesis of PHA occurs in a wide variety of organisms. Light microscopic investigation of the cells, staining of cells with Sudan black (for example, see ref. 86), or infrared spectroscopy of cells or cell extracts (for example, see ref. 87), provided easy means for the detection of PHA in cells. PHA biosynthesis is not restricted to certain taxonomic groups and it seems to be absent from only a very few taxa. In contrast to earlier assumptions[88], the presence or absence of PHA in the cell is therefore only a poor criterion for the classification of bacteria. It occurs in Gram-negative as well as in Gram-positive bacteria. PHB biosynthesis has been detected not only in aerobic bacteria but also in many anaerobic heterotrophic bacteria such as *Clostridium botulinum*[89], *C. perfringens*[90], *Desulfovibrio sapovorans*[91,92], *Desulfococcus multivorans*, *Desulfonema limicola*, *Desulfonema magnum*, *Desulfosarcina variabilis*[92] and *Syntrophomonas wolfei*[93–95]. Formation of PHB under anaerobic conditions has been detected in samples from complex bacterial communities such as those of anaerobic thermophilic digesters which produce methane[96]. In addition, PHB biosynthesis was shown to occur in anaerobic photosynthetic bacteria[97,98]. PHB has been detected in most nonsulphur and sulphur purple bacteria (for a recent survey, see ref. 99) and even in the green gliding bacterium *Chloroflexus auranthiacus*[100–102] and some unidentified *Chloroflexus* species[100]. Whereas nonsulphur and sulphur purple bacteria accumulated PHA in amounts of up to a high portion of the biomass (>50%), *Chloroflexus auranthiacus* accumulated only a little PHA (about 2.5% of biomass). There is so far no report of PHA synthesis in green sulphur bacteria.

There is an increasing number of reports on the synthesis of PHA in the aerobic photosynthetic bacteria, the cyanobacteria. PHA has been detected in *Chlorogloea fritschii*[103,104,105], in *Aphanothece* sp.[106], *Aphanocapsa* sp.[107], in four different species of the genus

Spirulina[108,109] as well as in one unidentified cyanobacterium[106]. However, PHA accumulated to only a very low portion of the biomass (≤5%).

PHB has been found even in enterobacteria like *E. coli*. However the PHB is not present in granules but in the cytoplasmic membrane (see above). Furthermore, PHB accumulates in *Staphylococcus aureus*, *S. xylosus* and *S. epidermis*[110,111]. Demonstration of PHB in various halobacteria[55,112,113] indicates that archaebacteria are not generally impaired in PHA biosynthesis.

Accumulation of PHA has not yet been detected in a few taxonomic or physiological groups of bacteria such as the methanogenic bacteria, the lactic acid bacteria and the green sulphur bacteria. It has yet to be determined whether these bacteria are impaired in the synthesis of these polyesters for physiological, biochemical, genetic or evolutionary reasons.

Commercial Significance of PHA

As *A. eutrophus* was considered as a candidate for the production of single-cell protein[14], early research on the metabolism of PHB in this bacterium focused on the understanding of the biosynthetic pathway[115–118], on the regulation of PHB-biosynthesis in the cell[119–120] and on the elimination of PHB in order that the efficiency of cell production was maximized. *A. eutrophus* was grown in pilot plants, and the nutritional value of cells of the wild type and of mutants of *A. eutrophus* lacking PHB was studied in feeding experiments in rats[122–124], broilers[125–127] and pigs[128]. However, when it became obvious that the development of processes for the production of single-cell protein would not be continued, the same pilot plants were considered for the production of PHB from cells of *A. eutrophus*[129].

The thermoplastic properties of poly(3HB-*co*-3HV) (Fig. 5) and of PHB were recognized in the early 1960s and formed the basis for patents assigned to W. R. Grace & Co.[130,131]. Later, ability to make copolyesters, and an application of their biodegradability, led to development of these polymers in several applications[25,27]. Although PHB and other PHA are also available from chemical synthesis[132–134], only biotechnological processes produce them in sufficient amounts for technical applications at the moment. Poly(3HB-*co*-3HV) is being produced by Imperial Chemical Industries in the United Kingdom on

145

an industrial scale from cells of a glucose-using mutant derived from *A. eutrophus* strain H16 with a glucose/propionic acid substrate mixture as a carbon source and under phosphate-limiting conditions[72,135,136].

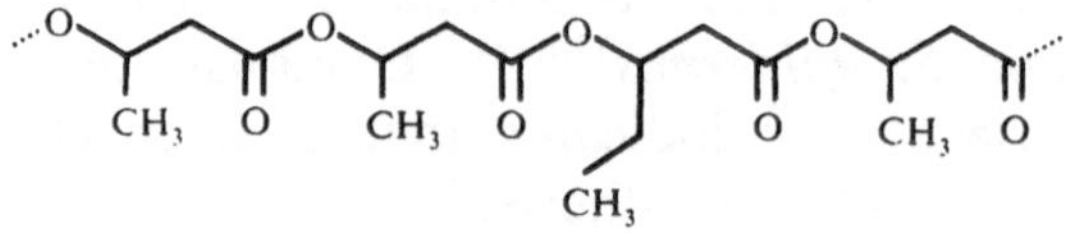

Fig. 5   Structural formula of poly(3HB-*co*-3HV).

The copolyester is distributed under the trade name Biopol and is already used for the manufacture of biodegradable films and bottles[26,137,138]. Shampoo bottles made from Biopol are already available in test markets in Germany[36]. In addition, the Biotechnologische Forschungsgesellschaft mbH in Austria produces PHB on a technical scale from cells of *Alcaligenes latus* with sucrose as a carbon source[139].

PHB and poly(3HB-*co*-3HV) are nontoxic and biodegradable, giving rise to several applications of the polymers for controlled drug release (for example, see ref. 140), such as the release of buserelin[141]. Other possible medical applications are as surgical sutures, surgical swabs, wound dressings or lubricants for surgeons' gloves[27]. As PHAs are optically active, the monomers derived from the polyesters have also been considered as a source for the synthesis of enantiomeric pure chemicals[142–145]. In addition, PHAs have been tested as carbon sources in denitrification processes for the elimination of nitrate from drinking water[146]. Whether the piezoelectric properties of PHB and poly(3HB-*co*-3HV)[147,148] are of any technological relevance has still to be investigated.

Mutants in the Accumulation of PHA

PHA-negative mutants, which were completely impaired in the accumulation of PHA, and PHA-leaky mutants, which accumulate less PHA than the wild type, have been isolated from *Alcaligenes eutrophus*[149–152], *Bacillus megaterium*[153], *Paracoccus denitrificans*[154], *Pseudomonas putida*[155], *Chromatium vinosum* (M. Liebergesell and

A. Steinbüchel, manuscript in preparation) as well as from *Rhodospirillum rubrum* and *Rhodobacter spheroides* (E. Hustede, A. Steinbüchel and H. G. Schlegel, manuscript in preparation). For enrichment and selection of these mutants, mutagenized cells were in most cases separated by centrifugation in sucrose[150] or Percoll[63] density gradients using the different buoyant densities of PHA-containing and PHA-free cells[63,65]. PHA-negative and PHB-leaky mutants of *B. megaterium* occurred spontaneously if the cells were repeatedly transferred in mineral salt medium containing glucose and 2,3-butanediol[153]; the selective principle was unknown. As colonies of PHA-negative mutants from most bacteria appear translucent and do not stain with Sudan black, identification of the mutants was readily achieved.

PHA-negative mutants were not only considered for the production of single-cell protein (see above), but were also used to study the metabolism of the cell. PHA-negative mutants of *A. eutrophus*, for instance, excreted large amounts of pyruvate if the cells were cultivated aerobically on gluconate, fructose or lactic acid under conditions that allowed PHA accumulation in the wild type[14]. If cells of PHA-negative mutants of *A. eutrophus* or *P. denitrificans* were cultivated under conditions of restricted oxygen supply, 3-hydroxybutyrate was excreted in appreciable amounts[154,156,157]. In addition, these mutants, especially those obtained by transposon mutagenesis[67,152,158], were valuable tools for the analysis of gene loci essential for PHA synthesis or of gene loci affecting the accumulation of the polyester (see below).

PHA Biosynthesis

Four different basic routes for the synthesis of PHA are known to exist in bacteria. These pathways, or part of them, account for the synthesis and incorporation of all the known constituents of bacterial PHA, which were mentioned above. However, the pathway for the synthesis of poly(β-malic acid), which occurs in some fungi, is still unknown. It has yet to be determined whether malyl-CoA or malic acid is the substrate of polymalate synthase.

In a three-step PHB-biosynthetic pathway, acetyl-CoA is converted to PHB (Fig. 6). A biosynthetic β-ketothiolase (EC 2.3.1.9) catalyses the formation of a carbon-carbon bond by a biological Claisen condensation of two acetyl-CoA moieties. An NADPH-dependent aceto-acetyl-CoA reductase (EC 1.1.1.36) catalyses the stereoselective

147

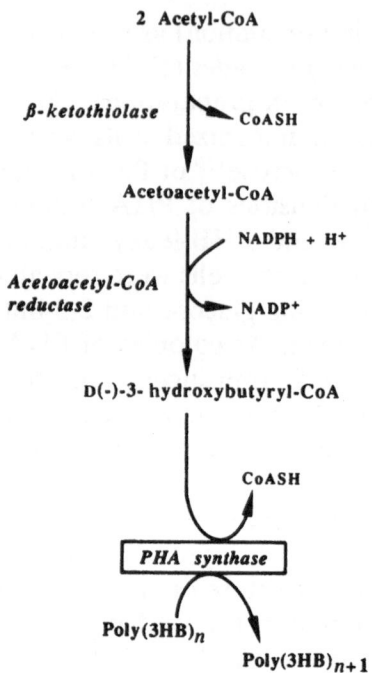

2 Acetyl-CoA

β-ketothiolase — CoASH

Acetoacetyl-CoA

NADPH + H⁺

Acetoacetyl-CoA
reductase — NADP⁺

D(-)-3- hydroxybutyryl-CoA

CoASH

PHA synthase

Poly(3HB)ₙ

Poly(3HB)ₙ₊₁

Fig.6 *Alcaligenes eutrophus* PHA-biosynthetic pathway.

reduction of acetoacetyl-CoA formed in the first reaction to D(−)-3-hydroxybutyryl-CoA. The third reaction of this pathway is catalysed by the enzyme PHB synthase and links the D(−)-3-hydroxybutyryl-moiety to an existing polyester molecule by an ester bond. Neither the reaction nor the enzyme has been studied in detail, and an EC number has not been assigned. As this pathway has been most extensively studied in *Alcaligenes eutrophus*, it is referred to as the *Alcaligenes eutrophus* PHA-biosynthetic pathway.

A modification of the *A. eutrophus* PHA-biosynthetic pathway is the five-step PHA-biosynthetic pathway which has been described only in *Rhodospirillum rubrum* and which is therefore referred to as the *Rhodosprillum rubrum* PHA-biosynthetic pathway (Fig. 7). The acetoacetyl-CoA formed by the β-ketothiolase is reduced by an NADH-dependent reductase to L(+)-3-hydroxybutyryl-CoA; the latter is then converted to the D(−)-3-hydroxybutyryl-CoA by two enoyl-CoA hydratases[159] (EC 4.2.1.17; no EC number has been assigned to the other hydratase). The fifth step of this pathway, which is the polymerization reaction, is catalysed by PHA synthase.

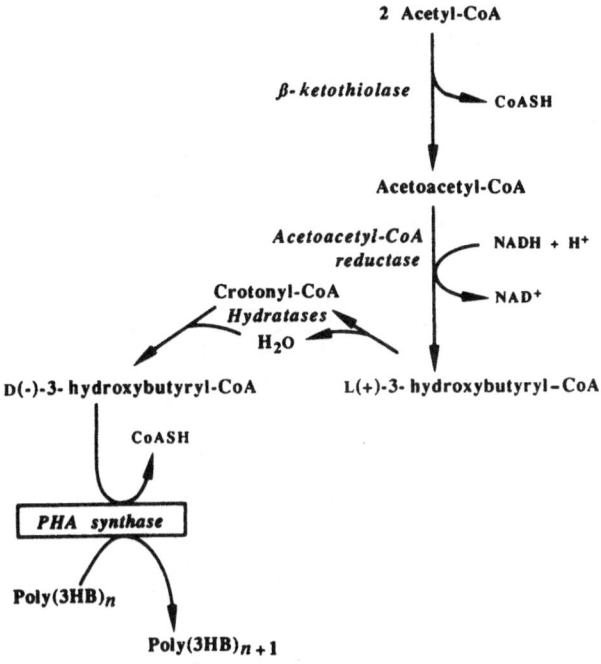

Fig. 7 *Rhodosprillum rubrum* PHA-biosynthetic pathway.

A third type of PHA-biosynthetic pathway seems to be active in most pseudomonads belonging to the ribosomal RNA homology group I. *Pseudomonas oleovorans*[43,46,54] and other bacteria belonging to this group[160–162] accumulate PHA consisting of 3-hydroxyalkanoic acids of medium chain length if the cells are cultivated on alkanes, alkanols or alkanoic acids[43,46,54]. During the cultivation on, for instance, octane or octanoic acid, a polyester is accumulated that contains 3-hydroxyoctanoate as the main constituent (about 90 mol%). The structural formula of poly(3-hydroxyoctanoic acid), or PHO, is shown in Fig. 8. Since evidence for this pathway was obtained first in *P. oleovorans*, it is referred to as the *Pseudomonas oleovorans* PHA-biosynthetic pathway (Fig. 9). It is likely that intermediates of the β-oxidation cycle, which occur during the oxidation of activated fatty acids derived from alkanes, alkanols or alkanoic acids, are directed to PHA biosynthesis in these bacteria. Whether octenoyl-CoA, L(+)-3-hydroxyoctanoyl-CoA or 3-ketooctanoyl-CoA are withdrawn from the cycle and are converted to D(−)-3-hydroxyoctanoyl-CoA by

Biomaterials

an enoyl-CoA hydratase, an epimerase or a 3-ketoacyl-CoA reductase, respectively, has still to be elucidated.

Almost all pseudomonads belonging to the rRNA homology group I, but not *P. oleovorans*, possess a fourth type of PHA-biosynthetic pathway for the synthesis of copolyesters consisting of medium-chain-length 3-hydroxyalkanoic acids from acetyl-CoA. The main constituent of the polyester is 3-hydroxydecanoate (3HD). This pathway

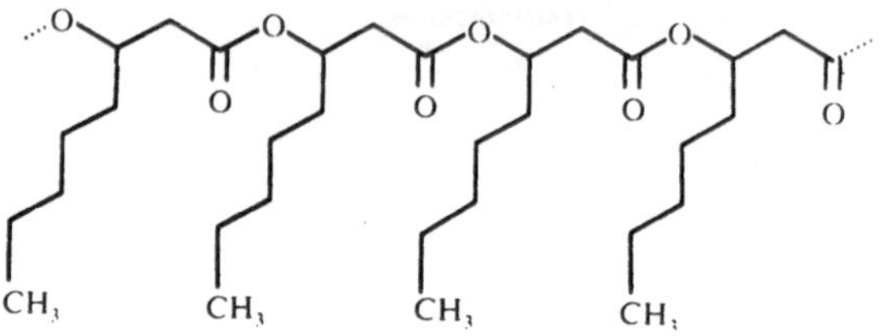

Fig. 8 Structural formula of poly(3-hydroxyoctanoic acid).

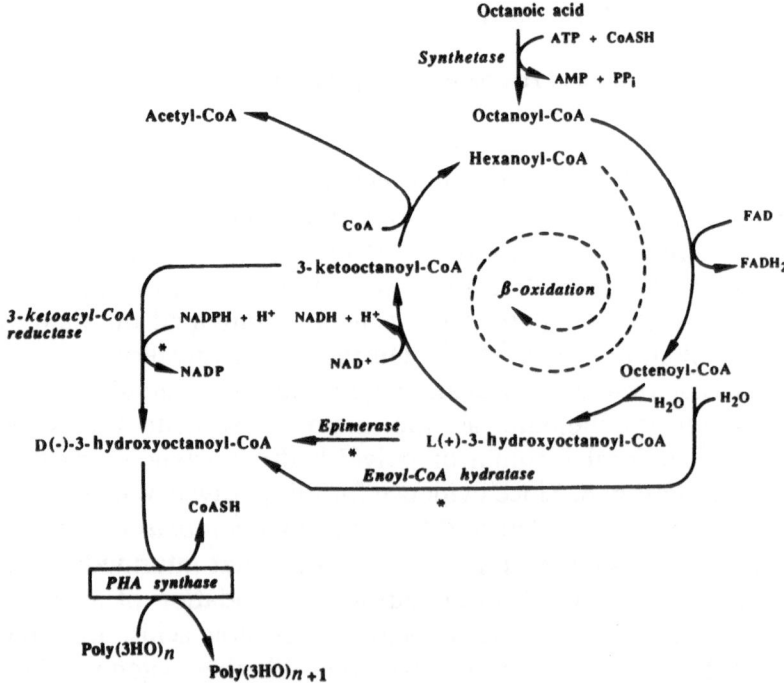

Fig. 9 *Pseudomonas oleovorans* PHA-biosynthetic pathway*.

150

has not yet been studied in detail. It is also not known which derivative of 3-hydroxydecanoate is the substrate of the PHA synthase in these bacteria and how this derivative is synthesized from acetyl-CoA. As the physiology of poly(3-hydroxydecanoate) or PHD, biosynthesis has been investigated mainly in *P. aeruginosa*, the pathway is referred to as the *P. aeruginosa* PHA-biosynthetic pathway. *P. aureofaciens*, *P. citronellolis*, *P. chlororaphis*, *P. marginalis*, *P. mendocina*, *P. putida* and *Pseudomonas* sp. strain DSM 1650 also accumulated this polymer at significant levels during cultivation on gluconate[162,163,164]. It remains an enigma why PHD synthesis from gluconate or from other carbohydrates was not detected earlier in *P. aeruginosa*, as many aspects of the metabolism of this bacterium have been studied thoroughly. Failure to detect PHA granules may be due to the application of unsuitable culture conditions and most probably due to the presence of a highly active PHA depolymerizing enzyme system.

Physiological Functions of PHA and Conditions for its Synthesis

Earlier reviews on the physiological function and relevance of PHA have been recently updated very comprehensively and do not therefore need to be repeated in detail here. The interested reader is referred to the excellent review of Anderson and Dawes[18] and to the references cited therein. Only a brief summary will be given here; an extensive survey of the studies performed on the physiology of PHA metabolism is included in Table 2.

PHB and most other PHA are synthesized and accumulated in the cells if a carbon source is provided in excess, but cell proliferation is impaired owing to the lack of one nutrient essential for growth such as nitrogen, sulphate, phosphoros, iron, magnesium or potassium. In many aerobic bacteria, PHB synthesis is also triggered if maximum growth cannot occur owing to limited availability of oxygen. In some bacteria, such as *Azotobacter vinelandii* and probably also *Bradyrhizobium japonicum*, accumulation of PHB occurs to an even greater extent during oxygen limitation than, for example, during nitrogen limitation. As the *A. eutrophus* PHA-biosynthetic pathway and the *R. rubrum* PHA-biosynthetic pathway include a pyridine nucleotide-dependent reduction of acetoacetyl-CoA, PHB is a sink for reducing equivalents and the polymer can be considered to be a fermentation product. By contrast, in bacteria relying on the *P. oleovorans* PHA-biosynthetic pathway or on the *P. aeruginosa* PHA-biosynthetic pathway, oxygen-deficiency exerts a negative effect on the accumulation of PHA, because either these pathways include the formation of reducing equivalents or energy in the form of ATP is consumed. Intracellular deposition or metabolites as a reduced polymer provides the advantage that they become unavailable as a carbon source for competing organisms, and that they are osmotically inert and do not therefore affect the osmotic pressure of the cells.

If the environmental conditions again allow proliferation of the cells and/or if the extracellular carbon source is exhausted, the polyesters are mobilized. Under these conditions PHA functions as a carbon and energy-reserve material. Mobilization of PHA and the degradation of the monomers contributes to the survival of starved cells of many bacteria. The mobilization of intracellular PHA provides a carbon and energy source for endospore formation and encystment in species of *Bacillus* or *Azotobacter*, respectively. It has been suggested that oxidation of the monomers derived from PHB provides a means of respiratory protection of nitrogenase in the nitrogen-fixing *Rhizobium* sp. when it is in root nodules at an elevated concentration of oxygen. The putative function of the polyester in $PHB:CA^{2+}:PP_i$ complexes in the membranes of prokaryotes and eukaryotes for the uptake of DNA, calcium or phosphate has already been mentioned.

Molecular Organization of PHA-Biosynthetic Genes

Molecular data on the PHA-biosynthetic genes are only available for *A. eutrophus*[68,152,165–167], *Z. ramigera*[165] and *P. oleovorans*[155,168,169]. Considerable efforts have been made to isolate and characterize the PHB-biosynthetic genes of *A. eutrophus*. The different cloning strategies employed in three different laboratories have been reviewed very recently[22,23] and so only the results of these studies and the conclusions will be given here. Parallel studies were recently carried out to investigate the molecular organization of the *P. oleovorans* PHA-biosynthetic genes[155,169,170].

A. eutrophus

In *A. eutrophus*, the genes for biosynthetic β-ketothiolase (*phbA*), NADPH-dependent acetoacetyl-CoA reductase (*phbB*) and PHB synthase (*phbC*) are clustered (Fig. 10). They are organized in one single operon (*phbCAB*) with 84-base-pair (bp) or 74-bp intergenic regions[67,68,166,171]. The transcription start site, which is preceded by a σ[70]-like promoter sequence, was identified 307 bp upstream of *phbC*[68], and a potential transcription terminator was localized downstream of *phbB*. The function of the unusually long leader sequence in this messenger RNA is not known. Northern blots revealed a length of approximately 4.1 kb for the putative primary transcript[172]. Whereas *phbC* is preceded by a relatively weak Shine-Dalgarno sequence, the corresponding sequences upstream of *phbA* and *phbB* are stronger[23]. The *A. eutrophus phbCAB* operon was not only readily expressed in *E. coli*[67,152,167] and in most pseudomonads tested[162,173,174], but was also functionally active resulting in the synthesis and accumulation of PHB in the recombinant strains.

The structural gene for a second, degradative β-ketothiolase (see below) has been probably cloned in the laboratory of Dennis[167]. By

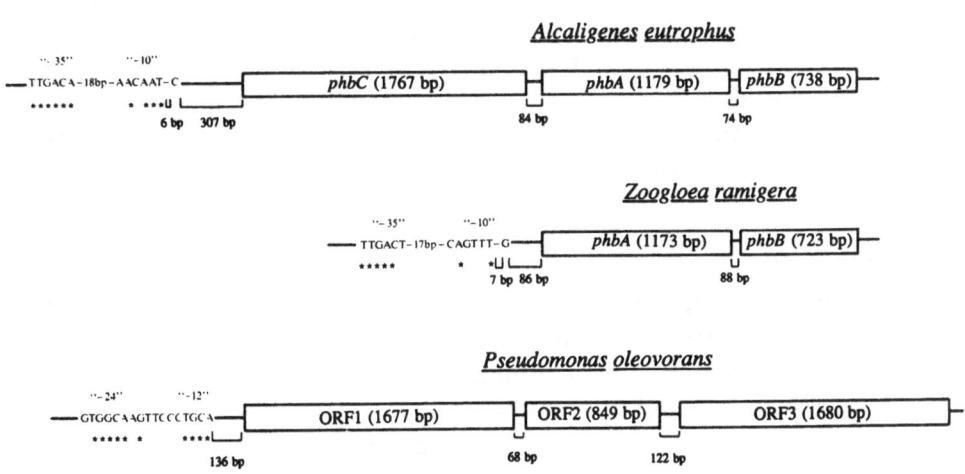

Fig. 10 Molecular organization of PHA-biosynthetic genes in bacteria. The molecular data were compiled from refs 67, 68, 152, 166, 167, 171, 172, 223 and 277 for *A. eutrophus*, from refs 165, 176 and 604 for *Z. ramigera* and from ref. 155 for *P. oleovorans*.

contrast, the gene for an NADH-dependent acetoacetyl-CoA reductase, which is present in addition to the NADPH-dependent enzyme in *A. eutrophus* (see below), has not yet been identified.

Insertions of Tn5 in *phbC* exerted a negative polar effect on the expression of *phbA* and *phbB*[152]. Insertions of Tn5 resulting in a PHB-negative phenotype have only been isolated within *phbC*. This indicates that the presence of active biosynthetic β-ketothiolase and NADPH-dependent acetoacetyl-CoA reductase is not essential for PHB synthesis in *A. eutrophus*. These functions are duplicated by other enzymes present in the organism. This is also indicated by restoration of the ability for PHB synthesis in a PHB-negative mutant of *A. eutrophus* upon transferring to it a genomic DNA fragment from *Thiocystis violacea*, which seems to encode only PHB synthase and β-ketothiolase but not an acetoacetyl-CoA reductase (M. Liebergesell, unpublished results). As a second degradative β-ketothiolase was found to contribute to PHB synthesis *in vitro* in reconstituted enzyme systems[175], insertions of Tn5 in *phbA* will probably have no effect on the synthesis of PHB *in vivo*. However, the reason for the failure to detect mutants harbouring Tn5 insertions in *phbB* is not known, because the NADH-dependent acetoacetyl-CoA reductase did not contribute to PHB biosynthesis *in vitro*[175]. Therefore, Steinbüchel and Schlegel[22] postulated the presence of isoenzymes of the NADP-dependent acetoacetyl-CoA reductase or of an alternative route, which circumvents the reaction catalysed by the NADPH-dependent reductase in *A. eutrophus*. This may be achieved by, for example, reactions similar to those in the *R. rubrum* PHA-biosynthetic pathway.

Z. ramigera

In *Z. ramigera*, only the structural genes for the biosynthetic β-ketothiolase (*phbA*) and for the NADPH-dependent acetoacetyl-CoA reductase (*phbB*) have been cloned and characterized[176]. Both genes are probably organized in one operon (*phbAB*) and are separated by an 88-bp intergenic region (Fig. 10). The transcriptional start site was identified 85 bp upstream of *phbA* and is preceded by a σ^{70}-like promoter. A potential transcriptional terminator was identified downstream of *phbB*[165]. The structural gene for PHB synthase has not yet been identified; obviously the synthase gene is not localized upstream of *phbA* as it is in *A. eutrophus*. The nucleotide sequences of the *phbA* genes from *Z. ramigera* and *A. eutrophus* were 70% identical. This explains why the use of the *Z. ramigera phbA* gene as a probe was successfully used for the detection of the *A. eutrophus phbA* gene[166]. The structural genes for the degradative β-ketothiolase and for the NADH-dependent acetoacetyl-CoA reductase of *Z. ramigera*[177,178] have not yet been cloned.

P. oleovorans

To clone the *P. oleovorans* PHA-biosynthetic genes, a gene library in a broad-host-range cosmid vector was screened for complementation of the mutant GPp104, a derivative of *P. putida* unable to accumulate PHA. One of the hybrid cosmids harboured a 6.4-kb *Eco*RI restriction fragment. Three open reading frames (ORF1, ORF2, ORF3) involved in the metabolism of PHA in *P. oleovorans* (Fig. 10) were identified on this fragment[155,169,170]. The primary structures of the putative gene products deduced from the nucleotide sequences of ORF1 (relative molecular mass (M_r 62,400) and ORF3 (M_r 62,600) exhibited 37.8 and 39.5% identity, respectively, with the *A. eutrophus* PHA synthase, indicating that both encode PHA polymerases. Each synthase gene was able to complement GPp104 and restore PHA synthesis. Although the transcription start site was not identified, an *ntrA*-dependent promoter consensus sequence was identified approximately 140 bp upstream of ORF1[155]. It has to be demonstrated in further studies whether single primary transcript derives from all three genes and whether the transcription is terminated at stem-loops downstream of ORF2 and/or ORF3.

The amino-acid sequence of the putative 31.5K protein deduced from ORF2, which is localized between ORF1 and ORF3 (Fig. 10), contained a decapeptide which comprised the lipase consensus sequence present in all known lipases. From this and from the complementation of a mutant of *P. oleovorans* impaired in the mobilization of PHA (see below), it was concluded[155] that ORF2 encodes a PHA depolymerase. Interestingly, no significant homology to the extracellular PHA depolymerase of *Alcaligenes faecalis* T1 was found in the ORF2 sequence.

Other bacteria

Two PHA synthase structural genes have been cloned from *Pseudomonas aeruginosa* PAO (Timm and Steinbüchel, manuscript in preparation). The primary structures of the synthases deduced from the nucleotide sequence and the organization of the genes revealed striking homologies with the corresponding genes of *P. oleovorans*.

We also cloned the PHB-biosynthetic genes from the anaerobic photosynthetic bacterium *Chromatium vinosum*. In this bacterium, the structural genes for PHB synthase, β-ketothiolase and for an acetoacetyl-CoA reductase, which exhibited preference to NADH rather than to NADPH as cofactor, are clustered on a 9-kb DNA restriction fragment. This fragment restored the wild-type phenotype in PHB-negative mutants of *A. eutrophus* and conferred the ability to synthesize PHB to *E. coli* (Liebergesell and Steinbüchel, manuscript in preparation). The analysis of the PHB-biosynthetic genes from other purple sulphur bacteria confirmed that those genes are frequently clustered in bacteria.

Enzymology of PHA Synthesis

The enzymes involved in the synthesis of PHA from acetyl-CoA and/or from intermediates of the β-oxidation cycle have not been studied in *P. aeruginosa* or *P. oleovorans* or in any other bacterium possessing the PHA-biosynthetic pathways of the types represented by these bacteria. Some properties of these enzymes may be deduced from the results of physiological studies or are available from the molecular data deduced from the nucleotide sequences of the structural genes and will be discussed later. Detailed information is only available for PHB-biosynthetic enzymes of bacteria possessing the *A. eutrophus* PHA-biosynthetic pathway.

Little information is available for the enzymes involved in the PHB biosynthesis in *R. rubrum* or in other photosynthetic non-sulphur purple bacteria. The only enzymatic studies were done in the laboratories of Merrick and of Stern more than 20 or 35 years ago, respectively. These studies resulted in the purification of L(+)-3-hydroxybutyryl-CoA-forming[179] as well as of D(−)-3-hydroxybutyryl-CoA-forming[159] enoyl-CoA hydratases from *R. rubrum* and provided data on some properties of these enzymes. In addition, reconstituted enzyme systems, which employed purified enoyl-CoA hydratases and native PHB granules, showed that PHB was only synthesized if D(−)-3-hydroxybutyryl-CoA was formed[159]. The composition of the PHA synthesized by these bacteria during cultivation on various fatty acids[99,180,181] allowed the conclusion that the substrate specificities of the PHB-biosynthetic enzymes of not only *R. rubrum*, but also of many other non-sulphur purple bacteria, are not restricted to four-carbon intermediates. The substrate specificities of these enzymes resemble most probably those of the *A. eutrophus* PHA-biosynthetic enzymes. The exceptions are the enzymes of *Rhodospirillum rubrum*[180] and *Rhodocyclus gelatinosus*[99], which seem to be slightly more nonspecific and which also allow the incorporation of significant amounts of 3-hydroxyheptanoate into the polymer. Moskowitz and Merrick[159] showed that the D(−)-3-hydroxybutyryl-CoA-forming enoyl-CoA hydratase hydrated not only crotonyl-CoA (100% relative activity) but also *trans*-2-hexanoyl-CoA (38% relative activity).

β-ketothiolase

A purification procedure for the β-ketothiolase of *A. eutrophus* has been described by Oeding and Schlegel[119,182]. However, the authors were not aware that two different β-ketothiolases exist in *A. eutrophus*. Evidence for the existence of two different enzymes was obtained later by the analysis of mutants impaired in the synthesis of PHB[152]. Dennis and collaborators succeeded in cloning two different *A. eutrophus* genomic DNA fragments that encode β-ketothiolase and which were able to express the enzymes in *E. coli*[167]. Subsequently, the separation and characterization of both thiolases was described by Dawes and collaborators[183]. The activity of β-ketothiolase A was (in the thiolytic direction) restricted to acetoacetyl-CoA (100% relative activity) and 3-ketovaleryl-CoA (3% relative activity), whereas β-ketothiolase B used not only these two substrates (100 or 39% relative activities, respectively) but also 3-ketohexanoyl-CoA (17%), 3-ketoheptanoyl-CoA (19%), 3-ketooctanoyl-CoA (10%) and 3-ketodecanoyl-CoA (12%) as substrates[183]. From this and from the location of the corresponding structural genes, β-ketothiolase A was referred to as the biosynthetic thiolase functioning primarily in PHB biosynthesis, whereas β-ketothiolase B was referred to as the degradative thiolase functioning primarily in fatty-acid catabolism. However, in reconstituted enzyme systems it was demonstrated that the degradative thiolase contributed to PHB synthesis almost as efficiently as the biosynthetic thiolase[175]. Thiolases A and B are tetrameric enzymes of identical subunits with M_rs 44K and 46K, respectively, as revealed by SDS-PAGE.

Two different thiolases were also detected in *Zoogloea ramigera*. The primary structures of the biosynthetic β-ketothiolases as deduced from the nucleotide sequences of the structural genes from *A. eutrophus* and *Z. ramigera* exhibited 63% identity. Homologies were significant but less with the β-ketothiolases from rat liver mitochondria[184] and rat liver peroxisomes[185]. In all four β-ketothiolases a cysteine residue (Cys 89 in *Z. ramigera* thiolase) is conserved that is involved in the formation of the acyl-enzyme intermediate. A second cysteine residue (Cys 378 in *Z. ramigera* thiolase) is also conserved in all four β-ketothiolases[166]. Mechanistic studies and site-directed mutagenesis at this residue[186,187] indicated that Cys 378 is involved in deprotonation in the condensation reaction.

Acetoacetyl-CoA reductase

As with the thiolases, two different acetoacetyl-CoA reductases are present in *A. eutrophus*. An NADPH-dependent acetoacetyl-CoA reductase (EC 1.1.1.36) catalyses the stereoselective reduction of acetoacetyl-CoA formed in the ketothiolase reaction to D(−)-3-hydroxybutyryl-CoA. The second reductase, in contrast, is NADH-dependent and will only form L(+)-3-hydroxyacyl-CoA but not D(−)-β-hydroxyacyl-CoA, although in the reverse direction the enzyme can also oxidize the latter[175]. Whereas the NADPH-dependent reductase is only active with D(−)-3-hydroxybutyryl-CoA (100% relative activity), D(−)-3-hydroxyvaleryl-CoA (48%) and D(−)-3-hydroxyhexanoyl-CoA (3.6%), the NADH-dependent reductase uses the L(+) enantiomers of these compounds (100, 35 or 50%, respectively) and also L(+)-3-hydroxyheptanoyl-CoA (99%), L(+)-3-hydroxyoctanol-CoA (146%) and L(+)-3-hydroxydecanoyl-CoA (60%) as substrates. The NADPH- and the NADH-dependent acetoacetyl-CoA reductase of *A. eutrophus* are tetrameric enzymes of identical subunits with M_rs of 23K and 30K, respectively, as estimated from SDS-PAGE.

Whereas the primary structure of the *Z. ramigera* NADPH-dependent acetoacetyl-CoA reductase exhibited 52% identity with corresponding enzyme from *A. eutrophus*, no significant homology was obtained with the L(+)-3-hydroxyacyl-CoA dehydrogenase from pig heart muscle[188]. Primary structures of other acetoacetyl-CoA reductases are not available. Near the *N* terminus of the *Z. ramigera* acetoacetyl-CoA reductase, a putative nucleotide-binding fold

(TGGXXG) for the binding of the ADP-moiety of NAD was identified[165].

PHA synthase

Little is known about the PHB synthase of *A. eutrophus*. The enzyme
is associated with the PHA granules during accumulation of the polyester. Neither the reaction nor the enzyme have been studied in detail
and an EC number has not been assigned. This enzyme links 3-
hydroxybutyryl moieties (100% relative activity) or 3-hydroxyvaleryl
moieties (7.5%) to existing polyester molecules by ester bonds. It is
absolutely stereospecific for the D($-$) stereoisomers and does not react
with the L($+$) stereoisomers. Whereas an M_r of 160K was determined
for the native enzyme[66], the sequence data predict an M_r of 63K for
the subunits. Lack of information is mainly due to the failure of
procedures for the purification of the enzyme. It will be a challenge to
determine the tertiary and quarternary structure of this enzyme.
Recombinant DNA technology will assist in providing cells that over-
produce this protein, allowing easier purification, and it will also
provide PHB synthases with modified characteristics resulting from
genes altered by site-directed mutagenesis.

The primary structures of the two putative PHA synthases from *P.
oleovorans* exhibited 53.6% identity with each other. Two regions of
54 (residues 180 to 234) and 155 (residues 337 to 492) amino acids,
which showed approximately 75% identity, were identified[155]. Each
region included one smaller segment of particular homology to the *A.
eutrophus* PHA synthase (almost 60% identity). This indicates that
these segments are important for the enzyme structure and for the
catalysts of the polymerization reaction. Witholt and collaborators
assumed that they are involved in substrate binding and that they
define the specificity of the synthases for hydroxyacyl-CoA thioesters[155].
Only two cysteine residues are conserved in the PHA synthases of *A.
eutrophus* (Cys 319 and Cys 459) and of *P. oleovorans* (Cys 296 and
Cys 430). Although both cysteine residues were not located in either
highly conserved region[155], Peoples and Sinskey suggested that they
are important for the catalysis of the polymerization reaction[67]. This
reaction is assumed to occur in two steps: after an acyl-S-enzyme
intermediate has been formed in the first reaction, the acyl moiety is
transferred to the polyester primer as suggested previously by Griebel
and Merrick[189]. That the N-terminal region of the PHA synthases,

159

which exhibit only weak homologies, are not essential for enzyme activity is indicated by the isolation of two truncated *A. eutrophus* PHA synthases that had lost 36 or 100 amino acids, respectively, but that were still at least partially active[68].

Biosynthesis of PHA from Related Substrates

Hydroxyalkanoic acids

Bacteria incorporate a large variety of 3-hydroxyalkanoates as well as 4-hydroxybutyrate and 5-hydroxyvalerate into the polyester. When Wallen and Rohwedder discovered the presence of poly(3HB-*co*-3HV) in activated sewage sludge almost 20 years ago, and when they failed to produce this polyester *in vitro* from bacterial strains isolated from the sewage treatment plant, they made a very important conclusion: 'possibly PHA production depends upon formation of essential precursors by more than one microorganism in a mixed bacterial population'[40]. It is known that most of the building blocks of PHA shown in Table 1 are incorporated into the polymer only if the cells are cultivated on a carbon source whose carbon skeleton is related to the structure of the constituent monomer. Exceptions to this rule, which allow the synthesis of PHA from unrelated substrates, will be discussed later. In addition, the hydroxyl group, which becomes part of the ester bond, must be present. Alternatively, a different functional group may be present, which is converted to a hydroxyl group during the degradation of the compound. Double bonds and additional functional groups must be already present in the carbon source. There is no evidence that the introduction of functional groups or of double bonds in R-pendant groups occur after the polyester has been synthesized.

Only the known enzymes of the four basic PHA-biosynthetic pathways are known to be involved in the incorporation of these hydroxyalkanoic acids into polymer in bacteria. There is no evidence that different PHA synthases are involved in the synthesis of polyesters consisting of 4- or 5-hydroxyalkanoic acids or of functional 3-hydroxyalkanoic acids. However, auxiliary enzymes must also be present in order to convert the carbon source into an acyl-CoA thioester intermediate that is used as a substrate by the PHA-biosynthetic enzymes (Fig. 11).

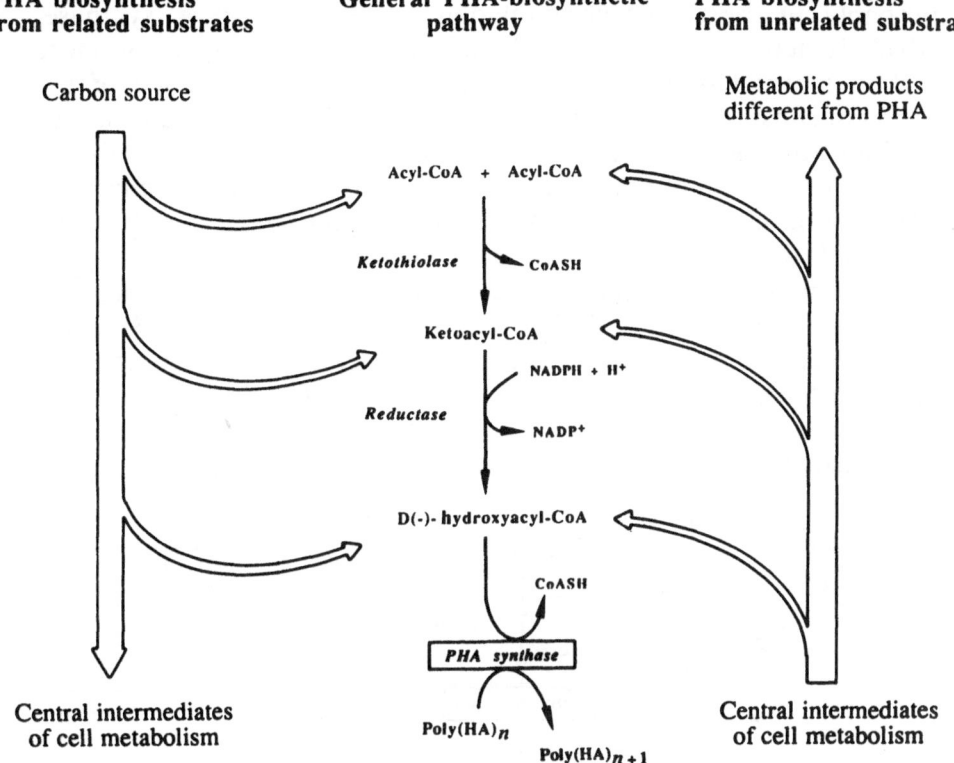

PHA biosynthesis
from related substrates

General PHA-biosynthetic
pathway

PHA biosynthesis
from unrelated substrates

Carbon source

Metabolic products
different from PHA

Acyl-CoA + Acyl-CoA

Ketothiolase CoASH

Ketoacyl-CoA

NADPH + H⁺

Reductase NADP⁺

D(-)- hydroxyacyl-CoA

CoASH

PHA synthase

Central intermediates
of cell metabolism

Poly(HA)ₙ

Poly(HA)ₙ ₊ ₁

Central intermediates
of cell metabolism

Fig. 11 General scheme for synthesis of PHA from related or from unrelated substrates.

A large flexibility with respect to the incorporation of hydroxy-alkanoic acids has been demonstrated for *A. eutrophus* and *P. oleovorans*. With one exception, all known constituents of biosynthetic PHA listed in Table 1 were detected in one or the other of these two bacteria. 3-Hydroxy-4-pentenoic acid has been demonstrated only in the polymer from *Rhodospirillum rubrum*. Whereas in *A. eutrophus*, hydroxyalkanoic acids with three to five carbon atoms were detected, hydroxyalkanoic acids with six to 14 carbon atoms were detected in *P. oleovorans*. This reflects two different types of PHA synthases in *A. eutrophus* and *P. oleovorans*: one PHA synthase dependent on short-chain-length hydroxyacyl-CoA thioesters and the other dependent on medium-chain-length hydroxyalkanoyl-CoA

161

Biomaterials

thioesters, respectively. One or the other type of PHA synthase occurs probably in most-PHA-accumulating bacteria. A PHA synthase, which is active with hydroxyalkanoic acids possessing 15 or more carbon atoms has not yet been detected. The significance of the different substrate specificities of both types of PHA synthases is also obvious from the polyesters accumulated by octanoate-grown cells of a recombinant strain of *P. oleovorans*, which expressed the *A. eutrophus* PHA synthase operon[173,174]. Although the polymer accumulated by these cells was composed of 3HB and 3HO as main constituents and of 3HHx as a minor constituent, evidence was obtained that the polymer consisted of a blend of PHB homopolyester and of poly(3HO-*co*-3HHx) copolyester[174]. There was no evidence for the formation of a terpolyester containing all three constituents. The situation is most probably similar in the corresponding recombinant strains of *P. aeruginosa* and other fluorescent pseudomonads[162].

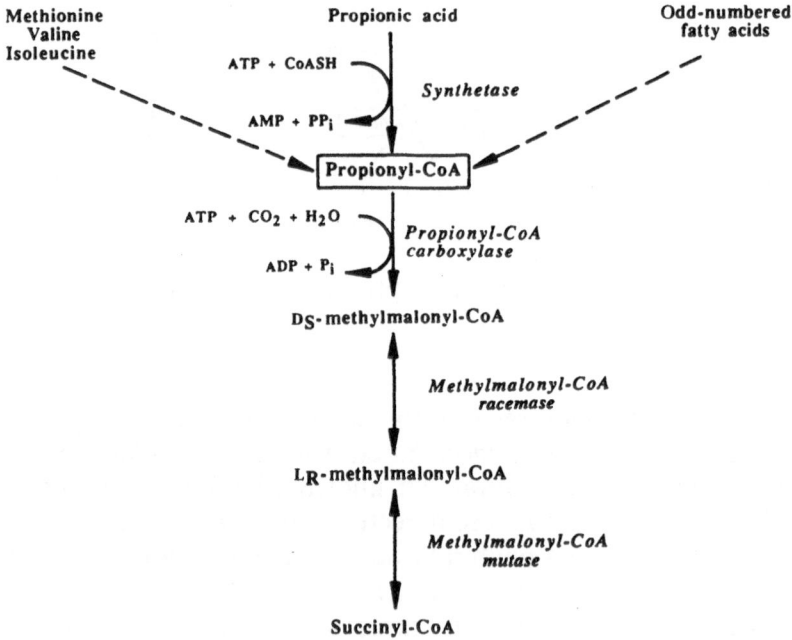

Fig. 12   Formation of propionyl-CoA and its conversion to succinyl-CoA

3-hydroxyvalerate

Synthesis of poly(3HB-*co*-3HV) in *A. eutrophus* and in most other bacteria possessing the three-step PHA-biosynthetic pathway is achieved if propionic acid is used alone or in combination with a second carbon source. Propionic acid is converted by an ATP-consuming synthetase to propionyl-CoA, which is usually converted to the Krebs-cycle intermediate succinyl-CoA by subsequent reactions catalysed by propionyl-CoA carboxylase, methylmalonyl-CoA racemase and methylmalonyl-CoA mutase (Fig. 12). The biosynthetic β-ketothiolase catalyses the condensation of propionyl-CoA with acetyl-CoA to 3-ketovaleryl-CoA, which is reduced to D(−)-3-hydroxyvaleryl-CoA by the acetoacetyl-CoA reductase. The hydroxyvaleryl moiety is then covalently linked to the polyester primer by

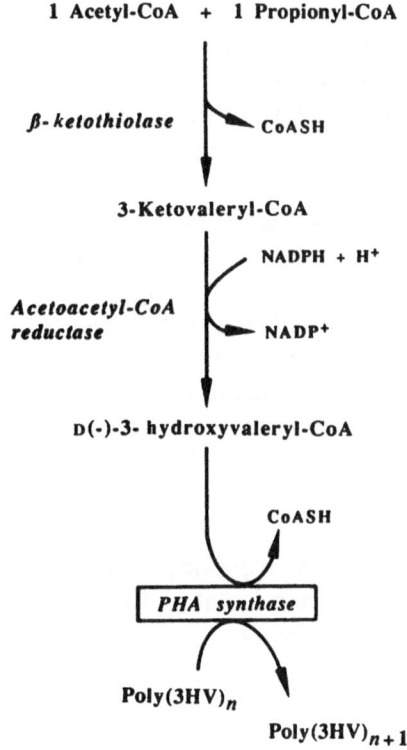

Fig. 13 Synthesis of D(−)-3-hydroxyvaleryl-CoA and incorporation of 3HV into PHA in *A. eutrophus*.

the PHA synthase (Fig. 13). By fast-atom-bombardment mass spectrometry of the oligomers formed by partial methanolysis of poly(3HB-*co*-3HV), random distribution of the comonomers was demonstrated in the copolyester[190,191]. Alternatively, other odd-numbered alkanoic acids, valine, isoleucine or methionine may be provided as a source for propionyl-CoA.

Non-3-hydroxyalkanoates

A. eutrophus incorporates 4-hydroxybutyric acid into the polyester only if 4-hydroxybutyric acid, γ-butyrolactone, 4-chlorobutyric acid, 1,4-butanediol[192,193] or ω-alkanediols with an even number of carbon atoms[194] are used as carbon sources (Fig. 14). A synthetase or a CoA transferase convert 4-hydroxybutyric acid, which derives also from 4-chlorobutyric acid by a dehalogenase, from 1,4-butanediol by dehydrogenases or from γ-butyrolactone, into 4-hydroxybutyryl-CoA. The latter is also an intermediate in the β-oxidation of derivatives of even-numbered ω-alkanediols (Fig. 14). In addition to the auxiliary enzymes mentioned above, PHA synthase is then the only enzyme of the basic PHA biosynthetic pathway required for the incorporation of 4-hydroxybutyryl-CoA into polymer.

5-hydroxyvaleric acid was detected in the polyester accumulated by *A. eutrophus* only if the cells were cultivated with 5-chlorovaleric acid as a carbon source[195]. Presumably, a dehalogenase plus a thiokinase catalyse the conversion of 5-chlorovaleric acid to 5-hydroxyvaleryl-CoA. The latter intermediate seems to be a substrate for the synthase in *A. eutrophus*. Again, only one enzyme of the basic *A. eutrophus* PHA-biosynthetic pathway is involved.

3-hydroxypropionate

A small amount of 3-hydroxypropionate ($\leqslant$7 mol%) was detected along with 3HB in the polymer isolated from cells of *A. eutrophus* when cultivated on 3-hydroxypropionic acid as the sole carbon source. Poly(3HB-*co*-3HP) was also synthesized and accumulated from odd-numbered ω-alkanediols[194]. This indicates that the PHA synthase of *A. eutrophus* can accept 3-hydroxypropionyl-CoA as a substrate. The pathway for the formation of this thioester is most probably similar to that of 4-hydroxybutyryl-CoA formation from 4-hydroxybutyric acid or from even-numbered ω-alkanediols (Fig. 14).

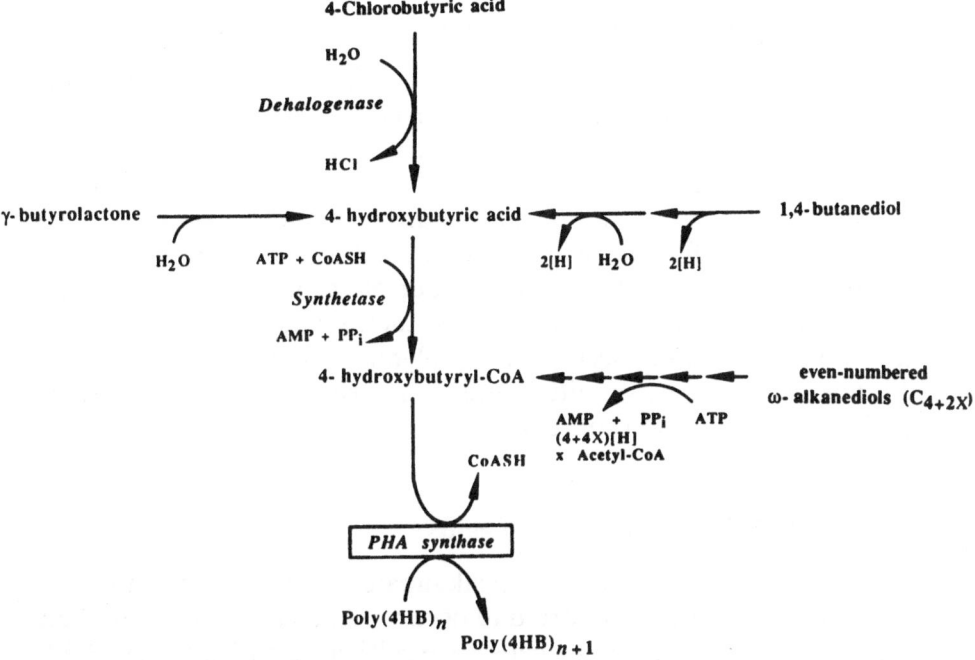

Fig. 14 Synthesis of 4-hydroxybutyryl-CoA and incorporation of 4HB into PHA in *A. eutrophus*.

3-hydroxyalkanoates of medium chain length

The accumulation of PHA containing 3-hydroalkanoates of medium chain length was first detected in *P. oleovorans*[54]. It has since become apparent that this capability occurs in all fluorescent pseudomonads belonging to the RNA homology group I[160–162]. Accumulation of these polyesters occurred under nitrogen limitation with alkanes, alkanols or alkanoic acids as carbon sources[46,54,160,196]. The composition of the accumulated polyester reflects the length of the carbon skeleton of the carbon source: 3-hydroxyoctanoate was the main constituent of PHA accumulated from octane, octanol or octanoic acid; 3-hydroxy-hexanoate occurred as a minor fraction. By contrast, 3-hydroxy-nonanoate was the main constituent of the accumulated polymer from nonane, nonanol or nonanoic acid; in addition, a minor fraction of 3-hydroxyheptanoate was also present[43,160,196].

Brominated 3-hydroxyalkanoic acids

Pseudomonas oleovorans accumulated PHA containing bromo-hydroxyalkanoic acids if the cells were cultivated with mixtures of nonanoic acid plus 6-bromohexanoic acid, 8-bromooctanoic acid or 11-bromoundecanoic acid[56]. It is likely that 6-bromo-3-hydroxyhexanoic acid, 8-bromo-3-hydroxyoctanoic acid and 11-bromo-3-hydroxyalkanoic acid, and presumably also other 3-hydroxy-bromoalkanoic acids with an odd number of carbon atoms, appeared in the polymer. The brominated constituents amounted to a molar fraction of 37.5% in the accumulated polymer. No polymer was detected if nonanoic acid was omitted and if the bromoalkanoic acids were supplied as single carbon sources.

Unsaturated 3-hydroxyalkanoic acids

The incorporation of 3-hydroxyalkanoates with unsaturated R-pendant groups was first reported to occur in *P. oleovorans* by Witholt and collaborators[43]. When the cells were cultivated on 1-alkenes, PHA accumulated to 5.1–13.4 mol% of the cellular dry matter. This polymer consisted of both 3-hydroxyalkanoates and terminally unsaturated 3-hydroxyalkenoates, the latter amounting to molar fractions from 39 to 55%. The monomers 3-hydroxy-7-octenoate and 3-hydroxy-5-hexenoate were detected in PHA of cells cultivated on 1-octene. Cells grown on 1-decene accumulated a polymer consisting of 3-hydroxy-9-decenoate and 3-hydroxy-7-octenoate. The only unsaturated compound detected in 1-nonene-cultivated cells was 3-hydroxy-8-nonenoate. Generally, the corresponding saturated hydroxyalkanoic acids were also present in the polymers[43].

Lenz and collaborators[197] observed the incorporation of 3-hydroxy-6-octenoic acid and 3-hydroxy-7-octenoic acid into PHA of *P. oleovorans* if the same compounds were used as carbon source. The PHA content of the cells amounted to approximately 10% of the cell mass with molar fractions of 76.3 or 61.9%, respectively, for the unsaturated constituents mentioned above. In addition to these constituents, the polyesters contained saturated 3-hydroxyalkanoic acids and also some minor amounts of 3-hydroxy-5-hexenoate and 3-hydroxy-4-hexenoate, respectively[197]. No PHA was accumulated from hexene and dodecene.

Although PHA was accumulated by *P. oleovorans* from *cis*-9-

octadecenoic acid (oleic acid), *trans*-9-octadecenoic acid (elaidic acid) and *cis,cis,cis*-9,12,15-octadecatrienoic acid (linolenic acid), Witholt and collaborators did not report unsaturated constituents in the polymer[170]. By contrast, Eggink and collaborators[198] detected unsaturated constituents in the polymer of *P. oleovorans* cells cultivated on *cis*-9-octadecenoic acid or *cis,cis*-9,12-octadecadienoic acid (linoleic acid). The total PHA content of those cells amounted to 33% of the cellular mass, and depending on the carbon source, the molar fraction of the unsaturated constituents, which were 3-hydroxy-6-dodecenoate, 3-hydroxy-5-tetradecenoate and 3-hydroxy-5,8-tetradecadienoate, were 13 or 17.2 mol%, respectively. The reason for these discrepancies remains to be resolved. It should be noted that the batches of oleic acid and linoleic acid used by Eggink and collaborators were only 76 or 62% (w/w) pure, respectively, and that technical substrates were obviously used.

Rhodospirillum rubrum incorporated 3-hydroxy-4-pentenoate into PHA if the cells were cultivated anaerobically in the light in a nitrogen-free medium with 4-pentenoic acid as the sole carbon source. Thermal analysis indicated that the polymer consisted probably of two different copolyesters, which together were composed of 30 mol% 3-hydroxy-4-pentenoate, 11 mol% 3HB and 59 mol% 3HV as building blocks[56]. *A. eutrophus*, the other bacterium tested in this study, synthesized a copolyester, which consisted of 56 mol% 3HB and 44 mol% 3HV from 4-pentenoic acid; 3-hydroxy-4-pentenoate was not detected.

Branched 3-hydroxyalkanoic acids

Evidence for the presence of an iso-branched 3-hydroxyalkanoic acid in the polymer was obtained first in a complex polyester isolated from *Bacillus megaterium*[199]. However, the chemical structure was not identified. PHA containing defined branched 3-hydroxyalkanoate constituents were only reported recently in *P. oleovorans*[52]. Whereas the cells accumulated a copolyester of 3-hydroxy-7-methyloctanoate and of 3-hydroxy-5-methylhexanoate from 7-methyloctanoic acid as the sole carbon source, no PHA at all accumulated if 5-methyloctanoic acid or 6-methyloctanoic acid was used as the sole carbon source. With the latter methyl-branched octanoic acids, PHA containing methyl-branched 3-hydroxyalkanoic acids accumulated only when octanoic acid was added to the carbon source. From a mixture of octanoic acid and 6-methyloctanoic acid, a polymer consisting of 3HO, 3HHx, 3-hydroxy-6-methyloctanoate and a small amount of 3-hydroxy-4-methylhexanoate accumulated. The main constituents of the polymer accumulated from a mixture of octanoic acid and 5-methyloctanoic acid were 3HO and 3-hydroxy-5-methyloctanoate.

Deviations from the standard PHA synthases

Probably only minor variations of both types of PHA synthases occur in bacteria. Bacteria, which synsthesize and accumulate PHA consisting of 3HB plus monomers larger than hydroxyvalerate, each in a significant amount, with only one single PHA synthase, have not been described. The polymer accumulated by cells of *P. aeruginosa* and by some other species of the fluorescent pseudomonads, when cultivated on valeric acid, also contained 3HV as constituent[162]. This indicates that the range of substrates for the PHA synthases of these bacteria is extended to 3-hydroxyalkanoyl-CoA thioesters with only five carbon atoms. On the other hand, a polyester was isolated from cells of *Bacillus megaterium*[199] and *Rhodospirillum rubrum* which contained 3HHp as well as the principle constituents 3HV and 3HB. This indicates an extension of the range of substrates for the PHA synthases of these bacteria to 3-hydroxyalkanoyl-thioesters with seven carbon atoms.

In cells of the strict anaerobe *Syntrophomonas wolfei*, McInnerny and collaborators detected a copolyester that contained 3HB plus a significant fraction (4 mol%) of 3HHx if the cells were cultivated on *trans*-3-hexenoate, *trans*-2-hexenoate or *trans*,*trans*-2,4-hexadienoate[93]. These carbon sources were fermented mainly to acetate, propionate, butyrate, valerate and hexanoate[94]. From *trans*-3-hexenoate or *trans*-2-pentenoate, PHB homopolyester or a copolyester of 3HB and 3HV was accumulated, respectively[238]. Unsaturated hydroxyalkanoic acids were not detected in the polymer. These results indicated that PHA derived mainly from the condensation of two acetyl-CoA molecules and from the subsequent reduction of acetoacetyl-CoA to give 3HB in the polymer and, to a much lesser extent, from intermediates of the β-oxidation without cleavage of the carbon-carbon bond which gave rise to the other monomers found. The latter occurred predominantly during the early stages of growth.

Biosynthesis of PHA from Unrelated Substrates

Recently, the synthesis of poly(3HB-*co*-3HV) and of copolyesters consisting of medium-chain-length 3-hydroxyalkanoic acids from fructose, gluconate and other unrelated substrates has been detected in several bacteria (Table 3). Dawes and collaborators referred to this capability as PHA biosynthesis from unrelated substrates[163]. In these bacteria, a central intermediate of the metabolism of the cell, which is used for the generation of energy or for anabolic purposes, must be directed to PHA biosynthesis (Fig. 11). These pathways are very important as they allow synthesis of complex PHA from simple substrates rather than from expensive or relatively toxic substrates.

Table 3 Synthesis and accumulation of PHA from unrelated substrates

Organism	Carbon source	PHA detected	References
Alcaligenes eutrophus strain R3	Fructose, gluconate	Poly(3HB-*co*-3HV)	Manuscript in preparation Steinbüchel & Pieper
Corynebacterium hydrocarboxydans	Fructose, glucose, succinate, acetate, lactate	Poly(3HB-*co*-3HV)	163, 200
Haloferax mediterranei	Starch	Poly(3HB-*co*-3HV)	201
Rhodococcus sp. ATCC 19070	Fructose, glucose, sucrose, acetate, lactate	Poly(3HB-*co*-3HV)	163, 200
Rhodococcus sp. NCIMB 40126	Fructose, glucose, molasses, succinate, acetate, lactate	Poly(3HB-*co*-3HV)	163, 200
Nocardia strain 107-322	Butane	Poly(3HB-*co*-3HBen)	38
Nocardia lucida	Fructose, glucose, acetate, succinate	Poly(3HB-*co*-3HB)	163, 200
Pseudomonas aeruginosa (and some other fluorescent pseudomonads)	Gluconate, fructose, glucose, ethanol	Poly(3HO-*co*-3HD-*co*-3HDD)	162, 163, 164, 223
Rhodobacter sphaeroides (and some other non-sulphur purple bacteria)	Acetate	Poly(3HB-*co*-3HV)	99

Abbreviations: PHA, polyhydroxyalkanoates; for other abbreviations, see Table 1.

3-hydroxyvalerate

Synthesis of poly(3HB-*co*-3HV) from single substrates like carbo-hydrates such as fructose or glucose or from organic acids such as succinate, acetate or lactate has been reported for several gram-positive bacteria. Using glucose as a carbon source and in the absence of a nitrogen source, *Rhodococcus sp.* NCIMB 401266, for instance, accumulated a polymer that amounted to 30% (w/w) of the biomass and that consisted of 76 mol% 3HV and 24 mol% 3HB[163,200]. Ter-polyesters, which contained small amounts of 4HB in addition to 3HB and 3HV, were accumulated from 4-hydroxybutyric acid or 1,4-butanediol as sole carbon sources[200]. The biosynthetic route which yields propionyl-CoA or, alternatively, one of the possible precursors with five carbon atoms, is not known.

Poly(3HB-*co*-3HV) formation has also been reported for *Haloferax mediterranei*[201] from starch (46% (w/w) of biomass, up to 17.7 mol% 3HV). In addition, several nonsulphur bacteria such as *Rhodobacter sphaeroides* (69.9% (w/w) of biomass, 7.3 mol% 3HV) and also *Rhodopseudomonas acidophila*, *Rhodopseudomonas blastica*, *Rhodospirillum rubrum* or *Rhodocyclus gelatinosus*, accumulated this copolyester from acetate or from a mixture of acetate plus succinate[99]. By contrast, accumulation of poly(3HB-*co*-3HV) from unrelated substrates was never detected in sulphur purple bacteria[99].

Small amounts of 3HV (3 to 5 mol%) were detected in addition to 3HB in the PHA accumulated by strain *A. eutrophus* R3 from fructose or gluconate (Steinbüchel and Pieper, manuscript in preparation). R3 is a spontaneously occurring prototrophic revertant from a threonine deaminase-defective isoleucine-auxotrophic mutant. R3 is still unable to synthesize catalytically active threonine deaminase, but synthesizes acetohydroxy acid synthase constitutively and at a high level in order to compensate for the metabolic defect[202]. During cultivation on fructose or on gluconate, the strain excreted large amounts of valine, and to a lesser extent also isoleucine, into the medium. The assumption must be that the 3HV building-blocks derive most probably from the use of propionyl-CoA formed during the degradation of these two amino acids. Propionyl-CoA is therefore formed by the cells themselves and need not be provided as an extra carbon source for the synthesis of poly(3HB-*co*-3HV).

3-hydroxyalkanoates of medium chain length

Pseudomonas aeruginosa PAO, and all other strains of this species tested, accumulated a copolyester with 3HD as main constituent and 3HO and 3HDD as minor constituents from gluconate and from the other carbon sources tested, all of which are catabolized to acetyl-CoA before an intermediate enters a PHA-biosynthetic pathway[162,164]. Nitrogen limitation was used to induce PHA accumulation. PHA was accumulated to almost 70% (w/w) of the biomass. This biosynthetic capability is shared by most other fluorescent pseudomonads that belong to the rRNA homology group I. *P. aureofaciens*, *P. citronellolis*, *P. mendocina*, *P. putida* and *Pseudomonas* sp. DSM 1650 accumulated abundant amounts of this polymer. The pathway, which synthesizes the 3-hydroxydecanoyl-derivative used for PHA synthesis from acetyl-CoA, is still unknown. In the bacteria listed above, this new pathway exists in addition to the *P. oleovorans* PHA-biosynthetic pathway. Curiously, *P. oleovorans* seems to be the only species belonging to this taxonomic group of pseudomonads that does not possess the *P. aeruginosa* PHA-biosynthetic pathway.

Competive Routes for Hydroxyacyl-CoA Thioesters

One other important aspect of polyester biosynthesis should be emphasized at this point. Although a large variety of hydroxyalkanoic acid monomers were detected in bacterial PHA, PHB and PHV are the only homopolyesters synthesized by bacteria. Whereas PHB homopolyester is accumulated by many bacteria, the formation of PHV homopolyester was detected only recently in *Chromobacterium violaceum* if the cells were cultivated with valerate as the sole carbon source under nitrogen-limiting conditions (A. Timm, unpublished data). *Rhodococcus* sp. NCIMB 40126 accumulated a polyester from valerate that contained 99 mol% 3HV and only traces of 3HB[163,200]. All other bacterial PHA are composed of at least two different constituents.

Although many CoA thioesters of hydroxyalkanoic acids are efficient substrates for PHA synthase, these thioesters are also further degraded to acetyl-CoA or to other central intermediates of metabolism. PHA synthase and the catabolic enzymes are competing for these thioesters, and the latter are continuously withdrawing a significant portion from the hydroxyacyl-CoA pool. They can convert these to different hydroxyacyl-CoA components or to acetyl-CoA. The competition between PHA synthase and catabolic enzymes for hydroxyalkanoyl-CoA thioesters as well as the lack of specificity of all PHA synthases investigated so far will therefore bias against the production of homopolyesters other than PHB or PHV.

This is illustrated in Fig. 15 for the synthesis of PHA containing 4-hydroxybutyrate (4HB). It has already been mentioned that a copolyester of 4HB and 3HB, poly(3HB-*co*-4HB), is accumulated if, for example, 4-hydroxybutyric acid is provided as a carbon source. Homopolyesters of 4HB have not been described, and the molar fraction of 4HB in the copolyester has never exceeded 50%[193,203,204]. This can be explained by the competition for 4-hydroxybutyryl-CoA between PHA synthase and catabolic enzymes which convert 4-hydroxybutyryl-CoA probably via intermediates of the Krebs cycle to acetyl-CoA. Only a small fraction of 4-hydroxybutyryl-CoA is directed to PHB synthase.

A. eutrophus synthesized a copolyester of 3HB and 3HV if propionate or valerate were provided as the sole carbon source. Under these conditions the molar fractions of 3HV amounted to 45 or 90 mol%, respectively[205,206]. The molar fractions of 3HV dropped if a second carbon source like acetate or butyrate was provided in addition to propionate or valerate. These data indicated that even with propionate

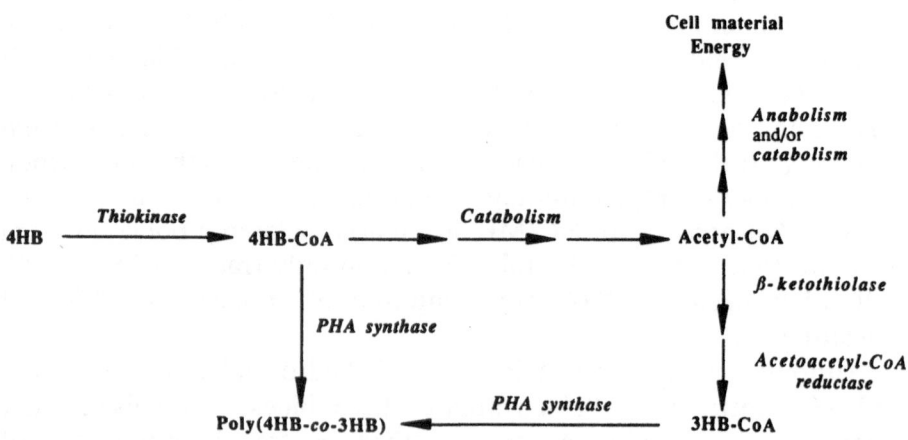

Fig. 15 Competition of PHA synthase and catabolic reactions for 4-hydroxybutyryl-CoA.

or valerate as the sole carbon source, a significant portion of the substrate was converted to acetyl-CoA. As the molar fraction of 3HV was higher with valerate than with propionate, 3-ketovaleryl-CoA was obviously reduced much more efficiently to D(-)-3-hydroxyvaleryl-CoA than it was cleaved into propionyl-CoA and acetyl-CoA[206].

When cells of *P. oleovorans* were cultivated on octane or octanoic acid, no PHO homopolyester was synthesized. The molar fraction of 3HO in the polyester amounted to about 90 to 95%; 3-hydroxyhexanoate was always present as an additional constituent at a significant fraction. This clearly indicated that some of the intermediates escaped from the first turn of the β-oxidation cycle without being converted to an intermediate of the PHA-biosynthetic pathway (Fig. 9) and entered a second turn of the cycle after one acetyl-CoA moiety had been removed. A recombinant strain of *P. oleovorans* harbouring multiple copies of the PHA synthase genes incorporated a slightly higher fraction of the substrate-derived monomers into the polyester and fewer monomers whose length was reduced by the cleavage of one acetyl-CoA moiety[155]. This indicated that in this strain, substrate-derived hydroxyacyl-CoA thioesters are more efficiently directed to PHA biosynthesis than to β-oxidation as compared to the wild type. No 3HB appeared in the polymer; this may be due to the fact that the *P. oleovorans* PHA synthase cannot use 3-hydroxybutyryl-CoA as a substrate.

Enlargement of the Substrate Spectrum of *Alcaligenes eutrophus* H16

The use of *A. eutrophus* for polymer production is impaired by its limited capability to grow on carbohydrates of technological interest. Carbon sources that give a low substrate cost per tonne of PHA produced[207] are not used by *A. eutrophus*. Fructose is the only sugar used by *A. eutrophus*, whereas glucose and disaccharides such as lactose, maltose or sucrose, are not used[208]. *A. eutrophus* is also unable to use methanol as carbon source. Glucose-using mutants of strain H16 were isolated many years ago[209,210]. Evidence was obtained that the use of glucose is not possible because of the lack of a transport system allowing passage of the cytoplasmic membrane, and that glucose is taken up by facilitated diffusion by mutant strain H16-G$^+$1. Glucose-utilizing mutants are now used by the chemical industry for the production of thermoplastic PHB on an industrial scale (see above).

Pathways for the use of lactose, galactose and sucrose were established in *A. eutrophus* by employing recombinant DNA technology. To confer the ability to use sucrose to the cells, the *Bacillus subtilis* levanase structural gene was transferred to *A. eutrophus*. However, growth of the recombinant strains on sucrose was rather poor; most probably the uptake of sucrose was not very efficient and/or the cells did not secrete the enzyme into the medium[211]. In a much more sophisticated approach, genomic libraries of *Paracoccus denitrificans* and *Rhodobacter capsulatus*, which are taxonomically much more related to *A. eutrophus*, were successfully screened for DNA fragments that conferred the ability to use sucrose on *A. eutrophus*[212].

To establish the use of lactose, a *phbCp'–'lacZ* fusion was ligated to Tn5 DNA in the suicide vector plasmid pSUP5011 and was inserted into the genomes of *A. eutrophus* H16 and the glucose-utilizing mutant H16-G^{+1} by transposition following conjugal transfer of the construct. In the recombinant cells the *lac*-operon was expressed under the control of the promoter of the PHB synthase operon. Although the recombinant strains grew on lactose, the cells only used lactose completely after the genes for the use of galactose had been transferred in addition[213,214].

Intracellular Degradation or Mobilization of PHA

In contrast to extensive studies on the biosynthesis of PHA, relatively little is known about the degradation of these polyesters. All naturally occurring bacteria accumulating PHA will most probably possess an enzyme system that mobilizes PHA intracellularly under certain conditions. These intracellular depolymeriases are different from the enzymes of microorganisms that use exogenous PHA in the environment as a sole carbon source for growth. The enzymes that degrade exogenous PHA are not only synthesized by bacteria but also by moulds[215] and are excreted by the cells to hydrolyse PHA extracellularly. Subsequently, the cells take up the hydrolysis products.

Experiments with isolated PHA granules

Mobilization of PHA has been frequently studied in *A. eutrophus*[216,217], *R. rubrum*[189,218,219] and *Bacillus megaterium*[220,221] employing native PHB granules and various cytoplasmic fractions from these organisms. Although no details of the proteins concerned were provided, it was evident that the PHA-mobilizing enzyme systems are rather complex and consisted of several components. Whereas a granule-associated heat-labile factor, a PHB depolymerase and an activator protein are required in *B. megaterium* for PHB mobilization[220], a granule-associated heat-labile factor, a PHB depolymerase, an oligomer hydrolase and a heat-stable activator protein were required in *R. rubrum*[189,219]. The action of the activator protein could be simulated by mild tryptic treatment in the samples derived from both bacteria.

3-Hydroxybutyric acid was identified as the sole or predominant product of PHB-hydrolysis, with the enzyme systems from *A. eutrophus*[217] and *R. rubrum*[219]. With the latter, dimeric esters of 3HB also occurred. Since cell-free extracts, which contained a mixture of enzyme, had been used for these experiments, it was not possible to determine the nature of the product resulting from the action of PHB depolymerase. However, the presence of an intracellular 3HB-dimer hydrolase in *R. rubrum*[219] may indicate that oligomers were an important and possibly obligate intermediate of PHB mobilization in this bacterium.

According to the studies of Hippe[216,217], the degradation of PHB was rather slow in *A. eutrophus* compared with other bacteria. By contrast, *P. oleovorans* and *P. aeruginosa* have probably a much more active PHA-mobilizing enzyme system. PHA disappeared at rather high rates after the extracellular carbon source was exhausted in these bacteria[43,162]. One can speculate whether the difference in M_rs of the polyesters of *A. eutrophus*[221] or of *P. oleovorans* and *P. aeruginosa*[196] are connected with these observations. A less active PHA-mobilizing enzyme system should be favourable with respect to the yield of PHA during fermentation, and therefore important for the production of PHA.

Physiology of PHA molibization

Doi and collaborators demonstrated intracellular degradation of PHA in *A. eutrophus* with whole cells[194,222]. PHB homopolyester was replaced by poly(3HB-*co*-3HV) if PHB-containing cells were incubated with valeric acid as the sole carbon source in a nitrogen-free medium. Further, the total content and the molar fraction of 3HV in poly(3HB-*co*-3HV) decreased drastically when cells, which had been cultivated on valeric acid, were transferred to a medium containing butyric acid as the sole carbon source. From this it can be concluded that degradation of the polyester, which accumulated in the first phase of the experiment, occurred in parallel with synthesis of the new polyester in the second phase. An average rate of approximately 60 mg 3HV perg protein perh was calculated for net PHA degradation for the initial phase. Later this rate decreased to approximately 12 mg 3HV perg protein perh[222]. These rates were in the same order of magnitude as calculated from the data provided by Hippe[216,217]. In contrast to the experiments mentioned above, no turnover of PHB was measured during the cultivation of *A. eutrophus* in a nitrogen-limited chemostat culture with glucose as the carbon source[66]. After ^{14}C-labelled glucose was replaced by nonlabelled glucose, the rate of the replacement of labelled polymer with nonlabelled polymer only paralleled the theoretical washout rate.

From the molecular analysis of PHB-leaky mutants of *A. eutrophus*, some evidence was obtained that two proteins, which were referred to as protein H and protein I and which exhibited a high degree of homology to the HPr protein and to enzyme I of the phosphoenolpyruvate:carbohydrate phosphotransferase system, are involved in the regulation of the mobilization of PHB in this bacterium (ref. 223; A. Pries, N. Krüger and A. Steinbüchel, manuscript in preparation). Inactivation of one or the other gene encoding these two proteins by transposon insertion resulted in an increased rate of PHB mobilization. Our working hypothesis suggests that in the wild type the intracellular concentration of phosphoenolpyruvate reflects the presence of an extracellular carbon source and that one component of the PHB-mobilizing enzyme system is inactivated by transfer of the phosphoryl group of PEP through protein H and protein I. As one consequence, synthesis and degradation of PHB would occur in parallel, and less PHB will accumulate in the cells of mutants unable to synthesize protein H or protein I.

Mutants defective in the mobilization of PHA

Mutants impaired in the degradation of accumulated polyesters were derived from *P. oleovorans*[155]. Cells that had accumulated PHA under nitrogen-limiting conditions retained their PHA if excess nitrogen source was added to restore growth. Mutants were distinguished from the wild type by the retention of Sudan black after the addition of nitrogen source[155].

Extracellular Degradation of PHA

Production of extracellular PHA hydrolases is usually demonstrated by zones of clearing around colonies on mineral salts agar medium which contains PHA as a carbon source. An alternative detection method is to incubate depolymerase-containing solutes with thin-cast polyester films and follow its dissolution and erosion (for example, see ref. 224). Where it has been studied in detail, degradation of PHB seems to occur in two steps. First, a PHB depolymerase releases oligomers, which are preferentially 3HB dimers. Second, an oligomer hydrolase cleaves the dimer into 3HB monomers. Whereas, for example, *P. lemoignei* has an intracellular dimer hydrolase[225], an extracellular oligomer hydrolase is synthesized by *A. faecalis* T1. It has not been demonstrated whether the dimers or the monomers are taken up by the cells.

Aerobic degradation

Although the 3HB dimer hydrolase[225] and isoenzymes of PHB depolymerase from *P. lemoignei*[226] were also studied, most detailed molecular data are available for the extracellular depolymerase system of *Alcaligenes faecalis* strain T1[227]. This bacterium synthesizes two different enzymes exhibiting activities that hydrolyse ester bonds. First, a 3-hydroxybutyrate oligomer hydrolase hydrolysed water-soluble as well as water-insoluble 3HB oligomers (for example, 3HB trimer); this enzyme did not hydrolyse PHB. The enzyme is monomeric with an M_r of 74K[228]. Second, a PHB depolymerase hydrolysed all three different types of polyesters mentioned above. Of various synthetic polyesters, only poly(3-propiolacetone) was hydrolysed at a significant rate; poly(5-caprolactone), polyglycolides, poly(ethylene-succinate), poly(ethyleneadipate) and poly(ethylenesebacate) were not hydrolysed[229]. The formation of both enzymes is induced in the presence of PHB, whereas they are absent from succinate- or 3-hydroxybutyrate-grown cells.

The PHB depolymerase from *A. faecalis* T1 was purified to homogeneity and has been thoroughly characterized[229]. Evidence was obtained that the enzyme hydrolyses 3HB trimers and tetramers from the hydroxyl terminus and that 3HB dimers are released, whereas it acted as an endo-type hydrolase towards higher-order oligomers. Monomers amounted to a significant fraction of the hydrolysis products only if oligomeric esters of 3HB with an odd number of monomer units were used as substrates[229,230]. Trypsin treatment removed part of the protein, and the M_r of the depolymerase decreased from 47K to 42K. Trypsin treatment removed a hydrophobic region of the depolymerase. The treated depolymerase lost its activity towards PHB, but not towards the water-soluble oligomers[231]. However, in the presence of 1 M ammonium sulphate, the activity of truncated depolymerase towards PHB was restored. It is assumed that the hydrophobic N-terminal region of PHB depolymerase binds to the polyester molecule[231].

Saito *et al.*[232] screened a gene bank of *A. faecalis* T1 genomic DNA in *E. coli* with antibodies raised against the depolymerase, and isolated the depolymerase structural gene. The nucleotide sequence predicted a precursor PHB depolymerase with 488 amino acids and an M_r of 49,934. This includes a signal peptide of 27 amino acids that is removed during the secretion of the protein, and the mature PHB depolymerase consisted of 461 amino acids, M_r 46,858. Recombinant strains of *E. coli* expressed the gene and the enzyme was fully active. However, most of the PHB depolymerase was located in the periplasmic fraction and only 10% was detected in the medium.

Solutions of cast films of different polyesters were incubated with an aqueous solution of purified PHB depolymerase from *A. eutrophus* T1[233]. The weight loss of these films was linear with time and was strongly dependent on the composition of polyesters: poly(3HB-*co*-4HB)>PHB>poly(3HB-*co*-3HV). The presence of 4HB units in the polyester accelerated the rates of degradation, whereas the rate was decreased with 3HB units. Doi and collaborators[233] assumed that the ester linkage at 4HB units is sterically more accessible for the depolymerase than at 3HB units. The M_rs of the polymer in the film remained constant during the course of the experiment. This is in agreement with scanning electron micrographs of the films which showed that the weight loss corresponded to the decrease of the initial thickness of the films. Therefore, the polymer erosion proceeds from the surface, whereas the inside of the film remains unchanged.

Anaerobic degradation

There is no *a priori* reason why the anaerobic degradation of PHA is not possible. Accumulation of PHA in anaerobic environments, which would happen if anaerobic degradation of PHA was not possible, has not been reported. Recently there was the first report on an anaerobic bacterium *Ilyobacter delafieldii* capable of degrading exogenous PHB[234]. *I. delafieldii*, *I. polytrophus* and *Clostridium homopropionicum* also used the monomer as a sole carbon source for growth and fermented 3-hydroxybutyrate to acetate, butyrate and molecular hydrogen[234,235,236]. *C. homopropionicum* seems to be a suitable candidate for the anaerobic degradation of copolyesters containing 4HB as this bacterium grew anaerobically on 4-hydroxybutyrate[235].

Outlook

In the last few years the number of constituents detected in bacterial PHA has increased dramatically. About 35 different monomers have already been identified and this number will probably increase further in the near future. The number of polyesters actually available is even higher, as these constituents occur in different combinations. These polymers will not only differ in their chemical composition but also in their physical and technical properties, which are almost unknown for many of the new polymers. The range of substrates that give rise to PHA is now large and it is no longer sufficient to assume that the presence of light-scattering granules in the cells or polymeric material absorbing at 5.7 μm indicates synthesis of PHB. Until detailed chemical analysis has been carried out to determine the composition of the polymer, it should be referred to as PHA.

It was generally assumed that bacterial PHA are isotactic only with D(−)-isomers of hydroxyalkanoic acids as constituents. When Haywood *et al.*[200] analysed the specific optical rotation of 3-hydroxy acid methyl esters liberated by methanolysis of poly(3HB-*co*-3HV), which was isolated from glucose-grown cells of *Rhodococcus* sp. NCIMB

40126, evidence was obtained that the polymer was optically impure. Although it consisted mainly of the D(−)-enantiomer, the specific optical rotation of the methyl esters indicated that L(+)-enantiomers of 3-hydroxyalkanoic acids were also present[200]. The authors also indicated that PHA synthase of *Rhodococcus* sp. is, in contrast to the synthase of *A. eutrophus*, active with both enantiomers of 3-hydroxyacyl-CoA. If this can be confirmed, it will greatly increase the number of potential constituents of bacterial PHA. Furthermore, the PHA-biosynthetic pathways have to be reconsidered.

An increasing number of PHA-biosynthetic genes will soon be available from various bacteria, in addition to the genes which have been already isolated from *A. eutrophus, Z. ramigera* and *P. oleovorans.* Recombinant DNA technology will allow any combination of these genes and their transfer and heterologous expression in other bacteria or in eukaryotic organisms. In addition, site-directed mutagenesis will be applied to alter, for example, the substrate specificity of the PHA-biosynthetic enzymes. Production of PHA with genetically engineered organisms will become possible. One of the challenges this technology presents is the generation of transgenic plants, which harbour and express PHA-biosynthetic genes[237]. If it is possible to direct the carbon flow in these transgenic plants towards acetyl-CoA, use of sunlight and of atmospheric carbon dioxide for the production of PHA would be possible. It would also provide new opportunities for agriculture, in the production of non-food-crops, allowing the industry to be maintained without over-production and the generation of food surpluses.

Acknowledgements

The studies performed in the laboratory of the author were supported by grants from the Bundesministerium für Forschung und Technologie (Forschungsschwerpunkt "Grundlagen der Bioprozeßtechnik"), the European Economic Community (ECLAIR AGRE.0006.C(H)), the Max Buchner Forschungsstiftung and Förderungsmittel des Landes Niedersachsen. The author is grateful to B. Witholt for providing a preprint of ref. 155.

References

1. Brandl, H., Gross, R.A., Lenz, R.W. & Fuller, R.C. (1990) Plastics from bacteria and for bacteria: Poly(β-hydroxyalkanoates) as natural, biocompatible, and biodegradable polyesters. In *Advances in Biochemical Engineering/Biotechnology* (eds Ghose, T.K. & Fiechter, A.), Vol. 41, Springer, Berlin.

2. Dawes, E.A. & Ribbons, D.W. (1964) Some aspects of the endogenous metabolism of bacteria. *Bacteriol. Rev.* **28**: 126–149.

3. Dawes, E.A. & Senior, P.J. (1973) Energy reserve polymers in microorganisms. *Arch. Microbiol. Physiol.* **14**: 135–266.

4. Dawes, E.A. & Senior, P.J. (1973) The role and regulation of energy reserve polymers in microorganisms. *Adv. Microbiol. Physiol.* **10**: 135–278.

5. Dawes, E.A. (1975) The role and regulation of poly-β-hydroxybutyrate as a reserve in microorganisms. In *Proceedings of the International Symposium on Macromolecules* (ed. Mano, E.B.), pp. 433–450. Elsevier, Amsterdam.

6. Dawes, E.A. (1986) *Microbial Energetics*. Blackie, Glasgow.

7. Dawes, E.A. (1988) Poly-β-hydroxybutyrate, an intriguing biopolymer. *Bioscience Reports* **8(6)**: 537–547.

8. Dawes, E.A. (ed.) (1990) *Novel Biodegradable Microbial Polymers*. Kluwer, Dordrecht.

9. Doi, Y. (1990) *Microbial Polyesters*. VCH, New York.

11. Merrick, J.M. (1978) Metabolism of reserve materials. In *The Photosynthetic Bacteria* (eds Clayton, R.K. & Sistrom, W.R.), pp. 199–219. Plenum, New York.

12. Schlegel, H.G. (1962) Bildung von Speicherstoffen durch Knallgas- und Purpurbakterien. *Veröffentlichung Deutsche Botanische Gesellschaft* **1**: 167–172.

13. Schlegel, H.G. & Gottschalk, G. (1962) Poly-β-hydroxybuttersäure, ihre Verbreitung, Funktion und Biosynthese. *Angewandte Chemie* **74**: 342–346.

14. Steinbüchel, A. & Schlegel, H.G. (1989) Excretion of pyruvate by mutants of *Alcaligenes eutrophus*, which are impaired in the synthesis of poly(β-hydroxybutyric acid), PHB, under conditions permissive for synthesis of PHB. *Appl. Microbiol. Biotechnol.* **31**: 168–175.

15. Tomita, K., Saito, T. & Fukui, T. (1983) Bacterial metabolism of poly-β-hydroxybutyrate. In *Biochemistry of the Metabolic Process* (eds Lennon, D.L.F., Stratman, F.W. & Zahlten, R.N.), pp. 353–366. Elsevier, New York.

16. Volova, T.G. (1990) Kalalewa: Polyoxybutyrate – ein thermoplastisch biodegradierbares Polymer. *USSR Acad. Sci. Preprint N 131B*.

17. Wilkinson, J.F. (1959) The problem of energy-storage compounds in bacteria. *Exp. Cell Res. Suppl.* **7**: 114–130.

18. Anderson, A.J. & Dawes, E.A. (1990) Occurrence, metabolism, metabolic role, and industrial uses of bacterial polyhydroxylalkanoates. *Microbiol. Rev.* **54**: 450–472.

19. Marchessault, R.H. (1988) History of polyalkanoate research. *Polym. Prepr.* **29**: 594–595. Toronto.

20. Schlegel, H.G. (1990) *Alcaligenes eutrophus* and its scientific and industrial career. In *Novel Biodegradable Microbial Polymers* (ed. Dawes, E.A.), pp. 133–141. Kluwer, Dordrecht.

21. Steinbüchel, A. (1989) Poly(hydroxyfettsäuren) – Speicherstoffe von Bakterien: Biosynthese und Genetik. *Forum Mikrobiologie.* **12**: 190–198.

22. Steinbüchel, A. & Schlegel, H.G. (1991) Physiology and molecular genetics of poly(β-hydroxyalkanoic acid) synthesis in *Alcaligenes eutrophus*. *Mol. Microbiol.* **5**: 535–542.

23. Steinbüchel, A. (1991) Recent advances in the knowledge of bacterial

poly(hydroxyalkanoic acid) metabolism and potential impacts on the production of biodegradable thermoplastics. *Acta Biotechnol.* (in the press).

24. Babel, W., Riis, V. & Hainich, E. (1990) Mikrobielle Thermoplaste: Biosynthese, Eigenschaft und Anwendung. *Plaste Kautschuk* **37**: 109–115.

25. Byrom, D. (1987) Polymer synthesis by microorganisms: technology and economics. *TIBTECH* **5**: 246–250.

26. Hartley, P. (1987) Abbaubare Polymere aus dem Fermenter. *Bioengineering* **3**: 66–68.

27. Holmes, P.A. (1985) Applications of PHB – a microbially produced biodegradable thermoplastic. *Phys. Technol.* **16**: 32–36.

28. Holmes, P.A. (1988) Biologically produced PHA polymers and copolymers. In *Developments in Crystalline Polymers* (ed. Bassett, D.C.), Vol. 2, pp. 1–65. Elsevier, London.

29. Lafferty, R.M. & Heinzle, E. (1977) Extraction of a thermoplastic from bacteria. *Chem. Rdsch.* **30**: 15–16.

30. Lafferty, R.M., Braunegg, G., Korneti, L., Strempfel, F., Bogensberger, B., Korsatko, W. & Wabnegg, B. (1984) PHB: Biotechnological production and polymer applications. Third European Congress on Biotechnology, Verlag Chemie, Weinheim, 521–527.

31. Lafferty, R.M., Korsatko, B. & Korsatko, W. (1988) Microbial production of poly-β-hydroxybutyric acid. In *Biotechnology* (eds Rehm, H.J. & Reed, G.), Vol. 66, pp. 136–176. Verlagsgesellschaft, Weinheim.

32. Steinbüchel, A. & Schlegel, H.G. (1988) Biologisch abbaubare Kunststoffe. Bakterielle poly-β-Hydroxyfettsäuren als Kunststoffe der Zukunft. Sonderpublikation Labor 2000, pp. 148–155.

33. Tomita, K., Fukui, T. & Saito, T. (1984) Microbial poly-β-hydroxybutyrate and its application. *Kobunshi* **33**: 370–373.

34. Uttley, N.L. (1986) Properties and possible applications of some biopolyesters. *Chimicaoggi* Juli/August-Heft, 71–72.

35. Witholt, B., de Smet, M.-J., Kingma, J., van Beilen, J.B., Kok, M., Lageveen, R.G. & Eggink, G. (1990) Bioconversions of aliphatic compounds by *Pseudomonas oleovorans* in multiphase bioreactors: background and economic potential. *Trends Biotechnol.* **8**: 46–52.

36. Anonymous (1990) Biodegradable plastic hits the production line. *New Scientist* **126**: 36.

37. Lemoigne. M. (1026) Produits de deshydration et de polymerisation de l'acide β-oxybutyric. *Bull. Soc. Chim. Biol. (Paris)* **8**: 770–782.

38. Davis, J.B. (1964) Cellular lipids of a *Nocardia* grown on propane and n-butane. *Appl. Microbiol.* **12**: 301–304.

39. Wallen, L.L. & Davis, E.N. (1972) *Environ. Sci. Technol.* **6** 161–164.

40. Wallen, L.L. & Rohwedder, W.K. (1974) Poly-β-hydroxyalkanoate from activated sludge. *Environ. Sci. Technol.* **8**: 576–579.

41. Marchessault, R.H., Morikawa, H., Revol, J.-F. & Bluhm, T.L. (1984) Physical properties of a naturally occurring polyester: poly(β-hydroxyvalerate)/poly (β-hydroxybutyrate). *Macromolecules* **17**: 1882–1884.

42. Morikawa, H. & Marchessault, R.H. (1981) Pyrolysis of bacterial poly-alkanoates. *Can. J. Chem.* **59**: 2306–2313.

43. Lageveen, R.G., Huisman, G.W., Preusting, H., Ketelaar, P., Eggink, G. & Witholt, B. (1988) Formation of polyesters by *Pseudomonas oleovorans*; effect of substrates on formation and composition of poly-(R)-3-hydroxyalkanoates and poly-(R)-3-hydroxyalkenoates. *Appl. Environ. Microbiol.* **54**: 2924–2932.

44. Nichols, P.D., Henson, J.M., Guckert, J.B., Nivens, D.E. & White, D.C.

(1985) Fourier transform-infrared spectroscopic methods for microbial ecology: analysis of bacteria, bacteria-polymer mixtures and biofilms. *J. Microbiol. Meth.* **4**: 79–94.

45. Apostolides, Z. & Potgieter, D.J.J. (1981) Determination of PHB (poly-β-hydroxybutyric acid) in activated sludge by a gas chromatographic method. *Eur. J. Appl. Microbiol. Biotechnol.* **13**: 62–63.

46. Brandl, H., Gross, R.A., Lenz, R.W. & Fuller, R.C. (1988) *Pseudomonas oleovorans* as a source of poly(β-hydroxyalkanoates) for potential applications as biodegradable polyesters. *Appl. Environ. Microbiol.* **54**: 1977–1982.

47. Braunegg, G., Sonnleitner, B. & Lafferty, R.M. (1978) A rapid gas chromatographic method for the determination of poly-β-hydroxybutyric acid in microbial biomass. *Eur. J. Appl. Microbiol. Biotechnol.* **6**: 29–37.

48. Comeau, Y., Hall, K.J. & Oldham, W.K. (1988) Determination of poly-β-hydroxybutyrate and poly-β-hydroxyvalerate in activated sludge by gas-liquid chromatography. *Appl. Environ. Microbiol.* **54**: 2325–2327.

49. Odham, G., Tunlid, A., Westerdahl, G. & Marden, P. (1986) Combined determination of poly-β-hydroxyalkanoic and cellular fatty acids in starved marine bacteria and sewage sludge by gas-chromatography with flame ionization or mass spectrometry detection. *Appl. Environ. Microbiol.* **52**: 905–910.

50. Riis, V. & Mai, W. (1981) Gas chromatographic determination of poly-β-hydroxybutyric acid in microbial biomass after hydrochloric and propanolysis. *J. Chromatogr.* **445**: 285–287.

51. Doi, Y., Kunioka, M., Nakamura, Y. & Soga, K. (1986) Nuclear magnetic resonance studies on poly(β-hydroxybutyrate) and a copolyester of β-hydroxybutyrate and β-hydroxyvalerate isolated from *Alcaligenes eutrophus* H16. *Macromolecules* **19**: 2860–2864.

52. Fritsche, K., Lenz, R.W. & Fuller, R.C. (1990) Bacterial polyesters containing branched poly(β-hydroxyalkanoate) units. *Int. J. Biol. Macromol.* **12**: 92–101.

53. Holmes, P.A., Wright, L.F. & Colins, S.H. (1985) β-Hydroxybutyrate polymers. European Patent Application EP 52,459.

54. De Smet, M.J., Eggink, G., Witholt, B., Kingma, J. & Wynberg, H. (1983) Characterization of intracellular inclusions formed by *Pseudomonas oleovorans* during growth on octane. *J. Bacteriol.* **154**: 870–878.

55. Fernandez-Castillo, R., Rodriguez-Valera, F., Gonzales-Ramos, J. & Ruiz-Berraquero, F. (1986) Accumulation of poly(β-hydroxybutyrate) by halobacteria. *Appl. Environ. Microbiol.* **51**: 214–216.

56. Lenz, R.W., Kim, B.-W., Ulmer, H.W. & Fritzsche, K. (1990) Functionalized poly-β-hydroxyalkanoates produced by bacteria. In *Novel Biodegradable Microbial Polymers* (ed. Dawes, E.A.), pp. 23–35. Kluwer, Dordrecht.

57. Shimida, K., Matsushima, K., Fukumoto, J. & Yamamoto, T. (1969) Poly-(L)-malic acid: a new protease inhibitor from *Penicillium cyclopium*. *Biochem. Biophys. Res. Commun.* **35**: 619–624.

58. Fischer, H., Erdmann, S. & Holler, E. (1989) An unusual polyanion from *Physarum polycephalum* that inhibits homologous DNA polymerase α in vitro. *Biochemistry* **28**: 5219–5226.

59. Gilding, D.K. & Reed, A.M. (1979) Biodegradable polymers for use in surgery. Polyglycolid/Polylactid homo- and copolymers. *Polymers* **20**: 1459–1464.

60. Holland, S.J., Tighe, B.J. & Gould, P.L. (1986) Polymers for biodegradable medical devices. 1. The potential of polyesters as controlled macromolecular release systems. *J. Contr. Rel.* **4**: 155–180.

61. Shively, J.M. (1974) Inclusion bodies of prokaryotes. *Ann. Rev. Microbiol.* **28**: 167–187.

62. Schlegel, H.G., Gottschalk, G. & Von Bartha, R. (1961) Formation and utilization of poly-β-hydroxybutyric acid by knallgas bacteria (*Hydrogenomonas*). *Nature* **191**: 463–465.

63. Pedros-Alio, C., Mas, J. & Guerrero, R. (1985) The influence of poly-β-hydroxybutyrate accumulation on cell volume and buoyant density in *Alcaligenes eutrophus*. *Arch. Microbiol.* **143**: 178–184.

64. Lundgren, D.G., Pfister, R.M. & Merrick, J.M. (1964) Structure of poly-β-hydroxybutyric acid granules. *J. Gen. Microbiol.* **34**: 441–446.

65. Mas, J., Pedros-Alio, C. & Guerrero, R. (1985) Mathematical model for determining the effects of intracytoplasmic inclusions on volume and density of microorganisms. *J. Bacteriol.* **164**: 749–756.

66. Haywood, G.W., Anderson, A.J. & Dawes, E.A. (1989) The importance of PHB-synthase substrate specificity in polyhydroxyalkanoate synthesis by *Alcaligenes eutrophus*. *FEMS Microbiol. Lett.* **57**: 1–6.

67. Peoples, O.P. & Sinskey, A.J. (1989) Poly-β-hydroxybutyrate biosynthesis in *Alcaligenes eutrophus* H16. Identification and characterization of the PHB polymerase gene (*phbC*). *J. Biol. Chem.* **264**: 15298–15303.

68. Schubert, P., Krüger, N. & Steinbüchel, A. (1991) Molecular analysis of the *Alcaligenes eutrophus* poly(3-hydroxybutyrate)-, PHB-, biosynthetic operon: identification of the N-terminus of PHB-synthase and identification of the promoter. *J. Bacteriol.* **173**: 168–175.

69. Barnard, G.N. & Sanders, J.K.M. (1988) Observation of mobile poly(β-hydroxybutyrate) in the storage granules of methylobacterium AM1 by *in vivo* ^{13}C-NMR spectroscopy. *FEBS Lett.* **231**: 16–18.

70. Barnard, G.N. & Sanders, K.M. (1989) The poly-β-hydroxybutyrate granule *in vivo*. *J. Biol. Chem.* **24**: 3286–3291.

71. Alper, R. Lundgren, D.G., Marchessault, R.H. & Cote, W.A. (1963) Properties of poly-β-hydroxybutyrate. I. General considerations concerning the naturally occurring polymer. *Biopolymers* **1**: 545–556.

72. Lundgren, D.G., Alper, R., Schnaitman, C. & Marchessault, R.H. (1965) Characterization of poly-β-hydroxybutyrate depolymerase of extracted from different bacteria. *J. Bacteriol* **89**: 245–251.

73. Reusch, R.N. & Sadoff, H.L. (1983) D-(−)-poly-β-hydroxybutyrate in membranes of genetically competent bacteria. *J. Bacteriol.* **156**: 778–788.

74. Reusch, R.N., Hiske, T.W. & Sadoff, H.L. (1986) Poly-β-hydroxybutyrate membrane structure and its relationship to genetic transformability in *Escherichia coli*. *J. Bacteriol.* **168**: 553–562.

75. Reusch, R., Hiske, T., Sadoff, H., Harris, R. & Beveridge, T. (1987) Cellular incorporation of poly-β-hydroxybutyrate into plasma membranes of *Escherichia coli* and *Azotobacter vinelandii* alters native membrane structure. *Can. J. Microbiol.* **33**: 435–444.

76. Gilbert, P. & Brown, M.R.W. (1978) Effect of R-plasmid RP1 and nutrient depletion on the gross cellular composition of *Escherichia coli* and its resistance to some uncoupling phenols. *J. Bacteriol.* **133**: 1062–1065.

77. Reusch, R.N. & Sadoff, H.L. (1988) Putative structure and functions of a poly-β-hydroxybutyrate/calcium polyphosphate channel in bacterial plasma membranes. *Proc. Natl. Acad. Sci. U.S.A.* **85**: 4176–4180.

78. Cornibert, J. & Marchessault, R.H. (1972) Physical properties of poly-β-hydroxybutyrate. IV. Conformational analysis and crystalline structure. *J. Mol. Biol.* **71**: 735–756.

79. Marchessault, R.H., Okamura, K. & Su, C.J. (1970) Physical properties of poly(β-hydroxybutyrate). II. Conformational aspects in solution. *Macromolecules* **3**:

735–740.

80. Marchessault, R.H., Cornibert, J., Benoit, H. & Weill, G. (1970) Physical properties of poly(β-hydroxybutyrate). III. Folding of helical segments in 2,2,2-trifluoroethanol. *Macromolecules* 3: 741–746.

81. Reusch, R.N. (1989) Poly-β-hydroxybutyrate/calcium polyphosphate complexes in eukaryotic membranes. *Proc. Soc. Ex. Biol. Med.* 191: 377–381.

82. Syldatk, C., Lang, S., Wagner, R., Wray, V. & Witte, L. (1985) Chemical and physical characterization of four interfacial-active rhamnolipids from *Pseudomonas* spec. DSM2874 grown on n-alkanes. *Z. Naturforsch.* 40c: 51–60.

83. Syldatk, C., Lang, S., Matulovic, U. & Wagner, F. (1985) Production of four interfacial active rhamnolipids from n-alkanese or glycerol by resting cells of *Pseudomonas* species DSM2874. *Z. Naturforsch.* 40c: 61–67.

84. Hollingsworth, R.I., Abe, M., Dazzo, F.B. & Hallenga, K. (1984) Identification of 3-hydroxybutanoic acid as a component of the acidic extracellular polysaccharide of *Rhizobium trifolii* 0403. *Carbohyd. Res.* 134: C7-C11.

85. Hollingsworth, R.I., Dazzo, F.B., Hallenga, K. & Musselman, B. (1988) The complete structure of the trifoliin, a lectin-binding capsular polysaccharide of *Rhizobium trifolii* 843. *Carbohydr. Res.* 172: 97–112.

86. Burdon, K.L. (1946) Fatty material in bacteria and fungi by staining dried, fixed slide preparations. *J. Bacteriol.* 52: 665–678.

87. Haynes, W.C., Melvin, E.H., Locke, J.M., Glass, C.A. & Senti, F.R. (1958) Certain factors affecting the infrared spectra of selected microorganisms. *Appl. Microbiol.* 6: 298–304.

88. Hayward, A.C. (1959) Poly-β-hydroxybutyrate inclusions in the classification of aerobic gram-negative bacteria. *J. Gen. Microbiol.* 21: 2–3.

89. Emeruwa, A.C. & Hawirko, R.Z. (1973) Poly-β-hydroxybutyrate metabolism during growth and sporulation of *Clostridium botulinum*. *J. Bacteriol.* 116: 989–993.

90. Nuti, M.P., De Bertoldi, M. & Lepidi, A.A. (1972) A simple method of extraction of poly-β-hydroxybutyrate from aerobic and anaerobic soil bacteria. *Tipolitografia 'Editrice Giardini' Pisa*, pp. 1–5.

91. Nanninga, H.J. & Gottschal, J.C. (1987) Properties of *Desulfovibrio carbinolicus* sp. nov. and other sulfate-reducing bacteria isolated from an anaerobic-purification plant. *Appl. Environ. Microbiol.* 53: 802–908.

92. Widdel, F. (1980) Anaerober Abbau von Fettsäuren und Benzoesäure durch neu isolierte Arten sulfatreduzierender Bakterien. Dissertation, Universität Göttingen.

93. Amos, D.A. & McInerney, M.J. (1989) Poly-β-hydroxyalkanoate in *Synthrophomonas wolfei*. *Arch. Microbiol.* 152: 172–177.

94. Amos, D.A. & McInerney, M.J. (1990) Growth of *Syntrophomonas wolfei* on unsaturated short chain fatty acids. *Arch. Microbiol.* 154: 31–36.

95. McInerney, M.J., Bryant, M.P., Hespell, R.B. & Costerton, J.W. (1981) *Syntrophomonas wolfei* gen. nov., sp. nov., an anaerobic syntrophic fatty acid-oxidizing bacterium. *Appl. Environ. Microbiol.* 41: 1029–1039.

96. Henson, J.M., Smith, P.H. & White, D.C. (1985) Examination of thermophilic methane-producing digesters by analysis of bacterial lipids. *Appl. Environ. Microbiol.* 50: 1428–1433.

97. Gaffron, H. (1933) Über den Stoffwechsel der schwefelfreien Purpurbakterien. *Biochem. Z.* 260: 1–17.

98. Gaffron, H. (1935) Über den Stoffwechsel der schwefelfreien Purpurbakterien II. *Biochem. Z.* 275: 301–319.

99. Liebergesell, M., Hustede, E., Timm, A., Steinbüchel, A., Fuller, R.C., Lenz, R.W. & Schlegel, H.G. (1991) Formation of poly(3-hydroxyalkanoates) by

phototrophic and chemolithotrophic bacteria. *Arch. Microbiol.* **155**: 415–421.

100. Krasil'nikova, E.N., Keppen, O.I. & Kondrat'yeva, E.N. (1986) *Chloroflexus aurantiacus* growth in media with different organic compounds and the pathways of their metabolism. *Mikrobiologiya* **55**: 425–430.

101. Pierson, B.K. & Castenholz, R.W. (1974) A phototrophic gliding filamentous bacterium of hot springs, *Chloroflexus aurantiacus, gen.* and *sp.* nov., *Arch. Microbiol.* **100**: 5–24.

102. Sirevag, R. & Castenholz, R. (1979) Aspects of carbon metabolism in *Chloroflexus. Arch. Microbiol.* **120**: 151–153.

103. Carr, N.G. (1966) The occurrence of poly-β-hydroxybutyrate in the blue-green alga, *Chlorogloea fritschii. Biochim. Biophys. Acta* **120**: 308–310.

104. Jensen, T.E. & Sicko, L.M. (1971) Fine structure of poly-β-hydroxybutyric acid granules in a blue-green alga, *Chlorogloea fritschii. J. Bacteriol.* **106**: 683–686.

105. Jensen, T.E. & Sicko, L.M. (1973) *Cytologica* **38**: 381–391.

106. Capon, R., Dunlop, R., Ghisalberti, E. & Jeffries, P. (1983) Poly-3-hydroxy-alkanoates from marine and freshwater cyanobacteria. *Phytochemistry* **22**: 1181–1184.

107. Allen, M.M. (1984) Cyanobacterial cell inclusions. *Ann. Rev. Microbiol.* **38**: 1–25.

108. Campbell, J. III, Stevens, S.E. Jr. & Balkwill, D.L. (1982) Accumulation of poly-β-hydroxybutyrate in *Spirulina platensis. J. Bacteriol.* **149**: 361–363.

109. Vincenzini, M., Sili, C., de Philipps, R., Ena, A. & Materassi, R. (1990) Occurrence of poly-β-hydroxybutyrate in *Spirulina species. J. Bacteriol.* **172**: 2791–2797.

110. Mikucki, J., Szarapinska-Kwaszweska, J. and Surewicz, K. (1979) Metabolizm endogenny a chorobotwarczosc gronkowcow koagulazoujemnyck. *Med. Dosw. Mikrobiol.* **31**: 65–75.

111. Szeweczyk, E. & Mickucki, J. (1989) Poly-β-hydroxybutyric acid in staphylococci. *FEMS Microbiol. Lett.* **61**: 279–284.

112. Altekar, W. & Rajagopalan, R. (1990) Ribulose bisphosphate carboxylase activity in halophilic archaebacteria. *Arch. Microbiol.* **153**: 169–174.

113. Kirk, R.G. & Ginzburg, M. (1972) Ultrastructure of two species of *Halobacterium. J. Ultrastruct. Res.* **41**: 80–94.

114. Schlegel, H.G. (1969) From electricity via water electrolysis to food. In *Fermentation Advances* (ed. Perlmann, D.), pp. 807–832. Academic, New York.

115. Gottschalk, G. (1964) Die Biosynthese der Poly-β-hydroxybuttersäure durch Knallgasbakterien. I. Ermittlung der ^{14}C-Verteilung in Poly-β-hydroxybuttersäure. *Arch. Mikrobiol.* **47**: 225–229.

116. Gottschalk, G. (1964) Die Biosynthese der Poly-β-hydroxybuttersäure durch Knallgasbakterien. II. Verwertung organischer Säuren. *Arch. Mikrobiol.* **47**: 230–235.

117. Schindler, J. (1964) Die Synthese der Poly-β-hydroxybuttersäure durch *Hydrogenomonas* H16: Die zu β-Hydroxybutyryl-Coenzym A führenden Reaktionsschritte. *Arch. Mikrobiol.* **49**: 236–255.

118. Oeding, V. (1972) Regulation des Poly-β-hydrobuttersäure-Stoffwechsels bei *Hydrogenomonas eutropha* Stamm H16 und PHBS-freie Mutaten. Dissertation, Universität Göttingen.

119. Oeding, V. & Schlegel, H.G. (1972) Wirksamkeit von Effektoren auf die β-Ketothiolase von *Hydrogenomonas eutropha* H16. Nachr. Akad. Wiss. in Göttingen. II. Math.-Phys. Klasse 6.

120. Ruhr, E.M. & Schlegel, H.G. (1975) Synthesis of poly-β-hydroxybutyrate in vivo and kinetics of β-ketothiolase in vitro in *Alcaligenes eutrophus* H16. *Biochem. Soc. Transactions* **3**: 1093–1094.

121. Ruhr, E.M. (1977) Regulation der Biosynthese von Poly-β-hydroxybuttersäure in *Alcaligenes eutrophus* H16. Dissertation, Universität Göttingen.

122. Brune, H. & Niemann, E. (1977) Über den Einsatz und die Verträglichkeit von Bakterieneinweiß (*Hydrogenomonas*) mit unterschiedlichem Gehalt an Poly-β-hydroxibuttersäure in der Tierernährung. 1. Mitteiluing. Gewichtsentwicklung und N-Bilanz bei wachsenden Ratten. *Z. Tierphysiol. Tierernährg. Futtermittelk.* **38**: 13–22.

123. Greife, H.A., Molnar, S. & Günther, K.-D. (1978) Biologische Bewertung der Proteinqualität von H₂-oxidierenden Bakterienstämmen an Ratten. *Z. Tierphysiol. Tierern. Futtermittelk.* **40**: 135–148.

124. Greife, H.A., Molnar, S. & Günther, K.-D. (1981) N-Stoffwechsel wachsender Ratten bei Aufnahme steigender Mengen des H₂-oxidierenden Bakterienstammes *Alcaligenes eutrophus*. *Z. Tierphysiol. Tierern. Futtermittelk.* **45**: 91–100.

125. Greife, H., Molnar, S. & Günther, K.-D. (1979) Proteinqualität des H₂-oxidierenden Bakterienstammes *Alcaligenes eutrophus* in der Broilermast. 1. Mitteilung: Wachstum und Futterverwertung bei steigendem Austausch von Sojaextraktionsschrot durch die Bakterienmasse. *Arch. Geflügelk.* **43**: 129–138.

126. Greife, H., Molnar, S., Badawy-Hefez, N. & Günther, K.-D. (1979) Proteinqualität des H₂-oxidierenden Bakterienstammes *Alcaligenes eutrophus* in der Broilermast. 2. Mitteilung: N-Verwertung und N-Stoffwechsel bei steigendem Austausch von Sojaextraktionsschrot durch die Bakterienmasse. *Arch. Geflügelk.* **43**: 182–192.

127. Greife, H.A., Molnar, S. & Günther, K.-D. (1981) Weitere Untersuchungen zur N-Verwertung des H₂-oxidierenden Bakterienstammes *Alcaligenes eutrophus* und alimentärer Ribonukleinsäure durch wachsende Broiler. *Arch. Geflügelk.* **45**: 57–68.

128. Brune, H. & Niemann, E. (1977) Über den Einsatz und die Verträglichkeit von Bakterieneinweiß (*Hydrogenomonas*) mit unterschiedlichem Gehalt an Poly-β-hydroxibuttersäure in der Tierernährung. 2. Mitteilung. Untersuchungen zur Gewichtsentwicklung, N-Bilanz und zum Fettsäuremuster der Organe Leber, Muskel und Nierendepotfett beim wachsenden Schwein. *Z. Tierphysiol. Tierernährg. Futtermittelk.* **38**: 81–93.

129. King, P.P. (1982) Biotechnology. An industrial view. *J. Chem. Tech. Biotechnol.* **32**: 2–8.

130. Baptist, J.N. (1962) Process for preparing poly-β-hydroxybutyric acid. US Patent Application US 3044942.

131. Baptist, J.N., Werber, F.X. (1965) Plasticized poly-beta-hydroxybutyric acid and process. US Patent Application US 3182036.

132. Bloembergen, S., Holden, D.A. & Marchessault, R.H. (1988) Non-biochemical synthesis and characterization of PHB/PHV. *Polym. Prepr.* **29**: 594–595.

133. Marchessault, R.H. & Faure, A.J. (1974) Synthesis of configurationally controlled oligomers of poly-β-hydroxybutyrate (PHB). *Polym. Prepr.* **15**: 87–88.

134. Shelton, J.R., Agostini, D.E. & Lando, J.B. (1971) Synthesis and characterization of poly-β-hydroxybutyrate. II. Synthesis of D-poly-β-hydroxybutyrate and the mechanism of ring-opening polymerization of β-butyrolactone. *J. Polym. Sci., Part A-1* **9**: 2789–2799.

135. Byrom, D. (1990) Industrial production of copolymer from *Alcaligenes eutrophus*. In *Novel Biodegradable Microbial Polymers* (ed. Dawes, E.A.), pp. 113–117. Kluwer, Dordrecht.

136. Sherwood, M. (1983) Bacterial plastic comes to market. *BioTechnology* **1**: 388–389.

137. Brandl, H. & Püchner, P. (1990) The degradation of shampoo bottles in a lake. In *Novel Biodegradable Microbial Polymers* (ed. Dawes, E.A.), pp. 421–422. Kluwer, Dordrecht.

138. Westlake, R.P. (1987) Biopol – the contribution of a biodegradable thermoplastic to the recycling issue. *Kautschuk + Gummi Kunststoffe* **40**: 203–204.

139. Hänggi, U.J. (1990) Pilot scale production of PHB with *Alcaligenes latus*. In

Novel Biodegradable Microbial Polymers (eds Dawes, E.A.), pp. 65–70. Kluwer, Dordrecht.

140. Trathnigg, B., Weidmann, V., Lafferty, R.M., Korsatko, B. & Korsatko, W. (1988) Niedermolekulares PHB. *Angew. Makromol. Chemie* **161**: 1–8.

141. Fraser, H.M., Sandow, J., Seidel, H.R. & Lunn, S.F. (1989) Controlled release of a GnRH agonist from a polyhydroxybutyric acid implant: reversible suppression of the menstrual cycle in the macaque. *Acta. Endocrinol.* **121**: 841–848.

142. Seebach, D. & Zueger, M. (1982) Über die Depolymerisierung von poly(R)-3-hydroxy-buttersäureester (PHB). *Helv. Chim. Acta* **65**: 495–204.

143. Seebach, D. & Zueger, M.F. (1985) On the preparation of methyl and ethyl (R)-(−)-3-hydroxyvalerate by depolymerization of a mixed PHB/PHV biopolymer. *Tetrahedron Lett.* **25**: 2747–2750.

144. Seebach, D., Roggo, S. & Zimmermann, J. (1987) Biological–chemical preparation of 3-hydroxycarboxylic acids and their use in EPC-synthesis. In *Stereochemistry of Organic and Bioorganic Transformations* (eds Bartmann, W. & Sharpless, K.B.) pp. 85–125. VCH, Weinheim.

145. Seebach, D. (1988) β-Hydroxycarboxylic acids from the biopolymer PHB-PHV. Small molecules with great potential for EPC synthesis. *Polym. Prepr.* **29**: 173–174.

146. Biehler, M.J. (1989) A first practical use of specially constructed starter cultures in a new biotechnological nitrate elimination process for drinking water with degradable biopolymers as adhesion material. In *DECHEMA Biotechnology Conferences* (eds Behrens, D. & Driesel, A.J.), Vol. 3, Part B, pp. 997–1002.

147. Ando, Y. & Fukada, E. (1984) Piezoelectric properties and molecular motion of poly(β-hydroxybutyrate) films. *J. Polym. Sci., Polym. Phys. Ed.* **22**: 1821–1834.

148. Fukada, E. & Ando, Y. (1986) Piezoelectric properties of poly-beta-hydroxy-butyrate and copolymers of beta-hydroxybutyrate and beta-hydroxyvalerate. *Int. J. Biol. Macromol.* **8**: 361–366.

149. Schlegel, H.G., Krauss, I. & Lafferty, R. (1969) Poly-β-hydroxybuttersäure-arme und -reiche Mutanten von *Hydrogenomonas* H16. Nachr. Akad. Wiss. Göttingen. II. Math.-Physk. Klasse 159–168.

150. Schlegel, H.G., Lafferty, R. & Krauss, I. (1970) The isolation of mutants not accumulating poly-β-hydroxybutyric acid. *Arch. Microbiol.* **71**: 283–294.

151. Schlegel, H.G., Lafferty, R. & Krauss, I. (1970) Bacterial mutants of *Hydrogenomonas* lacking poly-β-hydroxybutyric acid. *Experienta* **26**: 554.

152. Schubert, P., Steinbüchel, A. & Schlegel, H.G. (1988) Cloning of the *Alcaligenes eutrophus* genes for synthesis of poly-β-hydroxybutyrate (PHB) and synthesis of PHB in *Escherichia coli. J. Bacteriol.* **170**: 5837–5847.

153. Bohlken, G. (1969) Zur Speicherung von Reservestoffen in *Bacillus megaterium*. II. Untersuchungen an Poly-β-hydroxybuttersäure-freien Mutanten. *Zbl. Bakt. Parasitenkde. Abt. II*, **123**: 16–29.

154. Vollbrecht, D. & El Nawaway, M.A. (1980) Restricted oxgen supply and excretion of metabolites. I. *Pseudomonas* spec. and *Paracoccus dentrificans. Eur. J. Appl. Microbiol. Biotechnol.* **9**: 1–8.

155. Huisman, G.W., Wonink, E., Meima, R., Kazemier, B., Terpstra, P. & Witholt, B. (1991) Metabolism of poly(3-hydroxyalkanoates) by *Pseudomonas oleovorans*: identification and sequences of genes and function of the encoded proteins in the synthesis and degradation of PHA. *J. Biol. Chem.* **266**: 2191–2198.

156. Vollbrecht, D. & Schlegel, H.G. (1979) Excretion of metabolites by hydrogen bacteria. III. D(−)-3-hydroxybutanoate. *Eur. J. Appl. Microbiol. Biotechnol.* **7**: 259–266.

157. Vollbrecht, D., Schlegel, H.G., Stoschek, G. & Janczikowski, A. (1979) Excretion of metabolites by hydrogen bacteria. IV. Respiration rate-dependent formation of primary metabolites and of poly-3-hydroxybutanoate. *Eur. J. Appl. Microbiol.*

Biotechnol. **7**: 267–276.

158. Schubert, P., Steinbüchel, A. & Schlegel, H.G. (1988) Synthesis of poly-β-hydroxybutyric acid in *Alcaligenes eutrophus*, its genetic localization and conjugational transfer of the genes. Nachr. Akad. Wissensch. Göttingen, II. Math.-Physik. Klasse, 4, 1–10.

159. Moskowitz, G.J. & Merrick, J.M. (1969) Metabolism of poly-β-hydroxybutyrate. Enzymatic synthesis of D-(−)-β-hydrobutyryl Coenzyme A by an enoyl hydrase from *Rhodospirillum rubrum*. *Biochemistry* **8**: 2748–2755.

160. Haywood, G.W., Anderson, A.J. & Dawes, E.A. (1989) A survey of the accumulation of novel polyhydroxyalkanoates by bacteria. *Biotechnol. Lett.* **11**: 471–476.

161. Huisman, G.W., Leeuw, O. de, Eggink, G. & Witholt, B. (1989) Synthesis of poly-3-hydroxyalkanoates is a common feature of fluorescent pseudomonads. *Appl. Environ. Microbiol.* **55**: 1949–1954.

162. Timm, A. & Steinbüchel, A. (1990) Formation of polyesters consisting of medium-chain-length 3-hydroxyalkanoic acids from gluconate by *Pseudomonas aeruginosa* and other fluorescent pseudomonads. *Appl. Environ. Microbiol.* **56**: 3360–3367.

163. Anderson, A.J., Haywood, G.W., Williams, D.R. & Dawes, E.A. (1990) The production of polyhydroxyalkanoates from unrelated carbon sources. In *Novel Biodegradable Microbial Polymers* (ed. Dawes, E.A.), pp. 119–129. Kluwer, Dordrecht.

164. Haywood, G.W., Anderson, A.J., Ewing, D.F. & Dawes, E.A. (1990) Accumulation of polyhydroxyalkanoate containing primarily 3-hydroxydecanoate from simple carbohydrate substrates by *Pseudomonas* sp. strain NCIMB 40135. *Appl. Environ. Microbiol.* **56**: 3354–3359.

165. Peoples, O.P. & Sinskey, A.J. (1989) Fine structural analysis of the *Zoogloea ramigera phbA-phbB* locus encoding β-ketothiolase and acetoacetyl-CoA reductase: nucleotide sequence of *phbB*. *Mol. Microbiol.* **3**: 349–357.

166. Peoples, O.P. & Sinskey, A.J. (1989) Poly-β-hydroxybutyrate biosynthesis in *Alcaligenes eutrophus* H16. Characterization of the genes encoding β-ketothiolase and acetoacetyl-CoA reductase. *J. Biol. Chem.* **264**: 15293–15297.

167. Slater, S.C., Voige, W.H. & Dennis, D.E. (1988) Cloning and expression in *Escherichia coli* of the *Alcaligenes eutrophus* H16 poly-β-hydroxybutyrate biosynthetic pathway. *J. Bacteriol.* **170**: 4431–4436.

168. Huisman, G.W., Meima, R., Wonink, E. & Witholt, B. (1990) Genetic analysis of polyester synthesis in *Pseudomonas oleovorans*. In *Novel Biodegradable Microbial Polymers* (ed. Dawes, E.A.), pp. 451–452. Kluwer, Dordrecht.

169. Peoples, O.P. & Sinskey, A.J. (1990) Poly-hydroxybutyrate (PHB): A model system for biopolymer engineering: II. In *Novel Biodegradable Microbial Polymers* (ed. Dawes, E.A.), pp. 191–202. Kluwer, Dordrecht.

170. Witholt, B., Huisman, G.W. & Preusting, H. (1990) Bacterial Poly(3-hydroxyalkanoates). In *Novel Biodegradable Microbial Polymers* (ed. Dawes, E.A.), pp. 161–173. Kluwer, Dordrecht.

171. Janes, B., Hollar, J. & Dennis, D. (1990) Molecular characterization of the poly-β-hydroxybutyrate biosynthetic pathway of *Alcaligenes eutrophus* H16. In *Novel Biodegradable Microbial Polymers* (ed. Dawes, E.A.), pp. 175–190. Kluwer, Dordrecht.

172. Oelmüller, U., Krüger, N., Steinbüchel, A. & Friedrich, C.G. (1990) Isolation of prokaryotic RNA and detection of specific mRNA with biotinylated probes. *J. Microbiol. Meth.* **11**: 73–81.

173. Steinbüchel, A. & Schubert, P (1989) Expression of the *Alcaligenes eutrophus* poly(β-hydroxybutyric acid) synthetic pathway in *Pseudomonas* sp. *Arch. Microbiol.* **153**: 101–104.

174. Timm, A., Byrom, D. & Steinbüchel, A. (1990) Formation of blends of various poly(3-hydroxyalkanoic acids) by a recombinant strain of *Pseudomonas oleovorans*. *App. Microbiol. Biotechnol.* **33**: 296–301.

175. Haywood, G.W., Anderson, A.J., Chu, L. & Dawes, E.A. (1988) The role of NADH- and NADPH-linked acetoacetyl-CoA reductases in the poly-3-hydroxy-butyrate synthesizing organism *Alcaligenes eutrophus*. *FEMS Microbiol. Lett.* **52**: 259–264.

176. Peoples, O.P., Masamune, S., Walsh, C.T. & Sinskey, A.J. (1987) Biosynthetic thiolase from *Zoogloea ramigera*. III. Isolation and characterization of the structural gene. *J. Biol. Chem.* **262**: 97–102.

177. Nishimura, T., Saito, T. & Tomita, K. (1978) Purification and properties of β-ketothiolase from *Zoogloea ramigera*. *Arch. Microbiol.* **116**: 21–27.

178. Shuto, H., Fukui, T. Saito, T., Shirakura, Y. & Tomita, K. (1981) An NAD-linked acetoacetyl-CoA reductase from *Zoogloea ramigera* I-16-M. *Eur. J. Biochem.* **118**: 53–59.

179. Stephenson, M.P., Jackson, F.A. & Dawes, E.A. (1978) Further observations on carbohydrate metabolism and its regulation in *Azotobacter vinelandii*. *J. Gen. Microbiol.* **109**: 89–96.

180. Brandl, H., Gross, R.A., Knee, E.J., Lenz, R.W. & Fuller, R.C. (1989) The ability of the phototrophic bacterium *Rhodospirillum rubrum* to produce various poly(β-hydroxyalkanoates): Potential sources for biodegradable polyesters. *Int. J. Biol. Macromol.* **11**: 49–56.

181. Brandl, H., Gross, R.A., Lenz, R.W., Lloyd, R. & Fuller, R.C. (1991) The accumulation of poly(β-hydroxyalkanoates) in *Rhodobacter sphaeroides*. *Arch. Microbiol.* **155**.

182. Oeding, V. & Schlegel, H.G. (1973) β-Ketothiolase from *Hydrogenomonas eutropha* H16 and its signifcance in the regulation of poly-β-hydroxybutyrate metabolism. *Biochem. J.* **134**: 239–248.

183. Haywood, G.W., Anderson, A.J., Chu, L. & Dawes, E.A. (1988) Characterization of two 3-ketothiolases possessing differing substrate specificities in the poly-hydroxyalkanoate synthesizing organism *Alcaligenes eutrophus*. *FEMS Microbiol. Lett.* **52**: 91–96.

184. Arakawa, H., Takiguchi, M., Amaya, Y., Nagata, S., Hayashi, H. & Mori, M.S. (1987) cDNA-derived amino acid sequence of rat mitochondria 3-oxoacyl-CoA thiolase with no transient resequence: structural relationship with the peroxisomal isoenzyme. *EMBO J.* **6**: 1361–1366.

185. Hijikata, M., Ishi, N., Kagamiyama, H., Osumi, T. & Hashimoto, T. (1987) *J. Biol. Chem.* **262**: 8151–8158.

186. Masamune, S., Walsh, C.T., Sinskey, A.J. & Peoples, O.P. (1989) Poly-(R)-3-hydroxybutyrate (PHB) biosynthesis: mechanistic studies on the biological Claisen condensation catalyzed by β-ketothiolase. *Pure Appl. Chem.* **61**: 303–312.

187. Masamune, S., Palmer, M.A.J., Gamboni, R., Thompson, S., Davis, J.T., Williams, S.F., Peoples, O.P. Sinskey, A.J. & Walsh, C.T. (1989) Bio-Claisen condensation catalyzed by thiolase from *Zoogloea ramigera* – active site cysteine residues. *J. Am. Chem. Soc.* **111**: 1879–1881.

188. Bitar, K.G., Perez-Aranda, A. & Bradshaw, R.A. (1980) Amino acid sequences of L-3-hydroxyacyl-CoA dehydrogenase from pig heart muscle. *FEBS Lett.* **116**: 196–198.

189. Griebel, R.J. & Merrick, J.M. (1971) Metabolism of poly-β-hydroxybutyrate: Effect of mild alkaline extraction on native poly-β-hydroxybutyrate granules. *J. Bacteriol.* **108**: 782–789.

190. Ballistreri, A., Garozzo, D., Giuffrida, M., Impallomeni, G. & Montaudo, G.

(1989) Sequencing bacterial poly(β-hydroxybutyrate-co-β-hydroxyvalerate) by partial methanolysis, high-performance liquid chromatography fractionation and fast atom bombardment mass spectrometry analysis. *Macromolecules* **22**: 2107–2111.

191. Ballistreri, A., Garozzo, D., Giuffrida, M. & Montaudo, G. (1990) Microstructure of bacterial poly(β-hydroxybutyrate-*co*-β-hydroxyvalerate) by fast atom bombardment mass spectrometry analysis of their partial degradation products. In *Novel Biodegradable Microbial Polymers* (ed. Dawes, E.A.), pp. 49–64. Kluwer, Dordrecht.

192. Doi, Y., Segawara, A. & Kunioka, M. (1990) Biosynthesis and characterization of poly-3-hydroxybutyrate-co-4-hydroxybutyrate in *Alcaligenes eutrophus*. *Int. J. Biol. Macromol.* **12**: 106–111.

193. Kunioka, M., Nakamura, Y. & Doi, Y. (1988) New bacterial copolyesters produced in *Alcaligenes eutrophus* from organic acids. *Polym. Commun.* **29**: 174–176.

194. Doi, Y., Segawa, A., Kakamura, S. & Kunioka, M. (1990) Production of biodegradable copolyesters by *Alcaligenes eutrophus*. In *Novel Biodegradable Microbial Polymers* (ed. Dawes, E.A.), pp. 37–48. Kluwer, Dordrecht.

195. Doi, Y., Tamaki, A., Kunioka, M. & Soga, K. (1987) Biosynthesis of terpolyesters of 3-hydroxybutyrate, 3-hydroxyvalerate and 5-hydroxyvalerate from 5-chlorpentanoic and pentanoic acids. *Makromol. Chem., Rapid Commun.* **8**: 631–635.

196. Gross, R.A., Demello, C., Lenz, R.W., Brandl, H. & Fuller, R.C. (1989) The biosynthesis and characterization of poly(β-hydroxyalkanoates) produced by *Pseudomonas oleovorans*. *Macromolecules* **22**: 1106–1115.

197. Fritzsche, K., Lenz, R.W. & Fuller, R.C. (1990) Production of unsaturated polyesters by *Pseudomonas oleovorans*. *Int. J. Biol. Macromol.* **12**: 85–91.

198. Eggink, G., van der Wal, H. & Huyberts, G. (1990) Production of poly-3-hydroxyalkanoates by *P. putida* during growth on long-chain fatty acids. In *Novel Biodegradable Microbial Polymers* (ed. Dawes, E.A.), pp. 441–444. Kluwer, Dordrecht.

199. Findlay, R.H.D. & White, D.C. (1983) Polymeric beta-hydroxyalkanoates from environmental samples and *Bacillus megaterium*. *Appl. Environ. Microbiol.* **45**: 71–78.

200. Haywood, G.W., Anderson, A.J., Williams, D.R. & Dawes, E.A. (1991) The accumulation of a polyhydroxyalkanoate copolymer containing primarily 3-hydroxyvalerate from simple carbohydrate substrates by *Rhodococcus* sp. NCIMB 40126. *Int. J. Biol. Mol.* **13**: 83–88.

201. Rodriguez-Valera, F. & Lillo, J.A.G. (1990) Halobacteria as producers of poly-β-hydroxyalkanoates. In *Novel Biodegradable Microbial Polymers* (ed. Dawes, E.A.), pp. 425–426. Kluwer, Dordrecht.

202. Reh, M. & Schlegel, H.G. (1969) Die Biosynthese von Isoleucin und Valin in *Hydrogenomonas* H16. *Arch. Microbiol.* **67**: 110–127.

203. Kunioka, M., Kawaguchi, Y. & Doi, Y. (1989) Production of biodegradable copolyesters of 3-hydroxybutyrate and 4-hydroxybutyrate by *Alcaligenes eutrophus*. *Appl. Microbiol. Biotechnol.* **30**: 569–573.

204. Kunioka, M. & Doi, Y. (1990) Thermal degradation of microbial copolyesters: poly(3-hydroxybutyrate-*co*-hydroxyvalerate)   and   poly(3-hydroxybutyrate-*co*-4-hydroxybutyrate). *Macromolecules* **23**: 1933–1936.

205. Doi, Y., Tamaki, A., Kunioka, M. & Soga, K. (1987) Biosynthesis of an unusual copolyester (10 mol% 3-hydroxybutyrate and 90 mol% 3-hydroxyvalerate units) in *Alcaligenes eutrophus* from pentanoic acid. *J. Chem. Soc., Chem. Commun.* **21**: 1635–1636.

206. Doi, Y., Tamaki, A., Kunioka, M. & Soga, K. (1988) Production of

co-polyesters of 3-hydroxybutyrate from butyric and pentanoic acids. *Appl. Microbiol. Biotechnol.* **28**: 330–334.

207. Collins, S.H. (1987) Choice of substrate in polyhydroxybutyrate synthesis. In *Carbon Substrates in Biotechnology* (eds Stowell, J. D. *et al.*), pp. 161–168. Soc. Gen. Microbiol., London.

208. Wilde, E. (1962) Untersuchungen über Wachstum und Speicherstoffsynthese von *Hydrogenomonas. Arch. Mikrobiol.* **43**: 109–137.

209. König, C., Sammler, J., Wilde, E. & Schlegel, H.G. (1969) Konstitutiv Glucose-6-phosphat-Dehydrogenase bei Glucose-verwertenden Mutanten von einem kryptischen Wildstamm. *Arch. Mikrobiol.* **67**: 51–57.

210. Schlegel, H.G. & Gottschalk, G. (1965) Verwertung von Glucose durch eine Mutante von *Hydrogenomonas* H16. *Biochem. Z.* **342**: 249–259.

211. Friehs, K. & Lafferty, R.M. (1989) Enlargement of the substrate spectrum of *Alcaligenes eutrophus* H16 by integration of the gene for levanase from *Bacillus subtilis*. In *DECHEMA Biotechnology Conferences* (eds Behrens, D. & Driesel, A.), Vol. 3, Part A, pp. 341–344.

212. Nordsiek, G. & Bowien, B. (1990) Construction of sucrose-utilizing strains of *Alcaligenes entrophus. Forum Mikrobiologie* **13**: 89.

213. Pries, A., Steinbüchel, A. & Schlegel, H.G. (1989) Construction of lactose-utilizing strains of the poly(β-hydroxyalkanoic acid) accumulating *Alcaligenes eutrophus*. In *DECHEMA Biotechnology Conferences* (eds Behrens, D. & Driesel, A.), Vol. 3, Part A, pp. 421–424.

214. Pries, A., Steinbüchel, A. & Schlegel, H.G. (1990) Lactose and galactose utilizing strains of poly(hydroxyalkanoic acid) accumulating *Alcaligenes eutrophus* and *Pseudomonas saccharophila* by recombinant DNA technology. *Appl. Microb. Biotechnol.* **33**: 410–417.

215. McLellan, D.W. & Halling, P.J. (1988) Acid-tolerant poly(3-hydroxybutyrate) hydrolases from moulds. *FEMS Microbiol. Lett.* **52**: 215–218.

216. Hippe, H. (1967) Abbau und Wiederverwertung von Poly-β-hydroxybuttersäure durch *Hydrogenomonas* H16. *Arch. Microbiol.* **56**: 248–277.

217. Hippe, H. & Schlegel, H.G. (1967) Hydrolyse von PHBS durch intracelluläre Depolymerase von *Hydrogenomonas* H16. *Arch. Mikrobiol.* **56**: 278–299.

218. Merrick, J.M., Lundgren, D.G. & Pfister, R.M. (1964) Morphological changes in poly-β-hydroxybutyrate granules associated with decreased susceptibility to enzymatic hydrolysis. *J. Bacteriol.* **89**: 234–239.

219. Merrick, J.M. & Yu, C. (1966) Purification and properties of a D(−)-β-hydroxybutyric dimer hydrolase from *Rhodospirillum rubrum. Biochemistry* **5**: 3563–3568.

220. Griebel, R., Smith, Z. & Merrick, J.M. (1968) Metabolism of poly-β-hydroxybutyrate. I. Purification, composition, and properties of native poly-β-hydroxybutyrate granules from *Bacillus megaterium. Biochemistry* **7**: 3676–3681.

221. Bradel, R., Kleinke, A. & Reichert, K.H. (1989) Molecular weight of bacterially produced poly-D-(−)-3-hydroxybutyrate. In *DECHEMA Biotechnology Conferences* (eds Behrens, D. & Driesel, A.J.), Vol. 3, Part A, pp. 207–210.

221. Merrick, J.M. & Doudoroff, M. (1964) Depolymerization of poly-β-hydroxybutyrate by an intracellular enzyme. *J. Bacteriol.* **88**: 60–71.

222. Doi, Y., Segawa, A., Kawaguchi, Y. & Kunioka, M. (1990) Cyclic nature of poly(3-hydroxyalkanoate) metabolism in *Alcaligenes eutrophus. FEMS Microbiol. Lett.* **67**: 165–170.

223. Steinbüchel, A., Schubert, P., Timm, A. & Pries, A (1990) Genetic analysis of the *Alcaligenes eutrophus* poly(hydroxyalkanoate)-synthetic genes and accumulation of PHA in recombinant bacterial strains. In *Novel Biodegradable Microbial Polymers*

(ed. Dawes, E.A.), pp. 143–159. Kluwer, Dordrecht.

224. Malik, K.A. & Claus, D. (1978) A method for the demonstration of extra-cellular hydrolysis of poly-β-hydroxybutyrate. *J. Appl. Bacteriol.* **45**: 143–146.

225. Delafield, F.P., Cooksey, K.E. & Doudoroff, M. (1965) β-Hydroxybutyric dehydrogenase and dimer hydrolase of *Pseudomonas lemoignei*. *J. Biol. Chem.* **240**: 4023–4028.

226. Lusty, C.J. & Doudoroff, M. (1966) Poly-β-hydroxybutyrate depolymerase of *Pseudomonas lemoignei*. *Proc. Natl. Acad. Sci. U.S.A.* **56**: 960–965.

227. Tanaka, Y. *et al.* (1981) Purification and properties of D(−)-3-hydroxy-butyrate dimer hydrolase from *Zoogloea ramigera* I-16-M. *Eur. J. Biochem.* **118**: 177–182.

228. Shirakura, Y., Fukui, T., Tanio, T., Nakayama, K., Matsuno, R. & Tomita, K. (1983) An extracellular D(−)-3-hydroxybutyrate oligomer hydrolase from *Alcaligenes eutrophus*. *Biochim. Biophys. Acta* **880**: 46–53.

229. Tanio, T., Fukui, T., Saito, T., Tomita, K., Kaiho, T. & Masamune, S. (1982) An extracellular poly(β-hydroxybutyrate) depolymerase from *Alcaligenes faecalis*. *Eur. J. Biochem.* **124**: 71–77.

230. Shirakura, Y., Fukui, T., Saito, T., Okamoto, Y., Narikawa, T., Koide, K., Tomita, K., Takemasa, T. & Masamune, S. (1986) Degradation of poly(3-hydroxy-butyrate) by poly(3-hydroxybutyrate) depolymerase from *Alcaligenes eutrophus* T1. *Biochim. Biophys. Acta* **880**: 46–53.

231. Fukui, T., Narikawa, T., Miwa, K., Shirakura, Y., Saito, T. & Tomita, K. (1988) Effect of limited tryptic modifications of a bacterial poly(3-hydroxybutyrate) depolymerase on its catalytic activity. *Biochim. Biophys. Acta* **952**: 164–171.

232. Saito, T., Suzuki, K., Yamamoto, J., Fukui, T., Miwa, K., Tomita, K., Nakanishi, S., Odani, S., Suzuki, J.-I. & Ishikawa, K. (1989) Cloning, nucleotide sequence and expression in *Escherichia coli* of the gene for poly(3-hydroxybutyrate) depolymerase from *Alcaligenes faecalis*. *J. Bacteriol.* **171**: 184–189.

233. Doi, Y., Kanesawa, Y. & Kunioka, M. (1990) Biodegradation of microbial copolyesters: Poly(3-hydroxybutyrate-co-3-hydroxyvalerate) and poly(3-hydroxybuty-rate-*co*-4-hydroxybutyrate). *Macromolecules* **23**: 26–31.

234. Janssen, P.H. & Harfoot, C.G. (1990) *Ilyobacter delafieldii* sp. nov., a meta-bolically restricted anaerobic bacterium fermenting PHB. *Arch. Microbiol.* **154**: 253–259.

235. Dörner, C. & Schink, B. (1990) *Clostridium homopropionicum* sp. nov., a new strict anaerobe growing with 2-, 3-, or 4-hydroxybutyrate. *Arch. Microbiol.* **154**: 342–348.

236. Stieb, M. & Schink, B (1984) A new 3-hydroxybutyrate fermenting anaerobe, *Ilyobacter polytrophus*, ge. nov. sp. nov., posseses various fermentation pathways. *Arch. Microbiol.* **140**: 139–146.

237. Pool, R. (1989) In search of the plastic potato. *Science* **245**: 1187–1189.

238. Amos, D.A. & McInerney, M.J. (1990) Composition of polyhydroxy-alkanoate from *Syntrophomonas wolfei* grown on unsaturated fatty acid substrates. *Arch. Microbiol.* **155**: 103–106.

239. Witholt, B., Lageveen, R.G., Huisman, G.W., Preusting, H., Nijenhuis, A., Kingma, J., Tijsterman, A. & Eggink, G. (1988) The production of polyalkanoates by *Pseudomonas oleovorans*. *Polym. Prepr.* **29**: 592–593.

240. Guerin, P., Braud, C., Girault, J.P., Vert, M., Holler, E., Fischer, H. & Windisch, C. (1990) Poly(malic acid), a functional poly(β-hydroxy acid)-type poly-ester available from chemical and biological synthesis. In *Novel Biodegradable Micro-bial Polymers* (ed. Dawes, E.A.), pp. 419–420. Kluwer, Dordrecht.

241. Deinema, M.H. (1972) Bacterial flocculation and production of poly-

β-hydroxybutyrate. *Appl. Microbiol.* **24**: 857–858.

242. Yamasoto, K., Akagawa, M., Oishi, U. & Kuraiski, H. (1982) Carbon substrate assimilation profiles and other taxonomic features of *Alcaligenes faecalis, A. ruhlandii,* and *Achromobacter xylosoxidans. J. Gen. Appl. Microbiol.* **28**: 195–213.

243. Lotter, L.H., Wentzel, M.C., Loewenthal, R.E., Ekama, G.A. & Marais, G.R. (1986) A study of selected characteristics of *Acinetobacter* spp. isolated from activated sludge in anaerobic/anoxic/aerobic and aerobic systems. *Water SA* **12**: 203–208.

244. Vierkant, M.A., Martin, D.W. & Stewart, J.R. (1990) Poly-β-hydroxybutyrate production in eight strains of the genus *Acinetobacter. Can. J. Microbiol.* **36**: 657–663.

245. Wentzel, M.C., Lotter, L.H., Loewenthal, R.E. & Marais, G.R. (1986) Metabolic behavior in *Acinetobacter* spp. in enhanced biological phosphorus removal – a biochemical model. *Water SA* **12**: 209–224.

246. James, L.A. & Stewart, J.R. (1989) The monomeric composition of poly-β-hydroxyalkanoates of eight strains of *Acinetobacter calcoaceticus.* Abstract Annual Meeting ASM, New Orleans, p. 256.

247. Kannan, L.V. & Rehacek, Z. (1970) Formation of poly-beta-hydroxybutyrate by *Actinomyces. Ind. J. Biochem.* **7**: 126–129.

248. Kiredjian, M.M., Popoff, M., Coynault, C., Lefevre, M. & Lemelin, M. (1981) Taxonomie du genre *Alcaligenes. Ann. Microbiol. (Inst. Pasteur)* **132B**: 337–374.

249. Baumann, L., Baumann, P., Mandel, M. & Allen, R.D. (1972) Taxonomy of aerobic marine eubacteria. *J. Bacteriol.* **110**: 402–429.

250. Pichinoty, F., Vernon, M., Mandel, M., Durand, M., Job, C. & Garcia, J.L. (1978) Étude physiologique et taxonomique du genre *Alcaligenes: A. denitrificans, A. odorans, A. faecalis. Can. J. Microbiol.* **24**: 743–753.

251. Steinbüchel, A., Kuhn, M., Niedrig, M. & Schlegel, H.G. (1985) Fermentation enzymes in strictly aerobic bacteria: comparative studies on strains of the genus *Alcaligenes* and on *Nocardia opaca,* and *Xanthomonas autotrophicus. J. Gen. Microbiol.* **129**: 2825–2835.

252. Davis, D.H., Stanier, R.Y., Doudoroff, M. & Mandel, M. (1970) Taxonomic studies on some gram negative polarly flagellated 'hydrogen bacteria' and related species. *Arch. Microbiol.* **70**: 1–13.

253. Doi, Y., Kunioka, M., Nakamura, Y. & Soga, K. (1986) Biosynthesis of polyesters by *Alcaligenes eutrophus:* incorporation of ^{13}C-labelled acetate and propionate. *J. Chem. Soc., Chem. Commun.* **23**: 1696–1697.

254. Doi, Y., Kunioka, M. & Soga, K. (1986) Biosynthesis of polyesters by *Alcaligenes eutrophus:* incorporation of carbon-13-labeled acetate and propionate. *J. Chem. Soc., Chem. Commun.* **23**: 1696–1697.

255. Kawaguchi, Y. & Doi, Y. (1990) Structure of native poly(3-hydroxybutyrate) granules characterized by X-ray diffraction. *FEMS Microbiol. Lett.* **70**: 151–156.

256. Bitar, A. & Underhill, S. (1990) Effect of ammonium supplementation on production of poly-β-hydroxybutyric acid by *Alcaligenes eutrophus* in batch culture. *Biotechnol. Lett.* **12**: 563–568.

257. Blackkolb, F. & Schlegel, H.G. (1968) Katabolische Repression und Enzymhemmung durch molekularen Wasserstoff bei *Hydrogenomonas. Arch. Mikrobiol.* **62**: 129–143.

258. Jüttner, R.R., Lafferty, R.M. & Knackmuss, H.J. (1975) A simple method for the determination of poly-β-hydroxybutyric acid in microbial biomass. *Eur. J. Appl. Microbiol.* **1**: 233–237.

259. Doi, Y., Kunioka, M. & Soga, K. (1986) Conformational analysis of poly(β-

hydroxybutyrate) in *Alcaligenes eutrophus* by solid-state carbon-13 NMR spectroscopy. *Makromol. Chem., Rapid Commun.* **7**: 661–664.

260. Walther-Mauruschat, A., Aragno, M., Mayer, F. & Schlegel, H.G. (1977) Micromorphology of Gram-negative hydrogen bacteria. II. Cell envelope, membranes, and cytoplasmic inclusions. *Arch. Microbiol.* **114**: 101–110.

261. Bowien, B., Cook, A.M. & Schlegel, H.G. (1974) Evidence for the in vivo regulation of glucose 6-phosphate dehydrogenase activity in *Hydrogenomonas eutropha* H16 from measurements of the intracellular concentrations of metabolic intermediates. *Arch. Microbiol.* **97**: 273–281.

262. Haywood, G.W., Anderson, A.J., Chu, L. & Dawes, E.A. (1988) Accumulation of polyhydroxyalkanoates by bacteria and the substrate specificity of the biosynthetic enzymes. *Biochem. Soc. Trans.* **16**: 1046–1047.

263. Ramsay, J.A., Berger, E., Ramsay, B.A. & Chavarie, C. (1990) Recovery of polyhydroxyalkanoic acid granules by a surfactant-hypochlorite treatment. *Biotechnol. Tech.* **4**: 221–226.

264. Doi, Y., Kunioka, M., Nakamura, Y. & Soga, K. (1987) Biosynthesis of copolymer in *Alcaligenes eutrophus* H16 from ^{13}C-labeled acetate and propionate. *Macromolecules* **20**: 2988–2991.

265. Cook, A.M. & Schlegel, H.G. (1978) Metabolite concentrations in *Alcaligenes eutrophus* H16 and a mutant defective in poly-β-hydroxybutyrate synthesis. *Arch. Microbiol.* **119**: 231–235.

266. Doi, Y. Kunioka, M., Tamaki, A., Nakamura, Y. & Soga, K. (1988) Nuclear magnetic resonance studies on bacterial copolyesters of 3-hydroxybutyric acid and 3-hydroxyvaleric acid. *Makromol. Chem.* **1898**: 1077–1086.

267. Doi, Y., Kunioka, M., Nakamura, Y. & Soga, K. (1988) Nuclear magnetic resonance studies on unusual bacterial copolyesters of 3-hydroxybutyrate and 4-hydroxybutyrate. *Macromolecules* **21**: 2722–2727.

268. Doi, Y., Kawaguchi, Y., Nakamura, Y. & Kunioka, M. (1989) Nuclear magnetic resonance studies of poly(3-hydroxybutyrate) and polyphosphate metabolism in *Alcaligenes eutrophus*. *Appl. Environ. Microbiol.* **55**: 2932–2938.

269. Schindler, J. & Schlegel, H.G. (1963) D(−)-β-Hydroxybuttersäure-Dehydrogenase aus *Hydrogenomonas* H16. *Biochem. Z.* **339**: 309–310.

270. Gottschalk, G. (1964) Die Biosynthese der Poly-β-hydroxybuttersäure durch Knallgasbak. III. Synthese aus Kohlendioxid. *Arch. Mikrobiol.* **47**: 236–250.

271. Gottschalk, G. & Schlegel, H.G. (1965) Preparation of ^{14}C-D(−)-β-hydroxybutyric acid from using 'Knallgas' bacteria (*Hydrogenomonas*). *Nature* **254**: 308.

272. Groom, C.A., Luong, J.H.T. & Mulchandani, A. (1988) On-line culture fluorescence measurement during the batch cultivation of poly-beta-hydroxybutyrate producing *Alcaligenes eutrophus*. *J. Biotechnol.* **8**: 271–278.

273. Heinzle, E. & Lafferty, R.M. (1980) Continuous mass spectrometric measurement of dissolved H_2, O_2, and and CO_2 during chemolithoautotrophic growth of *Alcaligenes eutrophus* strain H16. *Eur. J. Appl. Microbiol. Biotechnol.* **11**: 17–22.

274. Heinzle, E. & Lafferty, R.M. (1980) A kinetic model for growth and synthesis of poly-β-hydroxybutyric acid (PHB) in *Alcaligenes eutrophus* H16. *Eur. J. Appl. Microbiol. Biotechnol.* **11**: 8–16.

275. Kunioka, M., Tamaki, A. & Doi, Y. (1989) Crystalline and thermal properties of bacterial copolyesters: poly(3-hydroxybutyrate-*co*-3-hydroxyvalerate) and poly(3-hydroxybutyrate-co-4-hydroxybutyrate). *Macromolecules* **22**: 694–697.

276. Kesler, T.G., Voitovich, Ya.V., Anistratova, N.A., Trubachev, I.N. & Eroshin, N.S. (1972) Growth and the biochemical composition of hydrogen bacteria under conditions of biosynthesis blockage by biogenic elements. *Mikrobiologiya* **41**: 456–460.

277. Schubert, P., Steinbüchel, A. & Schlegel, H.G. (1989) Genes involved in the synthesis of poly(β-hydroxyalkanoic acid) in *Alcaligenes eutrophus*. In *DECHEMA Biotechnology Conferences* (eds Behrens, D. & Driesel, A.J.), Vol. 3, Part A, pp. 433–436.

278. Linton, J.D. (1990) Physiology of exopolysaccharide production. In *Novel Biodegradable Microbial Polymers* (ed. Dawes, E.A.), pp. 311–330. Kluwer, Dordrecht.

279. Schubert, P., Pries, A., Krüger, N. & Steinbüchel, A. (1990) Molecular analysis of the *Alcaligenes eutrophus* PHB-biosynthetic genes: indentification of the NH$_2$-terminus of PHB synthase and identification of the transcription start site of *phbC*. In *Novel Biodegradable Microbial Polymers* (ed. Dawes, E.A.), pp. 447–448. Kluwer, Dordrecht.

280. Morinaga, Y., Yamanaka, S., Ishizaki, A. & Hirose, Y. (1978) Growth and cell composition of *Alcaligenes eutrophus* in chemostate culture. *Agric. Biol. Chem.* **42**: 439–444.

281. Mulchandani, A., Luong, J.H.T. & Groom, C. (1989) Substrate inhibition kinetics for microbial growth and synthesis of poly-β-hydroxybutyric acid by *Alcaligenes eutrophus* ATCC 17697. *Appl. Microbiol. Biotechnol.* **30**: 11–17.

281. Ramsay, B.A., Lomaliza, K., Chavarzi, C., Dube, B., Bataille, P. & Ramsay, J.A. (1990) Production of poly-(β-hydroxybutyric-*co*-β-hydroxyvaleric) acids. *Appl. Environ. Microbiol.* **56**: 2093–2098.

282. Repaske, R. & Repaske, A.C. (1976) Quantitative requirements for exponential growth of *Alcaligenes eutrophus*. *Appl. Environ. Microbiol.* **32**: 585–591.

283. Repaske, R. & Mayer, R. (1976) Dense autotrophic cultures of *Alcaligenes eutrophus*. *Appl. Environ. Microbiol.* **32**: 592–597.

284. Schlegel, H.G. & Steinbüchel, A. (1981) Die relative Respirationsrate (RRR), ein neuer Belüftungsparameter. In *Fermantation* (ed. Lafferty, R.M.), pp. 10–26. Springer, Wien.

285. Schuster, E. & Schlegel, H.G. (1967) Chemolithotrophic growth of *Hydrogenomonas* H-16 using electrolytic production of hydrogen and oxygen in a chemostat. *Arch. Microbiol.* **58**: 380–409.

286. Siegel, R.S. & Ollis, D.F. (1984) Kinetics of growth of the hydrogen-oxidizing bacterium *Alcaligenes eutrophus* (ATCC 17707) in chemostat culture. *Biotechnol. Bioeng.* **26**: 764–770.

287. Sonnleitner, B., Heinzle, E., Braunegg, G. & Lafferty, R.M. (1979) Formal kinetics of poly-β-hydroxybutyric (PHB) production in *Alcaligenes eutrophus* H16 and *Mycoplana rubra* R14 with respect to the dissolved oxygen tension in ammonium limited batch cultures. *Eur. J. Appl. Mircobiol. Biotechnol.* **7**: 1–10.

288. Srienc, F., Arnold, B. & Bailey, J.E. (1984) Characterization of intracellular accumulation of poly-β-hydroxybutyrate (PHB) in individual cells of *Alcaligenes eutrophus* H16 by flow cytometry. *Biotechnol Bioeng.* **26**: 982–987.

289. Thiele, O.W., Dreysel, J. & Herman, D. (1972) The 'free' lipids of two different strains of hydrogen-oxidizing bacteria in relation to their growth phase. *Eur. J. Biochem.* **29**: 224–236.

290. Trubachev, I.N., Kalachev, G.S., Andreeva, R.I. & Voitovich, Ya.V. (1971) Effect of growth conditions on the biochemical composition of hydrogen bacteria. *Mikrobiologiya* **40**: 424–427.

291. Vedenina, I.Y. (1968) Principal cell components during autotrophic growth of *Hydrogenomonas* Z-1. *Mikrobiologiya* **37**, 5–9.

292. Voytovich, Y.V., Gitelson, I.I., Ponomaryev, P.I., Sidko, F.Y., Terskov, I.A. & Trubachov, I.N. (1972) Autotrophic growth of hydrogen bacteria in continuous culture. *Z. Allg. Mikrobiol.* **12**: 69–73.

293. Vollbrecht, D. (1980) Oxygen deficiency and excretion of metabolites by strictly aerobic bacteria. *Biotechnol. Lett.* **2**: 49–54.

294. Ti, S.Y., Fukui, T., Saito, T., Okamoto, Y., Narikawa, T., Koide, K., Tomita, K., Takemasa, T. & Masamune, S. (1986) Degradation of poly(3-hydroxybutyrate) by poly(3-hydroxybutyrate) depolymerase from *Alcaligenes faecalis*. *Biochem. Biophys. Acta* **880**: 46–53.

295. Bogensberger, B. & Braunegg, G. (1987) Kinetics of poly-D(−)-3-hydroxybutyrate (PHB) accumulation in *Alcaligenes latus* – electron microscopical studies of the granula. *Proc. 4th Eur. Congress Biotechnol. Vol. 3*, 430.

296. Palleroni, N.J. & Palleroni, A.V. (1978) *Alcaligenes latus*, a new species of hydrogen-utilizing bacteria. *Int. J. Syst. Bacteriol.* **28**: 416–424.

297. Braunegg, G. & Bogensberger, B. (1985) Zur Kinetik des Wachstums und der Speicherung von Poly-D(−)-3-hydroxybuttersäure bei *Alcaligenes latus*. *Acta Biotechnol.* **5**: 339–345.

298. Malik, K.A., Jung, C., Claus, D. & Schlegel, H.G. (1981) Nitrogen fixation by the hydrogen-oxidizing bacterium *Alcaligenes latus*. *Arch. Microbiol.* **129**: 254–256.

299. Yoneda, M. & Kondo, M. (1959) Studies on poly-β-hydroxybutyrate in bacterial spores. I. Existence of poly-β-hydroxybutyrate in mature spores of a strain of *Bacillus cereus* and its relation to the acid-fast stainability. *Biken J.* **2**: 247–258.

300. Vries, W. de, Ras, J., Stam, H., Van Vlerken, M.M., Hilgert, U., Bruijn, F.J. de & Stouthamer, A.H. (1988) Isolation and characterization of hydrogenase-negative mutants of *Azorhizobium caulinodans* ORS 571. *Arch. Microbiol.* **150**: 595–599.

301. Berg, R.H., Vasil, V. & Vasil, J. (1979) The biology of *Azospirillum sugarcane* association. II. *Protoplasma* **101**: 143–163.

302. Sadasivan, L. & Neyra, C.A. (1985) Flocculation in *Azospirillum brasilense* and *Azospirillum lipoferum*: exopolysaccharides and cyst formation. *J. Bacteriol.* **163**: 716–723.

303. Nur, I., Okon, Y. & Henis, Y. (1982) The role of oxygen concentration in the synthesis of carotenoids, poly-β-hydroxybutyrate and succinate oxidase in continuous cultures of *Azospirillum*. *Isr. J. Bot.* **31**: 221–227.

304. Nur, I., Okon, Y. & Henis, Y. (1982) Effect on dissolved oxygen tension on production of carotenoids, poly-β-hydroxybutyrate, succinate oxidase, and superoxide dismutase by *Azospirillum brasilense* Cd grown in continuous culture. *J. Gen. Microbiol.* **128**: 2937–2943.

305. Tal, S., Smirnoff, P. & Okon, Y. (1990) Purification and characterization of D(−)-β-hydroxybutyrate dehydrogenase from *Azospirillum brasilense*. *Can. J. Gen. Microbiol.* **136**: 546–649.

306. Tal, S., Smirnoff, P. & Okon, Y. (1990) The regulation of poly-β-hydroxybutyrate metabolism in *Azospirillum brasilense* during balanced growth and starvation. *J. Gen. Microbiol,* **136**: 1191–1196.

307. Papen, H. & Werner, D. (1980) Biphasic nitrogenase activity in *Azospirillum brasilense* in long lasting batch culture. *Arch. Microbiol.* **128**: 209–214.

308. Papen, H. & Werner, D. (1982) Organic acid utilization, succinate excretion and encystation during biphasic nitrogenase activity in *Azospirillum brasilense* under microaerobic conditions. *Experientia, Suppl.* **42**: 75–91.

309. Tal, S. & Okon, Y. (1985) Production of the reserve material poly-β-hydroxybutyrate and its function in *Azospirillum brasilense*. *Can J. Microbiol.* **31**: 608–613.

310. Tanaka, G. (1982) Intrinsic viscosity and friction coefficient of flexible polymers. *Macromolecules* **15**: 1028–1031.

311. Bleakley, B.H., Gaskins, M.H., Hubbell, D.H. & Zam, S.G. (1988) Floc

formation by *Azospirillum lipoferum* grown on poly-β-hydroxybutyrate. *Appl. Environ. Microbiol.* **54**: 2986–2995.

312. Volpon, A.G.T., De-Polli, H. & Doebereiner, J. (1981) Physiology of nitrogen fixation in *Azospirillum lipoferum* Br 17 (ATCC 29 709). *Arch. Microbiol.* **128**: 371–375.

313. Stockdale, H., Ribbons, D.W. & Dawes, E.A. (1965) A survey of the distribution of poly-β-hydroxybutyrate in *Azotobacter* and related genera. *J. Gen. Microbiol.* **41**: xviii.

314. Sillman, C.E. & Casida, L.E. Jr. (1986) Cyst formation versus poly-β-hydroxybutyric acid accumulation in *Azotobacter. Soil Biol. Biochem.* **18**: 23–28.

315. Stockdale, H., Ribbons, D.W. & Dawes, E.A. (1968) Occurrence of poly-β-hydroxybutyrate in the Azotobacteriaceae. *J. Bacteriol.* **95**: 1798-1803.

316. Sobek, J.M., Charba, J.F. & Foust, W.N. (1966) Endogenous metabolism of *Azotobacter agilis. J. Bacteriol.* **92**: 687–695.

317. Ward, A.C. & Dawes, E.A. (1973) Disk assay for poly-β-hydroxybutyrate. *Analyt. Biochem.* **52**: 607–613.

318. Carter, I.S. & Dawes, E.A. (1979) Effect of oxygen concentration and growth rate on glucose metabolism, poly-β-hydroxybutyrate biosynthesis and respiration of *Azotobacter beijerinckii. J. Gen. Microbiol.* **110**: 393–400.

319. Ritchie, G.A.F. & Dawes, E.A. (1969) The non-involvement of acyl-carrier protein in poly-β-hydroxybutyric acid biosynthesis in *Azotobacter beijerinckii. Biochem. J.* **112**: 803–805.

320. Ritchie, G.A.F., Dawes, E.A. & Senior, P.J. (1971) The purification and characterization of acetoacetyl-coenzyme A reductase from *Azotobacter beijerinckii. Biochem. J.* **121**: 309–316.

321. Jackson, F.A. & Dawes, E.A. (1976) Regulation of the tricarboxylic acid cycle and poly-β-hydroxybutyrate metabolism in *Azotobacter beijerinckii* grown under nitrogen or oxygen limitation. *J. Gen. Microbiol.* **97**: 303–312.

322. Senior, P.J. & Dawes, E.A. (1970) Glyceraldehyde-3-phosphate dehydrogenase of *Azotobacter beijerinckii* and its possible significance in poly-β-hydroxybutyrate biosynthesis. *Biochem. J.* **119**: 38.

323. Senior, P.J. & Dawes, E.A. (1973) The regulation of poly-β-hydroxybutyrate metabolism in *Azotobacter beijerinckii. Biochem. J.* **134**: 225–238.

324. Senior, P.J. & Dawes, E.A. (1971) Role and regulation of poly-β-hydroxybutyrate synthesis in *Azotobacter beijerinckii. Biochem. J.* **123**: 29.

325. Senior, P.J. & Dawes, E.A. (1971) PHB-biosynthesis and regulation of glucose metabolism in *Azotobacter beijerinckii. Biochem. J.* **125**: 55–66.

326. Senior, P.J., Beech, G.A., Ritchie, G.F. & Dawes, E.A. (1972) The role of oxygen limitation in the formation of poly-β-hydroxybutyrate during batch and continuous culture of *Azotobacter beijerinckii. Biochem. J.* **128**: 1193–1201.

327. Ward, A.C., Rowley, B.I. & Dawes, E.A. (1977) Effect of oxygen and nitrogen limitation on poly-β-hydroxybutyrate biosynthesis in ammonium grown *Azotobacter beijerinckii. J. Gen. Microbiol.* **102**: 61–68.

328. Lemoigne, M. & Girard, H. (1943) Reserves lipidiques β-hydroxybutyriques chez *Azotobacter chroococcum. C. r. Acad. Sci., Paris* **217**: 557–558.

329. Nuti, M.P., De Bertoldi, M. & Lepidi, A.A. (1972) Influence of phenylacetic acid on poly-β-hydroxybutyrate (PHB) polymerization and cell elongation in *Azotobacter chroococcum. Can. J. Microbiol.* **18**: 1257–1261.

330. Ostle, A.G. & Holt, J.G. (1982) Nile Blue A as a fluorescent stain for poly-β-hydroxybutyrate. *Appl. Environ. Microbiol.* **44**: 238–241.

331. Dalton, H. & Postgate, J.R. (1969) Growth and physiology of *Azotobacter chroococcum* in continuous culture. *J. Gen. Microbiol.* **56**: 307–319.

332. Forsyth, W.G.C., Hayward, A.C. & Roberts, J.B. (1958) Occurrence of poly-β-hydroxybutyric acid in aerobic gram-negative bacteria. *Nature* **182**: 800–801.

333. Giles, K.L. (1975) The transfer of nitrogen fixing ability to a eukaryote cell. *Cytobios* **14**: 49–61.

334. Giles, K.L. (1976) Uptake and continued metabolic activity of *Azotobacter* within fungal protoplasts. *Science* **193**: 1125–1126.

335. Lemoigne, M. (1946) Fermentation β-hydroxybutyrique. *Helv. Chim. Acta* **29**: 1303–1306.

336. Page, W.J. (1990) Production of poly-β-hydroxybutyrate by *Azotobacter vinelandii* UWD in beet molasses culture at high aeration. In *Novel Biodegradable Microbial Polymers* (ed. Dawes, E.A.), pp. 423–424. Kluwer, Dordrecht.

337. Reusch, R.N. & Sadoff, H.L. (1981) Lipid metabolism during encystment of *Azotobacter vinelandii*. *J. Bacteriol.* **145**: 889–895.

338. Barham, P., Keller, A., Otun, E. & Holmes, P. (1984) Crystallization and morphology of a bacterial thermoplastic: poly-3-hydroxybutyrate. *J. Material Sci.* **19**: 2781–2794.

339. Lin, L.P. & Sadoff, H.L. (1968) Encystment and polymer production by *Azotobacter vinelandii* in the presence of β-hydroxybutyrate. *J. Bacteriol.* **95**: 2336–2343.

340. Lin, P., Pankratz, S. & Sadoff, H.L. (1978) Ultrastructural and physiological changes occurring upon germination and outgrowth of *Azotobacter vinelandii* cysts. *J. Bacteriol.* **135**: 641–646.

341. Stevenson, L.H. & Socolofsky, M.D. (1966) Cyst formation and poly(β-hydroxybutyric acid) accumulation in *Azotobacter*. *J. Bacteriol.* **91**: 304–310.

342. Tsai, J.C., Aladegbami, S.L. & Vela, G.R. (1979) Phosphate-limited culture of *Azotobacter vinelandii*. *J. Bacteriol.* **139**: 639–645.

343. Aladegbami, S.L., Tsai, J.C. & Vela, G.R. (1979) Adenylate energy charge of *Azotobacter vinelandii* during encystment. *Curr. Microbiol.* **2**: 327–329.

344. De La Rubia, T., Gonzalez-Lopez, J., Moreno, J., Martinez-Toledo, M.V. & Ramos-Cormenzana, A. (1987) Adenine nucleotide contents and energy charge of *Azotobacter vinelandii* grown at low phosphate concentration. *Arch. Microbiol.* **147**: 354–357.

345. De La Rubia, T., Gonzalez-Lopez, J., Martinez-Toledo, M.V., Moreno, J. & Ramos-Cormenzana, A. (1986) Adenine nucleotide content and energy charge in dry cells and cysts of *Azotobacter vinelandii*. *FEMS Microbiol. Lett.* **36**: 111–114.

346. Jurtshuk, P., Manning, S. & Barrerea, C.R. (1968) Isolation and purification of the D(−)β-hydroxybutyric dehydrogenase of *Azotobacter vinelandii*. *Can. J. Microbiol.* **14**: 775–783.

347. Martinez-Toledo, M.V., Gonzalez-Lopez, J., Salermon, V., De La Rubia, T., Ballesteros, F. & Ramos-Cormenzana, A. (1986) Properties of *Azotobacter vinelandii* in phosphate-limited batch cultures. *Folia Microbiol.* **31**: 154–163.

348. Mola, A.H. de (1975) Molecular weight distribution of native poly(D-β-hydroxybutyric) acid. *Makromol. Chemie* **176**: 2655–2667.

349. Page, W.J. (1989) Production of poly-β-hydroxybutyrate by *Azotobacter vinelandii* strain UWD during growth on molasses and other complex carbon sources. *Appl. Microbiol. Biotechnol.* **31**: 329–333.

350. Stevenson, L.H. & Socolofsky, M.D. (1973) Role of poly-β-hydroxybutyric acid in cyst formation by *Azotobacter*. *Ant. v. Leeuwenhoek; J. Microbiol. Serol.* **39**: 341–350.

351. Murrell, W.G. (1967) The biochemistry of the bacterial endospore. *Adv. Microb. Physiol.* **1**: 133–251.

352. Lemoigne, M., Delaporte, B. & Groson, M. (1944) Valeur du test des lipides

β-hydroxybutyriques pour la caracterisation des especes. *Annal. Inst. Pasteur* **70**: 224–233.

353. Blackwood, A.C. & Epp, A. (1957) Identification of β-hydroxybutyric acid in bacterial cells by infrared spectrophotometry. *J. Bacteriol.* **74**: 266–267.

354. Ottaway, J.H. (1962) A preparation of D(−)-β-hydroxybutyric acid. *Biochem. J.* **84**: 11–12.

355. Dunlop, W. & Robards, A. (1973) Ultrastructural study of poly-β-hydroxy-butyrate granules from *Bacillus cereus*. *J. Bacteriol.* **114**: 1271–1280.

356. Clifton, C.E. & Sobek, J.M. (1961) Endogenous respiration of *Bacillus cereus*. *J. Bacteriol.* **82**: 252–256.

357. Thompson, E.D. & Nakata, H.M. (1973) Characterization and partial purification of β-hydroxybutyrate dehydrogenase from sporulating cells of *Bacillus cereus* T. *Can. J. Microbiol.* **19**: 673–677.

358. Macrae, R.M. & Wilkinson, J.F. (1958) Poly-β-hydroxybutyrate metabolism in washed suspensions of *Bacillus cereus* and *Bacillus megaterium*. *J. Gen. Microbiol.* **18**: 210–222.

359. Ram, B.P. & Rana, R.S. (1978) Effects of phenylglyoxal on the growth and sporulation of *Bacillus cereus* T. *Ind. J. Exp. Biol.* **16**: 170–173.

360. Kondo, M., Yoneda, M., Nishi, Y. & Fukai, K. (1961) Studies on poly-β-hydroxybutyrate in bacterial spores. II. Localization of poly-β-hydroxybutyrate in relation to the morphological structure of mature spores of *Bacillus cereus*. *Bikens J.* **4**: 41–49.

361. Kominek, L.A. & Halvorson, H.O. (1965) Metabolism of poly-β-hydroxy-butyrate and acetoin in *Bacillus cereus*. *J. Bacteriol.* **90**: 1251–1259.

362. Kennedy, R.S., Malveaux, F.J. & Cooney, J.J. (1971) Effects of glutamic acid on sporulation of *Bacillus cereus* and on spore properties. *Can. J. Microbiol.* **17**: 511–519.

363. Williamson, D.H. & Wilkinson, J.F. (1958) The isolation and estimation of poly-β-hydroxybutyrate inclusions of *Bacillus* species. *J. Gen. Microbiol.* **19**: 198–209.

364. Pfister, R.M. & Lundgren, D.G. (1964) Electron microscopy of polyribosomes within *Bacillus cereus*. *J. Bacteriol.* **88**: 1119–1129.

365. Remsen, C.C. (1966) The fine structure of frozen-etched *Bacillus cereus* spores. *Arch. Microbiol.* **54**: 266–275.

366. Mynbayeva, B.N., Abdrashitova, S.A. & Ilyaletdinov, A.N. (1987) The ultra-structural organization of *Pseudomonas putida* cells oxidizing arsenite. *Mikrobiologiya* **56**: 95–99.

367. Vogt, J.C., McDonald, W.C. & Nakata, H.M. (1967) Effect of intracellular poly-β-hydroxybutyrate on the ultraviolet sensitivity of *Bacillus cereus*. *Radiat. Res.* **30**: 140–147.

368. Nam, D.H. & Ryu, D.D.Y. (1985) Relationship between butirosin bio-synthesis and sporulation in *Bacillus circulans*. *Antimicrob. Agents Chemother.* **27**: 798–801.

369. Hofsten, W.B. & Baird, G.D. (1962) Fractionation of cell constituents of *Bacillus megaterium* in a polymer two-phase system. *Biotechnol. Bioeng.* **6**: 403–410.

370. Kepez, A. & Peaud-Lenoel, C. (1952) Sur les proprietes et la constitution des lipides β-hydroxybutyriques. *C. R. Acad. Sci.* **234**: 756–757.

371. Law, J.H. & Slepecky, R.A. (1961) Assay of poly-β-hydroxybutyric acid. *J. Bacteriol.* **82**: 33–36.

372. Lemoigne, M. & Roukhelman, N. (1940) Fermentation β-hydroxybutyrique. Caracterisation et evolution des produits de deshydratation et de polymerisation de l'acide β-hydroxybutyrique. *Annal. Ferment.* **5**: 527–536.

373. Lemoigne, M., Grelet, N., Crosan, U. & Le Treis, M. (1945) Formation de

lipide β-hydroxybutyrique aux epens du glucose par le *Bacillus megatherium.* Donnies quantitatives. *Bull. Soc. Chim. Biol. Paris* **27**: 90–95.

374. Lemoigne, M., Milhaud, G. & Croson, M. (1949) Sur le metabolisme lipidique du *Bacillus megaterium. Bull. Soc. Chim. Biol.* **31**: 1587–1591.

375. Merrick, J.M. & Doudoroff, M. (1961) Enzymatic synthesis of poly-β-hydroxybutyric acid in bacteria. *Nature* **189**: 890–892.

376. Norris, K.P. & Greenstreet, J.E.S. (1958) On the infrared absorption spectrum of *Bacillus megaterium. J. Gen. Microbiol.* **19**: 566–580.

377. Sakharova, Z.V. (1977) Chemical composition of the cells of a chemostat culture of *Bacillus megaterium* at alkaline pH values. *Mikrobiologiya* **46**: 580–582.

378. Scott, C.C.L. & Finnerty, W.R. (1976) A comparative analysis of the ultrastructure of hydrocarbon-oxidizing micro-organisms. *J. Gen. Microbiol.* **94**: 342–350.

379. Slepecky, R.A. & Law, J.H. (1960) A rapid spectrophotometric assay of alpha, beta-unsaturated acids and β-hydroxy acids. *Analyt. Chem.* **32**: 1697–1699.

380. Weibull, C. (1953) Characterization of the protoplasmic constituents of *Bacillus megaterium. J. Bacteriol.* **66**: 696–702.

381. White, P.J. & Gilvarg, C. (1977) A teichuronic acid containing rhamnose from cell walls of *Bacillus megaterium. Biochemistry* **16**: 2428–2435.

382. Yan, L.-P. & Hitchins, A.D. (1980) Comparative macromolecular composition of filaments and rods of a *Bacillus megaterium* thermoconditional morphological mutant. *J. Bacteriol.* **144**: 454–456.

383. Doi, Y., Tamaki, A., Kunioka, M. & Soga, K. (1986) Proton and carbon-13 NMR analysis of poly(β-hydroxybutyrate) isolated from *Bacillus megaterium. Macromolecules* **19**: 1274–1276.

384. Bohlken, G. (1969) Zur Speicherung von Reservestoffen in *Bacillus megaterium.* I. Untersuchungen an Wildstämmen. *Zbl. Bakt. Parasitenkde. Abt. II,* **123**: 7–15.

385. Ellar, D., Lundgren, D.G., Okamura, K. & Marchessault, R.H. (1968) Morphology of poly-β-hydroxybutyrate granules. *J. Mol. Biol.* **35**: 389–502.

386. Freer, J.H. & Levinson, H.S. (1967) Fine structure of *Bacillus megaterium* during microcycle sporogenesis. *J. Bacteriol.* **94**: 441–457.

387. Rabotnova, I.L., Shul'govskaya, E.M., Pozmogova, I.N., Kuznetsov, L.E., Ibragimova, S.I. & Ryabchuk, V.A. (1983) Use of rifampicin for directed modification of the composition of *Bacillus megaterium* and *Candida utilis* cells. *Mikrobiologiya* **52**: 87–93.

388. De La Rubia, T., Gonzalez-Lopez, J., Moreno, J., Martinez-Toledo, M.V., Ramos-Cormenzana, A. (1986) Adenine nucleotide and energy charge of *Bacillus megaterium* during batch growth in low-phosphate medium. *FEMS Microbiol. Lett.* **35**: 5–9.

389. De la Rubia, T., Gonzalez-Lopez, J., Ballesteros, F. & Ramos-Cormenzana, A. (1986) Growth of *Bacillus megaterium* in phosphate-limited medium. *Folia Microbiol.* **31**: 98–105.

390. Gonzalez-Lopez, J., De la Rubia, N.T. & Ramos-Cormenzana, A. (1985) Effect of phosphate limitation on the morphology of *Bacillus megaterium. Microbios. Lett.* **28**: 7–13.

391. Macrae, R.M. & Wilkinson, J.F. (1958) The influence of the cultural conditions on poly-β-hydroxybutyrate synthesis in *Bacillus megaterium. Proc. R. Phys. Soc. Edinburgh* **27**: 73–78.

392. Shaforostova, L.D., Ivanova, I.I., Shul'govskaya, E.M. & Rabotnova, I.L. (1973) Growth of microorganisms during exponential phase. *Biotechnol. Bioeng. Symp.* **4**: 175–187.

393. Shul'govskaya, E.M., Pozmogova, I.N. & Rabotnova, I.L. (1980) Effect of

chloramphenicol, an inhibitor of protein synthesis on the main growth characteristics for continuous and batch cultures of *Bacillus megaterium*. *Mikrobiologiya* **49**: 893–901.

394. Slepecky, R.A. & Law, J.M. (1961) Synthesis and degradation of poly-β-hydroxybutyric acid in connection with sporulation of *Bacillus megaterium*. *J. Bacteriol.* **82**: 37–42.

395. Tinelli, R. (1955) Etude de la biochimie de la sporulation chez *Bacillus megaterium*. Etude du comportemente d'une souche de *B. megaterium* asporogene mise dans les conditions de sporulation. *Ann. Inst. Pasteur* **88**, 642–649.

396. Wilkinson, J.F. (1963) Carbon and energy storage in bacteria. *J. Gen. Microbiol.* **32**: 171–176.

397. Wilkinson, J.F. & Munro, A.L.S. (1967) Influence of growth-limiting conditions on the synthesis of possible carbon and energy-storage polymers in *Bacillus megaterium*. In *Microbial Physiology and Continuous Culture* (eds Powell, C., Evans, G.T., Strange, R.E. & Tempest, D.W.), pp. 173–185. HMSO, London.

398. Yamakawa, T., Aida, K. & Uemura, T. (1966) Spectrophotometric studies on the sporulation of *Bacillus megaterium*. *J. Gen. Appl. Microbiol.* **12**: 353–359.

399. Nickerson, K.W. (1982) Purification of poly-β-hydroxybutyrate by density gradient centrifugation in sodium bromide. *Appl. Environ. Microbiol.* **43**: 1208–1209.

400. Nickerson, K.W., Zarnick, W.J. & Kramer, V.C. (1981) Poly-β-hydroxybutyrate parasporal bodies in *Bacillus thuringiensis*. *FEMS Microbiol. Lett.* **12**: 327–331.

401. Wakisaka, Y., Masaki, E. & Nishimoto, Y. (1982) Formation of crystalline δ-endotoxin or poly-β-hydroxybutyric acid granules by asporogenous mutants of *Bacillus thuringiensis*. *Appl. Environ. Microbiol.* **43**: 1473–1480.

402. Pringsheim, E.G. & Wiessner, W. (1963) Minimum requirements for heterotrophic growth and reserve substance in *Beggiatoa*. *Nature* **197**: 102.

403. Strohl, W.R. & Larkin, J.M. (1978) Enumeration, isolation and characterization of *Beggiatoa* from freshwater sediments. *Appl. Environ. Microbiol.* **36**: 755–770.

404. Strohl, W.R., Cannon, G.C., Shively, J.M. Guede, H., Hook, L.A., Lane, C.M. & Larkin, J.M. (1981) Heterotrophic carbon metabolism by *Beggiatoa alba*. *J. Bacteriol.* **148**: 572–583.

405. Guede, H., Strohl, W.R. & Larkin, J.M. (1981) Mixotrophic and heterotrophic growth of *Beggiatoa alba* in continuous culture. *Arch. Microbiol.* **129**: 357–360.

406. Becking, J.H. (1974) Nitrogen-fixing bacteria of the genus *Beijerinckia*. *Soil Sci.* **118**: 196–212.

407. Baumann, P., Baumann, L. & Mandel, M. (1971) Taxonomy of marine bacteria: the genus *Beneckea*. *J. Bacteriol.* **107**: 268–294.

408. Reichelt, J.L. & Baumann, P. (1973) Taxonomy of the marine, luminous bacteria. *Arch. Microbiol.* **94**: 283–330.

409. Ionescu, M.D., Petrovigi, A., Andreescu, V., Mihai, G., Burghellea, B., Marion, M. & Ivan, I. (1983) Poly-β-hydroxybutyrate-type (PHB) inclusions in *B. pertussis* and their possible implication in toxicity. *Arch. Roum. Pathol. Exp. Microbiol.* **42**: 297–303.

410. McDernott, T.R., Griffith, S.M., Vance, C.P. & Graham, P.H. (1988) Carbon metabolism in *Bradyrhizobium japonicum* bacteroids. *FEMS Microbiol. Rev.* **63**: 327–340.

411. Suzuki, T., Zahler, W.L. & Emerich, D.W. (1987) Acetoacetyl-CoA thiolase of *Bradyrhizobium japonicum* bacteroids: purification and properties. *Arch. Biochem, Biophys.* **254**: 272–281.

412. Provost, P.J. & Doetsch, R.N. (1962) An appraisal of *Caryophanon latum*. *J. Gen. Microbiol.* **28**: 547–557.

413. Poindexter, J.S. (1981) Oligotrophy. Fast and famine existence. *Adv. Microb. Ecol.* **5**: 63–89.

414. Poindexter, J.S. (1964) Biological properties and classification of the *Caulobacter* group. *Bacteriol. Rev.* **28**: 231–295.

415. Poindexter, J.S. & Eley, L.F. (1983) Combined procedure for assays of poly-β-hydroxybutyric acid and inorganic polyphosphate. *J. Microbiol. Meth.* **1**: 1–17.

416. Poindexter, J.S. (1984) The role of calcium in stalk development and in phosphate acquisition in *Caulobacter crescentus*. *Arch. Microbiol.* **138**: 140–152.

417. Riley, R.G. & Koldziej, B.J. (1976) Pathway of glucose catabolism in *Caulobacter crescentus*. *Microbios.* **16**: 219–226.

418. Sand, W. & Bock, E. (1982) Effect of carbon and nitrogen sources on the induction of the nitrite oxidizing system and on the PHB-content in *Nitrobacter agilis* K_1. *Mitt. Inst. Allg. Bot. Hamburg* **18**: 61–70.

419. Van Gemerden, H. (1968) ATP generation by *Chromatium* in darkness. *Arch. Microbiol.* **64**: 118–124.

420. Kran, G., Schlote, F.W. & Schlegel, H.G. (1963) Cytologische Untersuchungen an *Chromatium okenii* Perty. *Naturwissenschaften* **50**: 728–730.

421. Schlegel, H.G. (1962) Die Speicherstoffe von *Chromatium okenii*. *Arch. Mikrobiol.* **42**: 110–116.

422. Madigan, M.T. (1986) *Chromatium tepidum* sp. nov., a thermophilic photosynthetic bacterium of the family *Chromatiaceae*. *Int. J. Syst. Bacteriol.* **36**: 222–227.

423. Hurlbert, R.E. (1967) Effect of oxygen on viability and substrate utilization in *Chromatium*. *J. Bacteriol.* **93**: 1346–1352.

424. Nicolay, K., Van Gemerden, H., Hellingwerf, K.J., Konings, W.N. & Kaptein, R. (1983) In vivo ^{31}P and ^{13}C nuclear magnetic resonance studies of acetate metabolism in *Chromatium vinosum*. *J. Bacteriol.* **155**: 634–642.

425. Dias, F.F. & Bhat, J.V. (1963) Accumulation of poly-β-hydroxybutyric acid and iodophilic material by the dominant activated sludge bacteria. *Curr. Sci.* **32**: 501–502.

426. Delafield, F.P., Doudoroff, M., Palleroni, N.J., Lusty, C.J. & Contopoulos, R. (1965) Decomposition of poly-β-hydroxybutyrate by Pseudomonads. *J. Bacteriol.* **90**: 1455–1466.

427. Stanier, R.Y., Palleroni, N.J. & Doudoroff, M. (1966) The aerobic pseudomonads: a taxonomic study. *J. Gen. Microbiol.* **43**: 159–271.

428. Galinski, E.A. & Herzog, R.M. (1990) The role of trehalose as a substitute for nitrogen-containing compatible solutes (*Ectothiorhodospira halochloris*). *Arch. Microbiol.* **153**: 607–613.

429. Wang, W.S. & Lundgren, D.G. (1969) Poly-β-hydroxybutyrate in the chemilithotrophic bacterium *Ferrobacillus ferrooxidans*. *J. Bacteriol.* **97**: 947–950.

430. Tezuka, Y. (1969) Cation-dependent flocculation in a *Flavobacterium* species predominant in activated sludge. *Appl. Microbiol.* **17**: 222–226.

431. Lillo, J.G., Rodriguez-Valera, F. (1990) Effects of culture conditions on poly(β-hydroxybutyric acid) production by *Haloferax mediterranei*. *Appl. Environ. Microbiol.* **56**: 2517–2521.

432. Elsohly, M.A., Elsayed, A.M. & Soliman, F.M. (1985) β-Hydroxybutyric acid polymer from *Hydroclathrus clathratus*. *J. Nat. Prod.* **48**: 809–810.

433. Heptinstall, J., Rittenhouse, H.G., McFadden, B.A. & Shumway, L.K. (1972) Effect of growth conditions on morphology of *Hydrogenomonas facilis* and on yield of a phospholipoprotein. *J. Bacteriol.* **110**: 363–367.

434. Hirsch, P. & Conti, S.F. (1964) Biology of budding bacteria. *Arch. Mikrobiol.* **48**: 339–357.

435. Nikitin, D.I., Pitryuk, I.A., Zagreba, E.D., Ginovska, M.K., Yacobson, Y.O. & Fetisova, M.B. (1986) *Hyphomicrobium* cells growing for a long time in poor media and studied by infrared spectrophotometry. *Mikrobiologiya* **55**: 648–651.

203

436. Vedenina, I.Y. & Lebedinskii, A.V. (1983) Effect of copper on the composition of denitrification products in *Hyphomicrobium*. *Mikrobiologiya* **52**, 917–923.

437. Graezer-Lampart, S.D., Egli, T. & Hamer, G. (1986) Growth of *Hyphomicrobium* ZV620 in the chemostat: regulation of ammonium-assimiliating enzymes and cellular composition. *J. Gen. Microbiol.* **132**: 3337–3347.

438. Semenov, A.M., Ganzlikova, A. & Tenov, N. (1989) Poly-β-hydroxybutyrate accumulation by some oligotrophic polyprostecate bacteria. *Microbiology (Moscow)* **58**: 923–926.

439. Kuhn, D.A. & Starr, M.P. (1965) Clonal morphogenesis of *Lampropedia hyalina*. *Arch. Mikrobiol.* **52**: 360–375.

440. Chandler, F.W., Blackmon, J.A., Hieklin, M.D., Cole, R.M. & Callaway, C.S. (1979) Ultrastructure of the agent of *Legionella* disease in the human lung. *Am. J. Clin. Pathol.*

441. Adams, L.F. & Ghiorse, W.C. (1985) Influence of manganese on growth of a sheathless strain of *Leptothrix discophora*. *Appl. Environ. Microbiol.* **49**: 556–562.

442. Adams, L.F. & Ghiorse, W.C. (1986) Physiology and ultrastructure of *Leptothrix discophora* SS-1. *Arch. Microbiol.* **145**: 126–135.

442a. Alizade, M.A. & Gaede, K. (1977) Chirality of the hydrogen transfer of NAD catalyzed by (3R) hydroxybutyrate dehydrogenase from *Pseudomonas lemoignei*. *Z. Naturforsch.* **32c**: 874–876.

443. Stokes, J.L. & Parson, W.L. (1968) Role of poly-β-hydroxybutyrate in survival of *Sphaerotilus discophorus* during starvation. *Can. J. Microbiol.* **14**, 785–789.

444. Stokes, J.L. & Powers, M.T. (1967) Stimulation of poly(β-hydroxybutyrate) oxidation in *Sphaerotilus discophorus* by manganese and magnesium. *Arch. Mikrobiol.* **59**: 295–301.

445. Dunstan, P.M. & Anthony, C. (1973) Microbial metabolism of C1 and C2 compounds. The role of acetate during growth of *Pseudomonas* AM1 on C1 compounds, ethanol and β-hydroxybutyrate. *Biochem. J.* **132**: 797–801.

446. Suzuki, T., Yamane, T. & Shimizu, S. (1986) Mass production of poly-β-hydroxybutyric acid by fully automatic fed-batch culture of methylotroph. *Appl. Microbiol. Biotechnol.* **23**: 322–329.

447. Suzuki, T., Yamane, T. & Shimizu, S. (1986) Kinetics and effect of nitrogen source feeding on production of poly-β-hydroxybutyric acid by fed-batch culture. *Appl. Microbiol. Biotechnol.* **24**: 366–369.

448. Suzuki, T., Yamane, T. & Shimizu, S. (1986) Mass production of poly-β-hydroxybutyric acid by fed-batch culture with controlled carbon/nitrogen feeding. *Appl. Microbiol., Biotechnol.* **24**: 370–374.

449. Choi, J.H., Kim, J.H., Daniel, M. & Lebeault, J.M. (1989) Optimization of growth medium and poly-beta-hydroxybutyric acid production from methanol in *Methylobacterium organophilum*. *Korean J. Appl. Microbiol. Biotechnol.* **17**: 392–396.

450. Slabova, O.I., Nikitin, D.I. & Zagreba, E.D. (1990) Poly-β-hydroxybutyric acid utilization during the oxidation of gas substrates and their mixtures by the cells of *Methylobacterium organophilum*. *Microbiology* **59**: 938–941.

451. Malashenko, Y.R. (1976) Isolation and characterization of new species (thermophilic and thermotolerant ones) of methane-utilizers. In *Symposium on Microbial Production and Utilisation of Gases (H₂, CH₄, CO)* (eds Schlegel, G., Gottschalk, G. & Pfennig, N.), pp. 293–300. Göttingen, Akademie der Wissenschaften, Göttingen.

452. Asenjo, J.A. & Suk, J.S. (1986) Microbial conversion of methane into poly-β-hydroxybutyrate (PHB): growth and intracellular product accumulation in a type II methanotroph. *J. Ferment. Technol.* **64**: 271–278.

453. Asenjo, J.A. & Suk, J.S. (1986) Kinetics and models for the bioconversion of methane into an intracellular polymer, poly-β-hydroxybutyrate (PHB). *Biotechnol. Bioeng. Symp.* **15**: 225–234.

454. Whittenbury, R., Davies, S.L. & Davey, J.F. (1970) Exospores and cysts formed by methane-utilising bacteria. *J. Gen. Microbiol.* **61**: 219–226.

455. Kallio, R.E. & Harrington, A.A. (1960) Sudanophilic granules and lipid of *Pseudomonas methanica. J. Bacteriol.* **80**: 321–324.

456. Harrington, A.A. & Kallio, R.E. (1960) Oxidation of methanol and formaldehyde by *Pseudomonas methanica. Can. J. Microbiol.* **6**: 1–7.

457. Weaver, T.L., Patrick, M.A. & Dugan, P.R. (1975) Whole-cell and membrane lipids of the methylotrophic bacterium *Methylosinus trichosporium. J. Bacteriol.* **124**: 602–605.

458. Best, D.J. & Higgins, I.J. (1981) Methane-oxidizing activity and membrane morphology in a methanol-grown obligate methanotroph, *Methylosinus trichosporium* OB36. *J. Gen. Microbiol.* **125**: 73–84.

459. Thomson, A.W., O'Neill, J.G. & Wilkinson, J.F. (1976) Acetone production by methylobacteria. *Arch. Microbiol.* **109**: 243–246.

460. Hazeu, W. & Steenis, P.J. (1970) Isolation and characterization of two vibrio shaped methane oxidizing bacteria. *Ant. v. Leeuwenhoek, J. Microbiol. Serol.* **36**: 67–72.

461. Sierra, G. & Gibbons, N.E. (1963) Sodium requirement of poly-β-hydroxybutyric acid depolymerase of *Micrococcus denitrificans. Can. J. Microbiol.* **9**: 491–497.

462. Kates, M., Sehgal, S.N. & Gibbons, N.E. (1961) The lipid composition of *Micrococcus halodenitrificans* as influenced by salt concentration. *Can. J. Microbiol.* **7**: 427–435.

463. Smithies, W.R., Gibbons, N.E. & Bayley, S.T. (1955) The chemical composition of the cell and cell wall of some halophilic bacteria. *Can. J. Microbiol.* **1**: 605–613.

464. Sierra, G. & Gibbons, N.E. (1962) Production of poly-β-hydroxybutyric acid granules in *Micrococcus halodenitrificans. Can. J. Microbiol.* **8**: 249–253.

465. Sierra, G. & Gibbons, N.E. (1962) Role and oxidation pathway of poly-β-hydroxybutyric acid in *Micrococcus halodenitrificans. Can. J. Microbiol.* **8**: 255–269.

466. Braunegg, G. (1979) Speicherung von PHB in *Mycoplana rubra* Stamm R14. *Österr. Chem. Zeitschrift.* **80**: 217.

467. Tobback, P. and Landelout, H. (1965) Poly-β-hydroxybutyric acid in Nitrobacter. *Biochim. Biophys. Acta* **97**: 589–590.

468. Garretson, A.L. & San Clemente, C.L. (1977) Inability of *Nitrobacter agilis* to grow heterotrophically on acetate. *Dev. Ind. Microbiol.* **19**: 541–552.

469. Pope, L.M., Hoare, D.S. & Smith, A.J. (1969) Ultrastructure of *Nitrobacter agilis* growth under autotrophic and heterotrophic conditions. *J. Bacteriol.* **97**: 936–939.

470. Tsien, H.C., Lambert, R. & Laudelout, H. (1968) Fine structure and the localization of the nitrite oxidizing system in *Nitrobacter winogradsky. Ant. v. Leeuwenhoek* **34**: 483–494.

471. Van Gool, A.P., Lambert, R. & Laudelout, H. (1969) The fine structure of frozen etched *Nitrobacter* cells. *Arch. Microbiol.* **69**: 281–293.

472. Van Gool, A.P., Tobback, P.P. & Fischer, I. (1971) Autotrophic growth and synthesis of reserve polymers in *Nitrobacter winogradskyi. Arch. Microbiol.* **76**: 252–264.

473. Emeruwa, A.C. (1981) Isolation and metabolism of glycogen and poly-beta-hydroxybutyrate in *Nocardia asteroides* at different development stages. *Ann. Microbiol.* **132B**: 13–21.

Biomaterials

474. Stal, L.J., Heyer, H. & Jacobs, G. (1990) Occurrence and role of poly-hydroxyalkanoate in the cyanobacterium *Oscillatoria limosa*. In *Novel Biodegradable Microbial Polymers*, pp. 435–438. Kluwer, Dordrecht.

475. Kalacheva, G.S., Vysotskii, E.S., Rodicheva, E.K. & Fish, A.M. (1981) Lipids of the luminescent bacterium *Photobacterium mandapamensis*. *Mikrobiologiya* **50**: 79–83.

476. Ebergardt, A. & Ruoso, G. (1971) Quantitative analysis of the phospholipids of some marine bioluminescent bacteria. *Lipids* **6**: 410–414.

477. Doudoroff, M. & Stanier, R.Y. (1959) Role of poly-β-hydroxybutyric acid in the assimilation of organic carbon by bacteria. *Nature* **183**: 1440–1442.

478. Morris, M.B. & Roberts, J.B. (1959) A group of Pseudomonads able to synthesize poly-β-hydroxybutyric acid. *Nature* **183**: 1538–1539.

479. Jacob, G.S., Garbow, J.R. & Schaefer, J. (1986) Direct measurement of poly(β-hydroxybutyrate) in a pseudomonad by solid-state carbon-13 NMR. *J. Biol. Chem.* **261**: 16785–16787.

480. Timm, A. & Steinbüchel, A. (1990) Formation of poly(3-hydroxyalkanoates) by wild type and recombinant strains of *Pseudomonas aeruginosa* and other fluorescent pseudomonads. In *Novel Biodegradable Microbial Polymers* (ed. Dawes, E.A.), pp. 445–446. Kluwer, Dordrecht.

481. Braunegg, G. & Kornetti, L. (1984) *Pseudomonas* 2F: kinetics of growth and accumulation of poly-D(−)-3-hydroxybutyric acid (Poly-HB). *Biotechnol. Lett.* **6**: 825–829.

482. Jones, K.L. & Rhodes-Roberts, M.E. (1981) The survival of marine bacteria under starvation conditions. *J. Appl. Bacteriol.* **50**: 247–258.

483. Zevenhuizen, L.P.T.M. & Ebbink, A.G. (1974) Interrelations between glycogen, poly-β-hydroxybutyric acid, and lipids during accumulation and subsequent utilization in a *Pseudomonas*. *Ant. v. Leeuwenhoek J. Microbiol. Serol.* **40**: 103–120.

484. Calcott, P.H., Zaborowski, C., Levine, W.E. & Truong, N.-H. (1979) Drug resistance plasmid (pPL1) mediated changes in the susceptibility of *Pseudomonas aeruginosa* to stress. *FEMS Microbiol. Lett.* **6**: 75–80.

484a. Chowdhury, A.A. (1963) Poly-β-hydroxybuttersäure abbauende Bakterien und Exoenzym. *Arch. Mikrobiol.* **47**: 167–200.

484b. Hayward, A.C., Forsyth, W.G.C. & Roberts, J.B. (1959) Synthesis and breakdown of poly-β-hydroxybutyric acid by bacteria. *J. Gen. Microbiol.* **20**: 510–518.

485. Takahashi, J. (1980) Production of intracellular and extracellular protein from n-butane by *Pseudomonas butanovora* sp. nov. *Adv. Appl. Microbiol.* **26**: 117–128.

486. Takahashi, J., Ichikawa, Y., Sagae, H., Komura, I., Kanou, H. & Yamada, K. (1980) Isolation and identification of n-butane assimilating bacterium. *Agric. Biol. Chem.* **44**: 1835–1840.

487. Hayward, A.C. (1960) A method for characterizing *Pseudomonas solanacearum*. *Nature* **186**: 405–406.

488. Higham, D.P., Sadler, P.J. & Scawen, M.D. (1986) Gold-resistant bacteria: excretion of a cystine-rich protein by *Pseudomonas cepacia* induced by an antiarthritic drug. *J. Inorg. Biochem.* **28**: 253–261.

489. Ramsay, B.A., Ramsay, J.A. & Cooper, D.G. (1989) Production of poly-β-hydroxyalkanoic acid by *Pseudomonas cepacia*. *Appl. Environ. Microbiol.* **55**: 584–589.

490. Wilkinson, S.G. (1969) Lipids of *Pseudomonas diminuta*. *Biochim. Biophys. Acta* **187**: 492–500.

491. Stinson, M.W. & Merrick, J.M. (1974) Extracellular enzyme secretion by *Pseudomonas lemoignei*. *J. Bacteriol.* **119**: 152–161.

493. Nakayama, K., Saito, T., Fuku, T., Shirakura, Y. & Tomita, K. (1985)

Purification and properties of extracellular poly(3-hydroxybutyrate) depolymerases from *Pseudomonas lemoignei. Biochem. Biophys. Acta* **827**: 63–72.

494. Imada, A., Kitano, K., Kiutaka, K., Muroi, M. & Asai, M. (1981) Sulfazecin and isosulfazecin, novel beta-lactam antibiotics of bacterial origin. *Nature* **289**: 590–591.

495. Bowman, J.P., Sly, L.I. & Hayward, A.C. (1988) *Pseudomonas mixta* sp. nov., a bacterium from soil with degradative activity on a variety of complex polysaccharides. *System. Appl. Microbiol.* **11**: 53–59.

496. Brandl, H., Gross, R.A., Lenz, R.W. & Fuller, R.C. (1988) *Pseudomonas oleovorans* as a source for novel poly(β-hydroxyalkanoates). *Polym. Prepr.* **29**: 590–591.

497. Gross, R.A., Brandl, H., Ulmer, H.W., Posada, M.A., Fuller, R.C. & Lenz, R.W. (1989) The biosynthesis and characterisation of new poly(β-hydroxyalkanoates). *Polym. Prepr.* **30**: 492–493.

498. Knee, E.J., Wolf, M., Lenz, R.W. & Fuller, R.C. (1990) Influence of growth conditions on production and composition of PHA by *Pseudomonas oleovorans*. In *Novel Biodegradable Microbial Polymers* (ed. Dawes, E.A.), pp. 439–440. Kluwer, Dordrecht.

499. Preusting, H., Nijenhuis, A. & Witholt, B. (1990) Physical characteristics of poly(3-hydroxyalkanoates) and poly(3-hydroxyalkenoates) produced by *Pseudomonas oleovorans* grown on aliphatic hydrocarbons. *Macromolecules* **23**: 4220–4224.

500. Preusting, H., Nijenhuis, A. & Witholt, B. (1990) Production and characterization of poly(3-hydroxyalkanoates). In *Novel Biodegradable Polymers* (ed. Dawes, E.A.), pp. 453–454. Kluwer, Dordrecht.

501. Gagnon, K.D., Bain, D.B., Lenz, R.W. & Fuller, R.C. (1990) Yield study of the poly(beta-hydroxyalkanoate) produced by *Pseudomonas oleovorans* grown on sodium octanoate. In *Novel Biodegradable Microbial Polymers* (ed. Dawes, E.A.), pp. 449–450. Kluwer, Dordrecht.

502. Holmes, B., Owen, R.J., Evans, A., Malnik, H. & Willcox, W.R. (1977) *Pseudomonas paucinobilis*, a new species isolated from human clinical specimens, the hospital environment, and other sources. *Int. J. Syst. Bacteriol.* **27**: 133–146.

503. Bertrand, J.-L., Ramsay, B.A., Ramsay, J.A. & Chavarie, C. (1990) Biosynthesis of poly-β-hydroxyalkanoates from pentoses by *Pseudomonas pseudoflava. Appl. Environ. Microbiol.* **56**: 3133–3138.

504. Levine, H.B. & Wolochow, H. (1960) Occurrence of poly-β-hydroxybutyrate in *Pseudomonas pseudomallei. J. Bacteriol.* **79**: 305–306.

505. Hayward, A.C. (1962) Studies on bacterial pathogens on sugar cane. II. Differentiation, taxonomy and renomenclature of the bacteria causing red stripe and mottled stripe disease. *Mauritius Sugar Ind. Res. Inst. Occas. Pap.* **13**: 13–27.

506. Young, H.L., Chao, F.C., Turnbull, C. & Philpott, D.E. (1972) Ultrastructure of *Pseudomonas saccharophila* at early and late log-phase of growth. *J. Bacteriol.* **109**: 862–868.

507. Granatskaya, T.A., Dvornikova, T.P., Platsynda, V.A. & Il'insakaya, S.P. (1981) Complex isolation of biologically active substances from hydrogen bacteria. *Izv. Akad. Nauk Mold. SSR, Ser. Biol. Khim. Nauk* **6**: 41–45.

508. Craig, A.S., Greenwood, R.M. & Williamson, K.B. (1973) Ultrastructural inclusions of rhizobial bacteroids of lotus nodules and their taxonomic significance. *Arch. Microbiol.* **89**: 23–32.

509. Tombolini, R. & Nuti, M.P. (1989) Poly(β-hydroxyalkanoate) biosynthesis and accumulation by different *Rhizobium* species. *FEMS Microbiol. Lett.* **60**: 299–304.

510. Scandola, M., Pizzoli, M., Ceccorulli, G., Searo, A., Paoletti, S. & Navarini,

L. (1988) PHB in solid state: dynamic mechanical and calorimetric properties. *Polym. Prepr.* **29**: 613–614.

511. Bonartseva, G.A., Myshkina, V.L. & Zagreba, E.A. (1989) Relationship between poly-β-hydroxybutyrate content and nitrogenase and hydrogenase activity in some strains of *Rhizobium. Mikrobiologiya* **58**: 742–745.

511. Starr, M.P. (1981) The genus *Lampropedia.* In *The Prokaryotes* (eds Starr *et al.*) Vol. II, pp. 1530–1536. Springer.

512. Bonartseva, G.A., Myshkina, V.L. & Zagreba, E.A. (1989) Poly(β-hydroxy-butyrate accumulation as a function of nitrogenase and hydrogenase activities in some *Rhizobium* strains. *Microbiology* **58**: 923–926.

513. Patel, J.J. & Gerson, T. (1974) Formation and utilization of carbon reserves by *Rhizobium. Arch. Microbiol.* **101**: 211–220.

514. Stam, H., Van Verseveld, H.W., De Vries, W. & Stouthamer, A.H. (1986) Utilization of poly-β-hydroxybutyrate in free-living cultures of *Rhizobium* ORS571. *FEMS Microbiol. Lett.* **35**: 215–220.

515. Vries, W. de, Stam, H., Duys, J.G., Ligtenberg, A.J.M., Simons, L.H. & Stouthamer, A.H. (1986) The effect of the dissolved oxygen concentration and anabolic limitations on the behaviour of *Rhizobium* ORS571 in chemostat cultures. *Ant. v. Leeuwenhoek. J. Microbiol. Serol.* **52**: 85–96.

516. Wong, P.P. & Evans, H.J. (1971) Poly-β-hydroxybutyrate utilization by soybean (*Glycine max*) nodules and assessment of its role in maintenance of nitro-genase activity. *Plant Physiol.* **47**: 750–755.

517. Fottrell, P.F. & O'Ttara, A. (1969) Multiple forms of D(−)-3-hydroxybutyrate dehydrogenases in *Rhizobium. J. Gen. Microbiol.* **57**: 287–292.

518. Casella, S., Leporini, C., Corti, A., Picci, G., Chiellini, E., Casini, E. & Solaro, R. (1990) Culture substrate effect in the production of poly(β-hydroxybutyrate by *Rhizobium* 'hedysari'. In *Novel Biodegradable Microbial Polymers* (ed. Dawes, E.A.), pp. 73–80. Kluwer, Dordrecht.

519. Hahn, M., Meyer, L., Studer, D., Regensburger, B. & Hennecke, H. (1984) Insertion and deletion mutations within the nif region of *Rhizobium japonicum. Plant Mol. Biol.* **3**: 159–168.

520. Karr, D.B., Waters, J.K. & Emerich, D.W. (1983) Analysis of poly-β-hydroxybutyrate in *Rhizobium japonicum* bacteroids by ion-exclusion high-pressure liquid chromatography and UV detection. *Appl. Environ. Microbiol.* **46**: 1339–1344.

521. Werner, D. & Moerschel, E. (1978) Differentiation of nodules of *Glycine max.* Ultrastructural studies of plant cells and bacteroids. *Planta* **141**: 169–177.

522. Karr, D.B., Waters, J.K., Lee, S.-Y. & Emerich, D.W. (1983) Relationship between nitrogen fixation and carbon metabolism in *Rhizobium japonicum* bacter-oides: enzymes of the poly-β-hydroxybutyrate cycle, the citric acid cycle and the pyruvate dehydrogenase complex. *Curr. Top. Plant Biochem. Physiol.* **1**: 248.

523. Karr, D.B., Waters, J.K., Suzuki, F. & Emerich, D.W. (1984) Enzymes of the poly-β-hydroxybutyrate and citric acid cycles of *Rhizobium japonicum* bacteroids. *Plant Physiol.* **75**: 1158–1162.

524. Faizova, G.K., Borodulina, Yu.S. & Samsonova, S.P. (1971) Lipids of the nodule bacterial, *Rhizobium leguminosarum. Mikrobiologiya* **40**: 471–474.

525. Kretovich, V.L., Romanov, V.I., Yushkova, L.A., Shramko, V.I. & Fedulova, N.G. (1977) Nitrogen fixation and poly-β-hydroxybutyric acid content in bacteroids of *Rhizobium lupini* and *Rhizobium leguminosarum. Plant Soil* **48**: 291–302.

526. Scandola, M., Pizzoli, M., Ceccorulli, G., Cesaro, A., Paoletti, S. & Navarini, L. (1988) Viscoelastic and thermal properties of bacterial poly-D-levo-beta-hydroxy-butyrate. *Int. J. Biol. Macromol.* **10**: 373–377.

527. Tikhonovich, I.A., Romanov, V.I., Alisova, S.M., Chermenskaya, I.E., Chetkova, S.A. & Fedulova, N.G. (1985) Nitrogen fixation and photoassimilates in the nodules of chlorophyll mutants of peas. *Genetika* **21**: 1021–1025.

528. Zevenhuizen, L.P.T.M. (1981) Cellular glucogen, β-1,2-glycan, poly-β-hydroxybutyric acid and extracellular polysaccharides in fastgrowing species of *Rhizobium. Ant. v. Leeuwenhoek* **47**: 481–497.

529. Gerson, T., Patel, J.J. & Wong, M.N. (1978) The effects of age, darkness and nitrate on poly-β-hydroxybutyrate levels and nitrogen-fixing ability of *Rhizobium* in *Lupinus angustifolius. Physiol. Plant* **42**: 420–424.

530. Romanov, V.I., Yushkova, L.A. & Kretovich, W.L. (1975) Synthesis and decomposition of poly-β-hydroxybutyric acid in *Rhizobium lupini. Microbiologiya* **44**: 820–824.

531. Romanov, V.I., Fedulova, N.G., Shramko, V.I., Molchanov, M.I. & Kretovich, V.L. (1980) Metabolism of poly-β-hydroxybutyric acid in bacteroids of *Rhizobium lupini* in connection with nitrogen fixation and photosynthesis. *Plant Soil* **56**: 379–390.

532. Fedulova, N.G., Chermenskaya, I.E., Romanov, V.I. & Kretovich, V.L. (1980) Enzymes of poly(β-hydroxybutyrate) metabolism in *Rhizobium lupini* bacteroids. *Fiziol. Rast.* **27**: 544–550.

533. Bonartseva, G.A. (1985) Activity of nodule bacteria in terms of poly(β-hydroxybutyrate) accumulation during colony staining with phosphine 3R. *Mikrobiologiya* **54**: 461–464.

534. Bonartseva, G.A. & Myshkina, V.L. (1985) Fluorescence intensity of *Rhizobium meliloti* and *Rhizobium phaseoli* differing in activity and growing in the presence of the lipophilic vital dye phosphine 3R. *Microbiologiya* **54**: 661–667.

535. Herrera de Mola, A., Marx-Figini, M. & Figini, R.V. (1975) Molecular weight distribution of native poly(D-β-hydroxybutyric acid). *Makromol. Chem.* **176**: 2655–2667.

536. Carranza, M.M., Rosas, S.B. & Ghittoni, N.E. (1986) Molecular composition of *Rhizobium meliloti* when parathion was added at the start of incubation. *J. Appl. Bacteriol.* **60**: 9–13.

537. Carranza de Storani, M.M., Rosas, S.B. & Ghittoni, N.E. (1985) Effect of parathion on growth, polysaccharides, lipids and proteins of *Rhizobium meliloti. Antonie van Leeuwenhoek* **51**: 249–254.

538. Tombolini, R., Boesten, B., O'Gara, F. & Nuti, M.P. (1990) Poly-β-hydroxyalkanoate (PHA) accumulation in *Rhizobium meliloti* affected in the dicarboxylate transport (DCT) genes. In *Novel Biodegradable Microbiol. Polymers* (ed. Dawes, E.A.), pp. 431–434. Kluwer, Dordrecht.

539. Soberon, M., Membrillo-Hernandez, J., Aquilar, G.R. & Sanchez, F. (1990) Isolation of *Rhizobium phaseoli* Tn5-induced mutants with altered expression of cytochrome terminal oxidases o and aa₃. *J. Bacteriol.* **172**: 1676–1680.

540. Vincent, J.M., Humphrey, B. & North, R.J. (1962) Some features of the fine structure and chemical composition of *Rhizobium trifoli. J. Gen Microbiol.* **29**, 551–555.

541. Karayiannis, V.G. & Madigan, M.T. (1990) Acetone as a substrate for poly-β-hydroxybutyrate production by phototrophic purple bacteria. In *Novel Biodegradable Microbial Polymers* (ed. Dawes, E.A.), pp. 427–430. Kluwer, Dordrecht.

542. Chong, E.P. & Berger, L.R. (1967) Fatty acids of extractable and bound lipids of *Rhodomicrobium vannielii. J. Bacteriol.* **93**: 230–236.

543. Bloomfield, G., Sandu, G. & Carr, N.G. (1969) Activation by Hg²⁺ of acetoacetyl-CoA reductase in extracts of *Rhodopseudomonas spheroides* and *Rhodomicrobium vannielli. FEBS Lett.* **5**: 246–248.

209

544. Kondrateva, E.N., Krasil'nikova, E.N. & Novikova, L.M. (1968) Production of polysaccharides by green photosynthesizing bacteria. *Mikrobiologiya* **37**: 417–424.

545. Williamson, D.H., Mellanby, J. & Krebbs, H.A. (1962) Enzymic determination of D(−)-β-hydroxybutyric acid and acetoacetic acid in blood. *Biochem. J.* **82**: 90–96.

546. Dierstein, R. & Drews, G. (1974) Nitrogen-limited continuous culture of *Rhodopseudomonas capsulata* growing photosynthetically or heterotrophically under low oxygen tensions. *Arch. Microbiol.* **99**: 117–128.

547. Goebel, F. (1978) Quantum efficiencies of growth of photosynthetic bacteria. In *Photosynthetic Bacteria* (eds Clayton, R.K. & Sistrom, W.R.), pp. 907–925.

548. Wijbenga, D.J. & Van Gemerden, H. (1981) The influence of acetate on the oxidation of sulfide by *Rhodopseudomonas capsulata*. *Arch. Microbiol.* **43**: 109–137.

549. Oadri, S.M.H. & Hoare, D.S. (1968) Formic hydrogenlyase and the photoassimilation of formate by a strain of *Rhodopseudomonas palustris*. *J. Bacteriol.* **95**: 2344–2357.

550. Imai, Y., Morita, S. & Arata, Y. (1984) Proton correlation NMR studies of metabolism in *Rhodopseudomonas palustris*. *J. Biochem.* **96**: 691–699.

551. Nicolay, K., Hellingwerf, K.J., Kaptein, R. & Konings, W.N. (1982) Carbon-13 nuclear magnetic resonance studies of acetate metabolism in intact cells of *Rhodopseudomonas spheroides*. *Biochim. Biophys. Acta* **720**: 250–258.

552. Peters, G.A. & Cellarius, R.A. (1972) Photosynthetic membrane development in *Rhodopseudomonas spheroides*. II. Correlation of pigment incorporation with morphological aspects of thylakoid formation. *Bioenergetics* **3**: 345–356.

553. Carr, N.G. & Lascelles, J. (1961) Some enzymic reactions concerned in the metabolism of acetoacetic-coenzyme A in Athiorhodaceae. *Biochem. J.* **80**: 70–72.

554. Ohashi, A., Oshihara, N. & Kikuchi, G. (1967) Pyruvate metabolism in *Rhodospeudomonas spheroides* under light-anaerobic conditions. *J. Biochem.* **62**: 497.

555. Bergmeyer, H.U., Gawehn, K. & Klotzsch, H. (1967) Purification and properties of crystalline 3-hydroxybutyrate dehydrogenase from *Rhodopseudomonas spheroides*. *Biochem. J.* **102**: 423–431.

556. Reiss-Husson, F., De Klerk, H., Jolchine, G., Jauneau, E. & Kamen, M.D. (1971) Effects of iron deficiency on *Rhodopseudomonas spheroides* strain Y. *Biochim. Biophys. Acta* **234**: 73–82.

557. Hurst, R., Pincock, A. & Broekhoven, L.H. (1973) Model discrimination and nonlinear parameter estimation in the analysis of the mechanisms of action of β-hydroxybutyrate dehydrogenase from *Rhodopseudomonas spheroides*. *Biochim. Biophys. Acta* **321**: 1–26.

558. Preuveneers, M.J., Peacock, D., Crook, E.M., Clark, J.B. & Brocklehurst, K. (1973) D-3-hydroxybutyrate dehydrogenase from *Rhodopseudomonas spheroides*. Kinetic mechanism from steady-state kninetics of the reaction catalysed by the enzyme in solution and covalently attached to diethylaminoethyl-cellulose. *Biochem. J.* **133**: 133–157.

559. Preuveneers, M.J., Peacock, D., Crook, E.M., Clark, J.B. & Bricklehurst, K. (1973) D-3-hydroxybutyrate dehydrogenase from *Rhodopseudomonas spheroides*. Kinetics of radioisotope redistribution at chemical equilibrium catalysed by the enzyme in solution. *Biochem. J.* **133**: 159–164.

560. Giesbrecht, P. (1968) Zur Darstellung der DNS von Bakterien und plastischer biologischer Strukturen mit Hilfe der Gefrierätzung. *Zbl. Bakt. I. Abt. Orig.* **207**: 198–221.

561. Boatman, E.S. (1964) Observations on the fine structure of spheroplasts of *Rhodospirillum rubrum*. *J. Cell Biol.* **20**: 297–306.

562. Cohen-Bazire, G. Kunisawa, R. (1963) The fine structure of *Rhodospirillum rubrum. J. Cell. Biol.* **16**: 401–419.

563. Uffen, R.L., Sybesma, C. & Wolfe, R.S. (1971) Mutants of *Rhodospirillum rubrum* obtained after long-term, anaerobic dark growth. *J. Bacteriol.* **108**: 1348–1356.

564. Bosshard-Herr, E. & Bachofen, R. (1969) Synthese von Speicherstoffen aus Pyruvat durch *Rhodospirillum rubrum. Arch. Microbiol.* **65**: 61–75.

564. Stanier, R.Y., Doudoroff, M., Kunisawa, R. & Contopoulou, R. (1959) The role of organic substrates in bacterial photosynthesis. *Proc. Natl. Acad. Sci. U.S.A.* **45**: 1246–1260.

565. Merrick, J.M. (1965) Effect of polymyxin B, tyrocidine, gramicidine D and other antibiotics on the enzymatic hydrolysis of poly-β-hydroxybutyrate. *J. Bacteriol.* **90**: 965–969.

566. Shuster, C.W. & Doudoroff, M. (1962) A cold-sensitive D(−)-β-hydroxybutyric acid dehydrogenase from *Rhodospirillum rubrum, J. Biol. Chem.* **237**: 603–607.

567. Stern, J.R., Del Campillo, A. & Raw, I. (1956) *J. Biol Chem.* **218**: 971–978.

568. Mulder, E.G., Deinema, M.H., Van Neen, W.L. & Zevenhuizen, L.P.T.M. (1962) Polysaccharides, lipids and poly-β-hydroxybutyrate in microorganisms. *Rec. Trav. Chim.* **81**: 797–809.

569. Mulder, E.G. & Van Veen, W.L. (1963) Investigations of the *Sphaerotilus-Leptothrix* group. *Ant. v. Leeuwenhoek* **29**: 121–153.

570. Nakata, H.M. (1963) Effect of pH on intermediates produced during growth and sporulation of *Bacillus cereus. J. Bacteriol.* **86**: 577–581.

571. Rouf, M.A. & Stokes, J.L. (1962) Isolation and identification of the sudanophilic granules of *Sphaerotilus natans. J. Bacteriol.* **83**: 343–347.

572. Van Den Eynde, E., Vriens, L., Wynants, M. & Verachtert, H. (1984) Transient behaviour and time aspects of intermittently and continuously fed bacterial cultures with regard to filamentous bulking of activated sludge. *Appl. Microbiol. Biotechnol.* **19**: 44–52.

573. Van Veen, W.L., Mulder, E.G. & Deinema, M.H. (1978) The *Sphaerotilus-Leptothrix* group of bacteria. *Microbiol. Rev.* **42**: 329–356.

574. Matin, A., Veldhuis, C., Stegeman, V. & Veenhuis, M. (1979) Selective advantage of a *Spirillum* sp. in a carbon-limited environment. Accumulation of poly-β-hydroxybutyric acid and its role in starvation. *J. Gen. Microbiol.* **112**: 349–355.

575. Martinez, R.J. (1962) On the nature of the granules of the genus *Spirillum. Arch. Microbiol.* **44**: 334–343.

576. Ohou, Y., Albrecht, S.L. & Burris, R.H. (1976) Carbon and ammonia metabolism of *Spirillum lipoferum. J. Bacteriol.* **128**: 592–597.

577. Semenov, A.M., Hanzlikova, A. & Jasndera, A. (1989) Quantitative estimation of poly-3-hydroxybutyric acid in some oligotrophic polyprosthecate bacteria. *Folia Microbiol.* **34**: 267–270.

578. Kutty, M.R., Kannan, L.V. & Rehacek, Z. (1969) Effect of phosphate on biosynthesis of antimycin A and production and utilization of poly-β-hydroxybutyrate by *Streptomyces antibioticus. Ind. J. Biochem.* **6**: 230–231.

579. Packter, N.M. & Flatman, S. (1983) Characterization of acetoacetyl-CoA reductase (3-oxoreductase) from *Streptomyces coelicolor*: its possible role in poly-hydroxylbutyrate biosynthesis. *Biochem. Soc. Trans.* **11**: 598–599.

580. Zychlinsky, E. & Matin, A. (1983) Effect of starvation on cytoplasmic pH, protonmotive force, and viability of an acidophilic bacterium, *Thiobacillus acidophilus. J. Bacteriol.* **153**: 371–374.

581. Tabita, R. & Lundgren, D.G. (1971) Utilization of glucose and the effect of organic compounds on the chemolithotroph *Thiobacillus ferrooxidans. J. Bacteriol.* **108**: 328–333.

582. Claassen, P.A.M., Dijkema, C., Visser, J. & Zehnder, A.J.B. (1986) In vivo carbon-13 NMR analysis of acetate metabolism in *Thiobacillus versutus* under denitrifying conditions. *Arch. Microbiol.* **146**: 227–232.

583. Boon, J.J., de Leeuw, J.W. & Krumbein, W.E. (1985) Biogeochemistry of Gavish Sabkha sediments. II. Pyrolysis mass spectrometry of the laminated microbial mat in the permanently water-covered zone before and after the desert sheetflood of 1979. *Ecol. Stud.* **53**: 368–380; 440–470.

584. Tamura, Y., Fujino, T., Kondo, M. & Kotani, S. (1968) Occurrence of poly(β-hydroxybutyric acid) inclusions in *Vibrio parahaemolyticus* A55. *Biken J.* **11**: 225–234.

585. Tamura, T., Kato, K., Iwata, S., Kotani, S. & Kitaura, T. (1976) Studies of the cell envelope of *Vibrio parahaemolyticus* A55: isolation and purification of bag-shaped peptidoglycan (murein sacculus). *Biken J.* **19**: 93–113.

586. Wiegel, J., Wilke, D., Baumgarten, J., Opitz, P. & Schlegel, H.G. (1978) Transfer of the nitrogen-fixing hydrogen bacterium *Corynebacterium autotrophicum*, Baumgarten et al. to *Xanthobacter* gen. nov. *Int. J. Syst. Bacteriol.* **28**: 578–581.

587. Angelbeck, D.I. & Kirsch, E.J. (1969) Influence of pH and metal cations on aggregated growth of non-slime forming strains of *Zoogloea ramigera*. *Appl. Microbiol.* **17**: 435–440.

588. Friedman, B.A. & Dugan, P.R. (1968) Identification of *Zoogloea* species and the relationship to Zoogloeal matrix and floc formation. *J. Bacteriol.* **95**: 1903–1909.

589. Friedman, B.A., Dugan, P.R., Pfister, R.M. & Remsen, C.C. (1968) Fine structure and composition of the Zoogloeal matrix surrounding *Zoogloea ramigera*. *J. Bacteriol.* **96**: 2144–2153.

590. Crabtree, K., McCoy, E., Boyle, W.C. & Rohlich, G.A. (1965) Isolation, identification, and metabolic role of the sudanophilic granules of *Zoogloea ramigera*. *Appl. Microbiol.* **13**: 218–226.

591. Crabtree, K., Boyle, W., McCoy, E. & Rohlich, G.A. (1966) Mechanism of floc formation of *Zoogloea ramigera*. *J. Water Pollut. Contr. Fed.* **38**: 1968–1980.

592. Cooper, T.A., Flatt, J.H., Lightfoot, E.N. & Cameron, D.C. (1989) Exopolysaccharide production from lactose by wild-type and polyhydroxybutyrate minus strains of *Zoogloea ramigera*. *Abstr. PAP Am. Chem. Soc.* **198**:

593. Fukui, T., Yoshimoto, A., Matsumoto, M., Hosokawa, S., Saito, T., Nishikawa, H. & Kenkichi, T. (1976) Enzymatic synthesis of poly-β-hydroxybutyrate in *Zoogloea ramigera*. *Arch. Microbiol.* **110**: 149–156.

594. Parsons, A.G. & Dungan, P.R. (1971) Production of extracellular polysaccharide matrix by *Zoogloea ramigera*. *Appl. Microbiol.* **21**: 657–661.

595. Tomita, K. & Saito, T. (1976) Metabolism of poly-β-hydroxybutyrate in *Zoogloea ramigera*. *Seikagaku* **48**: 1045–1048.

596. Davis, J.T., Moore, R.N., Imperiali, B., Pratt, A.J., Kobayashi, K., Masamune, S., Sinskey, A.J. & Walsh, C.T. (1987) Biosynthetic thiolase from *Zooloea ramigera*. I. Preliminary characterization and analysis of proton transfer reaction. *J. Biol. Chem.* **262**: 82–89.

597. Davis, J.T., Chen, H.-H., Moore, R., Nishitani, Y., Masamune, S., Sinskey, A.J. & Walsh, C.T. (1987) Biosynthetic thiolase from *Zoogloea ramigera*. II. Inactivation with haloacetyl CoA analogs. *J. Biol. Chem.* **262**: 90–96

598. Fukui, T., Ito, M., Saito, T. & Tomita, K. (1987) Purification and characterization of NADP-linked acetoacetyl-CoA reductase from *Zoogloea ramigera* I-16-M. *Biochim. Biophys. Acta* **917**: 365–371.

599. Ito, M., Saito, T. & Tomita, K. (1987) Purification and characterization of NADP-linked acetoacetyl-CoA reductase from *Zoogloea ramigera* I-16-M. *Biochim. Biophys. Acta* **917**: 365–371.

600. Ploux, O., Masamune, S. & Walsh, C.T. (1988) The NADPH-linked acetoacetyl-CoA reductase from *Zoogloea ramigera*. Characterization and mechanistic studies on the cloned enzyme over-produced in *Escherichia coli. Eur. J. Biochem.* **174**: 177–182.

601. Saito, T., Fukui, T., Ikeda, F., Tanaka, Y. & Tomita, K. (1977) An NADP-linked acetoacetyl CoA reductase from *Zoogloea ramigera. Arch. Microbiol.* **114**: 211–217.

602. Saito, T., Takemura, N., Ito, M. & Tomita, K. (1990) An enzymatic assay method for D(−)-3-hydroxybutyrate and acetoacetate involving acetoacetyl-Coenzyme A synthetase from *Zoogloea ramigera. Chem. Pharmac. Bulletin* **38**: 1627–1629.

603. Thompson, S., Mayer, F., Peoples, P., Masamune, S., Sinskey, A.J. & Walsh, C.T. (1989) Mechanistic studies on β-ketoacyl thiolase from *Zoogloea ramigera*: identification of the active-site nucleophile as Cys 89, its mutation to Ser 89, and kinetic and thermodynamic characterization of wild-type and mutant enzymes. *Biochemistry* **28**: 5735–5742.

604. Walsh, C.T., Chen, W.H., Differding, E., Masamune, S., Peoples, O.P., Ploux, O., Sinskey, A.J. & Thompson, A.J. (1989) Enzymes in the biosynthesis of polyesters. *Proc. Robert A. Welsh Found. Conf. Chem. Res.* **31**: 184–203.

4 Microbial polysaccharides

J. D. Linton*, S. G. Ash* & L. Huybrechts†

*Shell Research Limited, Sittingbourne Research Centre,
    Sittingbourne, Kent, ME8 9AG, UK.
†International Bio Synthetics, Komvest 43, B-8000, Brugge, Belgium.

Introduction

Microorganisms are a rich and as yet underexploited source of exo-polysaccharides (EPS). These complex molecules serve as energy reserves, structural materials and stores of information, and are involved in molecular interaction and recognition. Although plants produce a wide range of polysaccharides, their diversity is considerably less than that of those produced by micro-organisms. The number of different sugars found in polysaccharides is an indicator of diversity of structure and some 200 different sugars have been found in microbial polysaccharides, compared to approximately 25 associated with those of plant origin[1]. As the structure and function of these biofunctional molecules become known, microorganisms will become important as sources of them and will in addition provide the means of producing them from other diverse sources using molecular genetic methods already available. In the short term, however, industrial applications of microbial polysaccharides will be confined to their use as rheological modifiers of aqueous solutions. The wide range of structures, their properties and potential applications have been the subject of a number of reviews[2-10]. This review will focus on the production and applications of microbial polysaccharides that are in commercial use and discuss the prospects for those approaching commercialization in the food, oilfield and industrial areas.

Physiology of EPS Production

Yields of EPS from the carbon source and oxygen

For a polysaccharide of known composition, the theoretical yield from the carbon source and oxygen can be calculated if the ATP/O quotient for the organism is known. The ATP/O quotient is defined as the number of moles of ATP synthesized per 0.5 moles of O_2 consumed. Several assumptions have to be made in order to calculate the ATP/O quotient and it should be stressed that the values obtained are not precise, but are nevertheless a useful comparative yardstick of the growth efficiency of different microorganisms [11].

Studies carried out at Shell Research Limited, Sittingbourne[11-18], have shown that all exopolysaccharide-producing organisms examined to date have ATP/O quotients that are considerably lower than that expected from the composition and properties of their respiratory chains (Table 1).

Table 1 A comparison of theoretical yields of EPS from oxygen and the carbon source with observed yields corrected for cellular requirements for carbon and oxygen

Exopolysaccharide and organism	$Y_{O_2}^{max}$	ATP/O quotient	Yield of EPS (g g^{-1})			
			Corrected for cell production		Theoretical	
			Y_{glc}	Y_{O_2}	Y_{glc}	Y_{O_2}
Succinoglycan *A. radiobacter*	40	1.4–1.6	0.74	5.7	0.83	8.1
Succinogalactan *E. herbicola*	31	1.1–1.25	0.69	8.15	0.83	7.8
Xanthan *X. campestris*	26.4	0.94–1.0	0.95	4.0	0.80	5.32
Alginate *P. mendocina*	25.0	0.9–1.0	0.52	0.85	0.45	0.91

The theoretical yields were calculated using ATP/O quotients derived from $Y_{O_2}^{max}$ obtained from carbon-limited cultures[11].

The scope for EPS yield improvement can be assessed by comparing the theoretical yield values calculated on the basis of the ATP/O quotient (determined experimentally) for a given organism with the observed yield. However, before this is done, the latter has to be corrected for the amount of glucose and oxygen required for cell production. This assessment has been carried out for a number of EPS-producing organisms and in all cases the corrected yields approach (>70%) the theoretical calculated on the basis of the experimentally determined ATP/O quotients[11,12] (Table 1). EPS production in these organisms is a very efficient process, as virtually all the carbon consumption and respiratory activity occurring in excess of growth requirements can be accounted for in terms of the EPS produced. To approach the theoretical yields under process conditions, it is necessary to isolate strains exhibiting high specific rates of EPS production so that the EPS-to-cell ratio is maximized and therefore the amount of substrate and oxygen consumed for cell production is a small fraction of the total consumed.

These studies indicate that although EPS production is an efficient process with respect to the ATP/O quotient in different organisms, the low ATP/O quotients of these organisms mean that the yields from oxygen are inherently poor and there is little scope for major improvements[11]. Thus *Xanthomonas campestris*, with an ATP/O quotient of 1, has a theoretical xanthan yield of approximately $5.3 g/g$ O_2, whereas a yield of $15.97 g/g$ O_2 would be possible at an ATP/O quotient of 3 (ref. 19). This is the primary reason why the productivity of EPS fermentation processes is limited by the rate of oxygen mass transfer. The high viscosity that develops in broths containing relatively low (2–4%) EPS concentrations exacerbates this problem by hindering good mixing in the reactor.

Effect of carbon source

The oxidation states of exopolysaccharides, even those of the most oxidized forms, like alginate, are very close to that of a hexose with respect to the CHO ratio. The yield of EPS and the rate of production decreases when carbon sources either more oxidized or reduced than a hexose are used for production[15]. Although the carbon-carbon yields of EPS may be acceptable from these substrates, the yield from oxygen is reduced significantly, and this is accompanied by a corresponding increase in the heat output of these fermentations (Table 2). As discussed above, the productivity of EPS processes from glucose is limited by O_2 mass transfer. With methanol and methane the yields of EPS from O_2 are much worse, and any process using these substrates will have significant additional costs with respect to O_2 supply and fermenter cooling. (Table 2)[20].

Closely related hexoses may be metabolized by metabolic pathways that differ in their efficiency of precursor and energy generation, and this can greatly influence EPS yields. This is observed with alginate production by *Pseudomonas* sp. that catabolize glucose and fructose via different pathways. Incorporation of fructose into alginate occurs directly, whereas glucose is converted into two C3 fragments, only one of which is incorporated[21]. The remaining C3 fragment is oxidized to CO_2 and therefore the yield of EPS from oxygen is also significantly higher with fructose than it is with glucose as the carbon source[11]. For alginate production, carbon and energy flux is better integrated with

Table 2 The influence of the carbon source on the yield of an unsubstituted hexose exopolysaccharide

Carbon source	Yield of EPS (g/g) from		Heat of fermentation (KJ/hexose in EPS)
	carbon source	oxygen	
Glucose	0.86	16.3	146
Methanol	0.82	1.58	1550
Methane	0.84	0.28	7760

The relative yields from oxygen and the heat released during EPS production illustrate the magnitude of oxygen transfer and cooling required with substrates considerably more reduced than glucose.

fructose than glucose as the carbon source and the maximum specific rate of production is considerably faster from the former[11]. Similarly, the highest EPS yields and rate of production from methanol occur via the RuMP pathway of C1 fixation, which of all the known pathways facilitates the best integration of carbon and energy flux during EPS synthesis[13].

Rates of production and stability

Rapid rates of EPS production necessitate rapid fluxes of carbon and energy through the metabolic system of an organism. The capacity to process these fluxes rapidly appears to be inversely related to the growth efficiency (ATP/O quotient) of the producing organism[11]. Moreover, the oxidation state and the specific rate of EPS production are also inversely related to the growth efficiency of the producing organism[11]. Thus, oxidized polysaccharides are produced by bacteria possessing relatively low growth efficiencies, whereas unsubstituted polysaccharides are produced by organisms possessing relatively higher growth efficiencies, such as yeasts and fungi[11]. Although EPS synthesis is an efficient process, the energy demand for synthesis is very high. For example, at a specific production rate routinely observed with *X. campestris*, the rate of ATP use for EPS production may be more than twice that required for cell production[11]. Nevertheless, between 30 and 70% of this energy is supplied during the synthesis of the oxidized moieties that comprise the polysaccharides. It appears that optimal EPS synthesis only occurs when carbon and energy fluxes are integrated. The extent to which this occurs will depend on the oxidation state of the EPS, the ATP/O quotient of the organism and the oxidation state of the carbon source. For further discussion of this aspect the reader is referred to ref. 11.

IMPACT OF MOLECULAR GENETICS

The genes involved in the synthesis of xanthan have been identified. In addition, a range of mutants altered in their capacity to produce xanthan have been produced and their biochemical lesions characterized[22,23]. This elegant work has shown that in principle various forms of xanthan gum can be produced. One particular modification, the poly-trimer, has been shown to have improved rheological properties. Similarly, the genetic loci involved in succinoglycan synthesis have been identified and various mutants generated[24]. Nevertheless, there have been no well-substantiated reports that indicate that genetic engineering has been used to significantly alter EPS composition or production rate on an industrial scale.

The cloning and amplification of EPS-associated loci without establishing the flux control points in its biosynthesis is the main reason for the apparent lack of success in achieving improvements in production rates. The elegant work of Cornish and co-workers[25,26] has shown unequivocally that substrate uptake is the major flux control point in succinoglycan synthesis. Using continuous selection under glucose limitation they were able to isolate a stable mutant in which the glucose uptake system was partially derepressed under conditions of glucose excess. This mutant exhibited a significantly higher rate of EPS production than the parent strain from which it was derived. The use of molecular genetics to deregulate substrate uptake will undoubtedly lead to further improvements in the specific production rate. However, repeated flux control analysis will be necessary to identify further flux control points and to achieve maximum impact from genetic manipulation. Moreover, as discussed earlier, any proposed modification to EPS composition should take into account the consequent effect on carbon and energy flux to EPS synthesis (see ref. 11).

STRAIN STABILITY

The strains of *Agrobacterium radiobacter* and *X. campestris* used for large-scale ($220\,m^3$) batch production of succinoglycan and xanthan respectively have proved to be very stable with respect to EPS production. In continuous culture, all EPS-producing organisms that have been examined have shown instability with respect to EPS production when grown under conditions of carbon excess. For example, in cultures of *E. herbicola*, the rate of EPS production is reduced from 0.37 g/g cells/h to 0.05 g/g cells/h after approximately 17 generations[17]. In *P. aeruginosa*, 10 generations in continuous culture resulted in the reduction of EPS producers in the population from 99.99% to 55%[27]. Similar observations have been made with most EPS producers, including *X. campestris*[28].

Commercial Production

The discussion regarding the production of microbial exopolysaccharides will be confined to the production of xanthan gum and succinoglycan. These polysaccharides are commercially significant, representative and the most familiar to the authors. For convenience the various steps involved in the commercial process will be discussed in chronological order as summarized in Fig. 1. It should, however, be

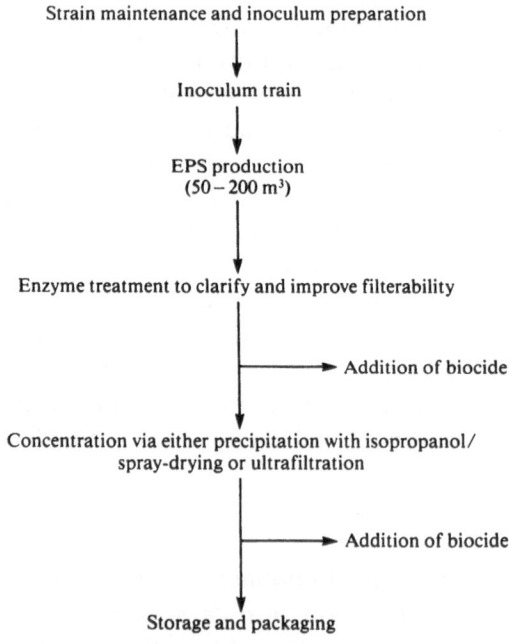

Strain maintenance and inoculum preparation

Inoculum train

EPS production
(50 – 200 m³)

Enzyme treatment to clarify and improve filterability

Addition of biocide

Concentration via either precipitation with isopropanol/
spray-drying or ultrafiltration

Addition of biocide

Storage and packaging

Fig. 1   A generalized scheme for the commercial production of microbial exopolysaccharides.

stressed that successful process development relies on the integration of the fermentation and downstream processing steps, and is an iterative process requiring a multidisciplinary approach.

Storage and production of inoculum

Exopolysaccharides are sold as performance chemicals, and so stored strains have to be assessed to ensure that the EPS they produce meets specifications. Many individual ampoules containing the required strain are freeze-dried or stored in liquid nitrogen and used as the source of future inocula. Great care must be taken to ensure that the stored cultures are pure and the viability is known, so that a standard inoculum can be provided on demand.

Inoculum train

Although a large inoculum comprising 10% of each inoculum stage may be beneficial in terms of efficiently producing the required biomass, it involves many transfers, with a concomitant increased risk of contamination. Thus a 10% inoculum train starting with the inoculation of a 200-ml shake flask with a freeze-dried culture, would require seven culture transfers to inoculate a 200,000-l fermentation. In practice, the number of inoculum steps is kept to a minimum (for example 3 or 4) to reduce the risk of contamination. The overall time required to produce the inoculum for the final production does not depend significantly on the number of transfers. Having fewer transfers leads to less capital investment and running costs, because considerably fewer vessels are required for development of the inoculum. It is usual to run two inoculum trains, either in parallel or staggered by 12-24h, so that, if necessary, one train can be aborted before the final transfer to the production vessel. Contamination of the production vessel leads to serious financial penalties that could threaten the economic viability of the process.

Each step of the inoculum train is monitored for contamination. However, to reduce the risk of contamination, sampling directly from the inoculum vessels is minimized. It is preferable to take a sample from the residue left in each vessel after its contents have been transferred to the next reactor. The detection of contamination is retrospective and therefore heavy reliance is placed on the characteristic growth kinetics of the given organism. Parameters such as growth rate, the rate of oxygen consumption and CO_2 evolution are generally sufficient to provide a reliable fingerprint of a given organism.

The medium used to produce the inoculum may be designed to facilitate rapid biomass production without concomitant EPS production. The latter causes the development of highly viscous cultures that are difficult to transfer from one reaction to another.

Fermentation

The capacity of the production vessel is usually in the 100–200-m³ range. Although various new fermenters [29] have been designed to cope with the high viscosity developed in EPS fermentations, in practice, economic pressures necessitate the use of existing fermentation facilities. Processes have to be developed to suit existing facilities rather than vice versa. Fermenters are usually conventional continuously stirred tank reactors (CSTR) fitted with Rushton turbine impellers, and these may be combined with other impellers to cause axial flow and improve bulk mixing. The productivity of EPS fermentations is limited by the viscosity of the EPS produced. In the case of xanthan, the viscosity developed at a concentration of 3–4% is approximately 20,000 cP/s, and this limits mixing and O_2 mass transfer. The prospects for overcoming these process limitations will be discussed later.

EPS PRODUCTION MEDIA

Xanthan gum is commercially the most important microbial EPS currently being manufactured. The discussion on media requirements will therefore be confined mainly to xanthan fermentations. The early patent literature[30,32] indicated that *X. campestris* required complex nutrients and vitamins to sustain an adequate growth and EPS production rate. However, these complex nutrient sources are expensive and variable in their composition and consequently it is difficult to maintain process reproducibility within the narrow window required to produce a product of consistent quality.

Strains have been developed that grow in a mineral salts medium without added vitamins or complex nutrients such as yeast extract[31]. Typical media for xanthan and succinoglucan production are given in Table 3. A complex nitrogen source such as glutamate that supports a significantly faster growth and EPS production rate than that observed with ammonia may be used (Table 4). Great care has to be taken to monitor water quality because the presence of certain metal salts, for example, calcium, copper and iron, can have an adverse effect on both the growth of the organism and the rheological properties of the EPS produced.

Table 3 Defined mineral salts medium for xanthan and succinoglucan production[12,16,31]

Constituent	*X. campestris* (g/l)	*A. radiobacter* (g/l)
KH_2PO_4	0.68	0.68
Glutamate	3.5–7.0	0.0
$(NH_4)_2SO_4$	0.0	0.79
$MgSO_4.7H_2O$	0.4	0.49
$CaCl_2.2H_2O$	0.012	0.007
$FeSO_4.7H_2O$	0.011	0.014
Glucose	25.0–40.0	25.0–40.0
	Concentrations (mg/l)	
$MnSO_4.4H_2O$	0.57	0.44
$ZnSO_4.7H_2O$	0.72	0.57
H_3BO_4	0.077	0.06
Na_2MoO_4	0.312	0.24
$CuSO_4.5H_2O$	0.625	0.25
$CoCl_2.6H_2O$	0.30	0.23
KI	0.20	0.16

Table 4 A comparison of the kinetics of growth and EPS production by *X. campestris* in glutamate- or ammonia-containing mineral salts production medium[31]

	Nitrogen source	
Kinetic parameter	Ammonia	Glutamate
u_{max} (h^{-1})	0.09	0.12
q_p (g xanthan/g cells/h)	0.27	0.36
Y_p (g xanthan/g glucose)	0.53	0.56
Productivity (g xanthan/l/h)	0.39	0.49

The characteristics of a typical fermentation production run for xanthan and succinoglucan are shown in Figs 2 and 3. The medium is formulated to limit the cell density to between 1.5 and 5 g/l as a result of nitrogen exhaustion. The cell production phase is completed in

15–20h and EPS is then produced for a further 30–50h depending on the specific EPS production rate (q_p) of the organism employed. In the case of xanthan production, the fermentation is completed in approximately 30–50h because of the high q_p and EPS production during growth of the organism (Fig. 2). The sugar concentration and viscosity of the fermentation is monitored at regular intervals and the production is considered to be complete when all the glucose is exhausted and the desired viscosity has been produced[34].

It should be emphasized that EPS is a performance product and, as

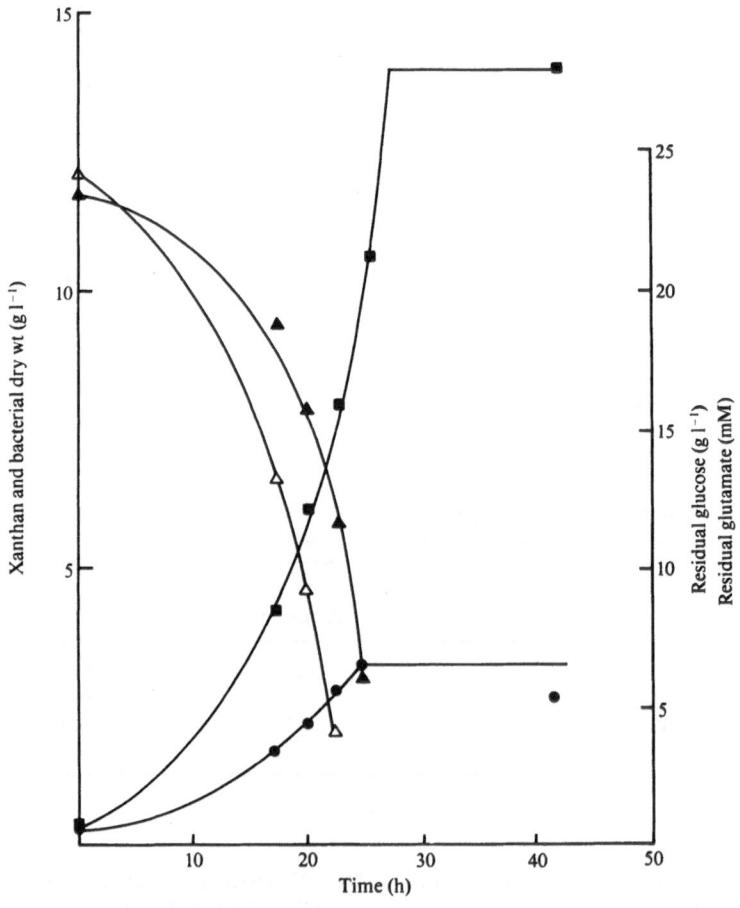

Fig. 2 Batch production process for xanthan gum using *Xanthomonas campestris* NCIB 11854. Bacterial dry wt., ●; xanthan gum ■; residual glucose, ▲; and glutamate, △[31].

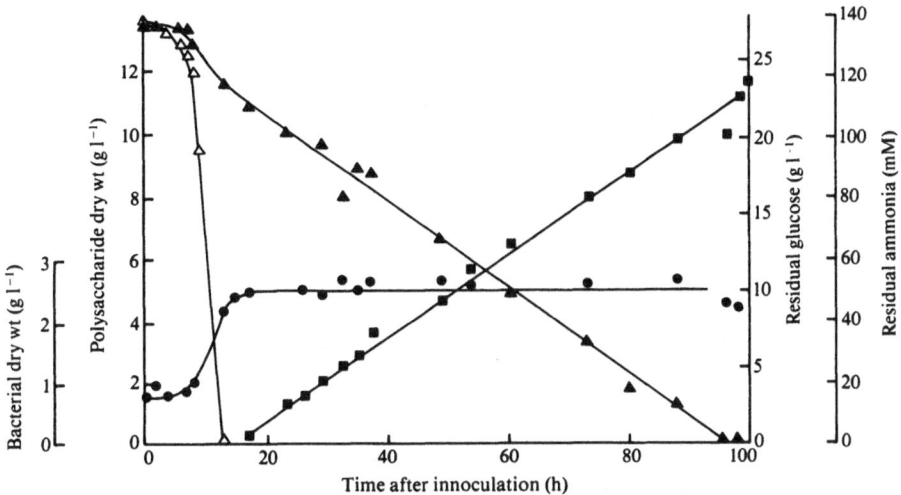

Fig. 3 Batch production of succinoglycan by *Pseudomonas* sp. NCIB 11592[33]. Bacterial dry wt., •; polysaccharide concentration, ■; residual glucose, ▲; and residual ammonia, △.

such, its rheological properties are as important as the concentration of the product at the end of the fermentation. In certain cases, the viscosity per unit weight may vary significantly, depending on the precise growth conditions. For example, although the concentration of succinoglycan produced by *A. radiobacter* is similar when the temperature is controlled at 30 or 35 °C, the viscosity of EPS produced at the latter temperature is considerably higher per unit weight than at 30 °C (Fig. 4). Unlike xanthan, succinoglycan does not exhibit a viscosity increase when heated. The increased viscosity per unit weight of succinoglycan produced at 35 °C is probably due to an increase in the average relative molecular mass (M_r) of the EPS produced at the higher temperature. As the EPS is sold on the basis of its viscosifying power, an increase in viscosity per unit weight is clearly economically advantageous.

In large-scale production, the yield of xanthan from glucose is approximately 0.6–0.7 g/g glucose[35]. If an allowance is made for the glucose requirement for cell production, these yields are close to the theoretical for *X. campestris*[11]. This has also been shown to be the case for the production of succinoglycan and alginate by *A. radiobacter* and *P. aeruginosa*, respectively[11]. In practice, a high specific production

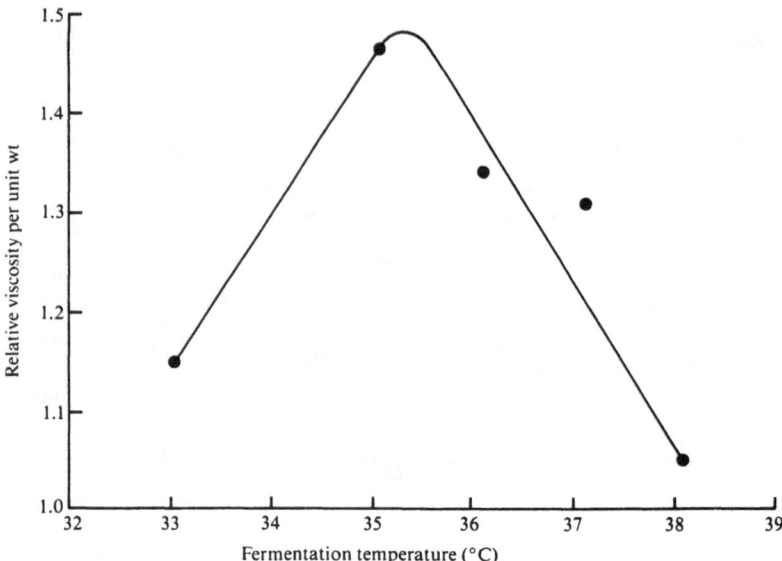

Fig. 4 The effect of fermentation temperature on the relative viscosity of succino-glycan produced by *Agrobacterium radiobacter* NCIB 11883 (ref. 14, and Rye, personal communication).

rate for an extended period in the absence of cell growth will ensure that observed yields approach the theoretical and will result in the most favourable polymer-to-cell ratio. As discussed earlier, there is little scope for improving yields significantly, either from the carbon source or, more importantly, oxygen.

Enzyme treatment

The primary aim of enzyme treatment is to improve the clarity and filterability of the product. Most patents claim that this is achieved via the removal of particulate material such as cells and cell debris. On a commercial scale, an enzyme-treatment step is used and this causes cell lysis and the release of soluble materials such as proteins, nucleic acids and polysaccharides. However, in most cases microscopic examination reveals the presence of virtually intact microbial cell walls. The complete removal of cells necessitates either filtration or centrifugation of relatively dilute solutions, which is not economically attractive.

There are many patents covering the enzyme-treatment step. In general, the step is combined with a heat-treatment step and may involve the addition of agents such as phenols or detergents to aid cell disruption[36]. It is claimed that heat treatment for 15–30 min at 80–100°C at pH 4.5–5 before enzyme treatment, greatly improves the clarification process and also results in a significant (40–50%) increase in the viscosity per unit weight of the treated xanthan gum [37–39]. The type of enzyme and chemical package used will be governed by the strict requirements of the particular application. For example, all chemicals and enzymes used in the treatment of EPS destined for the food market must be food-approved, whereas for industrial or oilfield application the requirements are not as stringent.

Enzyme treatment may involve the use of a single enzyme such as an alkaline protease[40] but most industrial proteases such as Maxatase (Gist) and Alkalase (Novo) contain additional enzymatic activities involved in the treatment. For example, Novozyme 234 (Novo) contains mutase (a 1,3–glucosidase), cellulase, laminarinase, xylanase, chitinase and protease activity and these activities are claimed to play an important role in the production of xanthan possessing good clarity and filterability[37]. Similarly, the 'enzyme' Kitalase (Kumiai Chem. Ind. Co. Ltd.) contains cellulase, carboxymethylcellulase, glucanase, chitinase, protease, hemicellulase and amylase[41]. The amount of enzyme used depends on its specific activity, but the dosage used is usually dependent on the protein content of the broth to be treated and falls within the range 0.4–1 g/g N for Novozyme[37]. The precise conditions employed in the enzyme-treatment step will be dependent on the particular enzyme being used and on the nature of the EPS-producing organism. For Novozyme 234, treatment of xanthan-containing broths is carried out at pH values between 4 and 5.8 at temperatures between 25 and 55°C for a duration of 4–24 h (ref. 37). The clarity of the xanthan solution is determined by optical-density measurements and is typically reduced by 70–95% by the enzyme-treatment step.

In addition to the general treatment discussed above, treatment with specific enzymes improves injectivity for enhanced oil recovery (EOR) applications (see later). In the case of succinoglycan production by *A. radiobacter* on a commercial scale, co-production of small amounts of cellulose causes adverse effects on broth filterability that can be reversed by treatment with the enzyme cellulase[42]. Filterability is also seriously impaired if the DNA and RNA released during cell lysis is not degraded. The addition of nucleases to these broths results in substantial improvement in filterability[43].

The enzyme-treatment step may be combined with the ultrafiltration step because the enzyme is retained by the membrane, but the hydrolysed fragments of contaminants pass through allowing the production of a cleaner product.

Concentration of EPS

On completion of the enzyme- and heat-treatment steps, the EPS is concentrated either by solvent precipitation followed by spray drying, or by ultrafiltration. Ultrafiltration results in the production of a concentrate containing 8–12% EPS by weight. The method used is governed by the market in which the EPS is to be sold. In general, food-related markets have a requirement for a dry powder, whereas concentrates are preferred for EOR. The advantages/disadvantages of concentrates versus powders are discussed later.

SOLVENT PRECIPITATION

Exopolysaccharides are readily precipitated from aqueous solutions by the addition of polar organic solvents such as acetone, methanol or isopropanol (IPA). The amount of solvent required for efficient precipitation is strongly influenced by the salt concentration of the aqueous solution[44]. A minimum of 2.3 volumes of IPA is required at a salt concentration $<10^{-2}$N, but the volume falls to 0.8 vols when the salt concentration is increased to 10^{-1}N (Fig. 5)[44]. Calcium nitrate was found to be more effective than either sodium sulphate or magnesium sulphate and a concentration of 1.82×10^{-1}N, which is equivalent to 3.3–4 times the normality of polyanions on xanthan, was shown to be optimal[44]. Further reduction in solvent requirement can be achieved by heat treatment of the EPS for 1–15 min at 100–130°C before solvent precipitation[44].

The precipitated EPS is dehydrated by washing with fresh anhydrous solvent before it is dried and packaged. Great care has to be taken to ensure that the solvent content in the dried EPS is below acceptable limits. For application of IPA-precipitated EPS in foods, the residual IPA level must be <750 mg/kg xanthan.

The disadvantage of precipitation with a water-soluble solvent is that relatively large volumes have to be used and this requires costly solvent recovery processes. In addition, this method precipitates

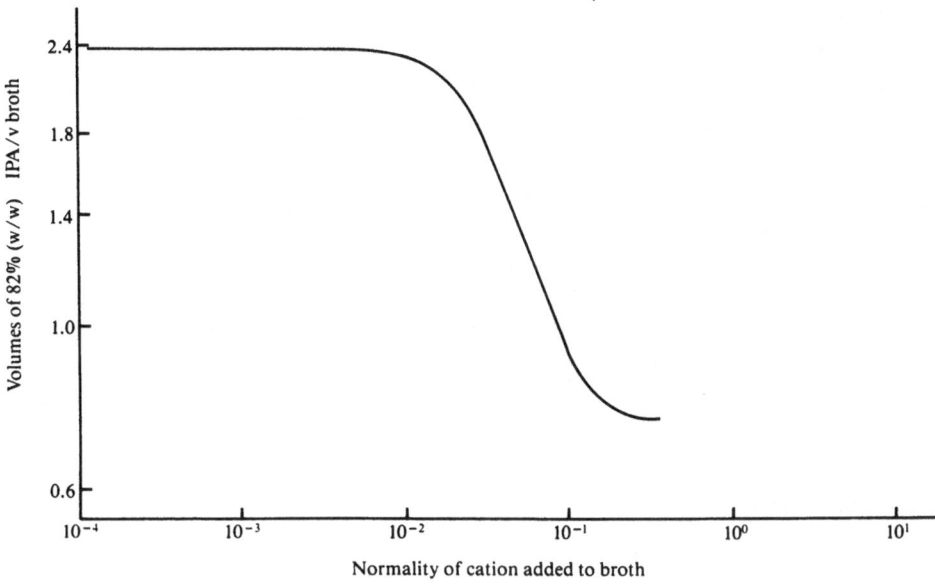

Fig. 5 The relationship between salt concentration (Ca^{2+}) and the volume of IPA required to precipitate xanthan gum from fermentation broths[44].

proteins, salts and in certain cases, for example pullulan production, pigments are also precipitated. The latter leads to the product being discoloured. Consequently, other methods of precipitation have been proposed to overcome these limitations.

Solvents with lower hydrophilicity, such as methylethylketone (MEK), have been claimed to be better for use than methanol or IPA, because only 22.6% (w/w) is sufficient to saturate the aqueous phase and precipitate EPS quantitatively[45]. Therefore, considerably smaller volumes of solvent are required. Moreover, as two distinct phases are generated, pigments are extracted into the organic phase and salts remain in the aqueous phase, resulting in the production of a pure white precipitate. The precipitate is washed twice with 0.2 vol. MEK (or methyl acetate) and roller-, fluidized-bed- or spray-dried.

ULTRAFILTRATION

Ultrafiltration is a process in which relatively high-M_r solutes, such as proteins and natural gums or colloidal substances, are separated and concentrated from their solvent. However, for concentration of viscous polysaccharides, ultrafiltration is only applicable if the polysaccharides are pseudoplastic. Thus pullulan, which is considerably less pseudoplastic than xanthan gum, can be concentrated only to a viscosity of 3000 centipoise, which is approximately 10% of that obtainable for xanthan gum. The flux rates achievable with pullulan are proportionally lower[46].

The ultrafiltration process is fairly simple and relies on the flow, under pressure, of the solution to be concentrated over the surface of a suitably supported membrane. Polysulphone and polyvinylidine fluoride membranes with an M_r cut-off between 20–60,000 are used in commercially available plate- and frame-type ultrafiltration equipment. The only energy requirement is for pumping the fluid through the equipment at the desired pressure. The pressure difference across the membrane is of the order 2–14 atmospheres and the linear flow rate in the region of 50–500 cm/s. Ultrafiltration has been used to concentrate xanthan to 10–20% (w/v)[46–48].

Preservation and storage

PRESERVATION OF CONCENTRATES

Preservation of the EPS has to be initiated immediately after the enzyme-treatment step. At this stage, addition of the minimum biocide level compatible with product protection is advisable because most of this material will remain in the aqueous effluent after the EPS concentration phase. This effluent will have to be treated before discharge. On completion of the concentration step (ultrafiltration), the biocide content is raised to a level that affords protection against microbial spoilage. However, the biocide level in the concentrate will not be sufficient to prevent microbial spoilage in the final formulation when diluted.

INDUSTRIAL AND OILFIELD APPLICATIONS

The type of biocide used is governed by the nature of the particular application. For example, the nature of the biocide is considerably more critical for cosmetic than it is for industrial and oilfield applications. Nevertheless, in each case the requirements of the customer have to be met and storage trials conducted in order to establish the efficacy of the biocide and to ensure that the rheological properties of the EPS remain unaffected. In the case of xanthan and succinoglucan,

0.4% Proxel CF (ICI) has been shown to be an effective preservative for concentrates to be used in industry and oilfields. However, this biocide is not acceptable for concentrates destined for use in cosmetics.

FOOD AND PHARMACEUTICAL APPLICATIONS

Xanthan gum is the major food-approved microbial exopolysaccharide and its use in foods is governed by existing statutory microbiological standards that are product-dependent. Nevertheless, for food and pharmaceutical applications, the microbial count must be reduced to less than 10,000 viable cells per g polysaccharide, preferably <250 per g (ref. 49). Propylene oxide treatment has been recommended as a means of reducing the count of viable microorganisms in xanthan powders. In a process patented by the Kelco Company[49], the viable count was reduced by 92% (<1800 per g) following propylene oxide treatment for 3 h at 105°F in a tumbling reactor at a dose of 1 oz propylene oxide/ft³. The process involves an initial evacuation step before propylene oxide treatment for the required time. After treatment, evacuation and tumbling are alternated and if necessary the contents flushed with sterile nitrogen gas to reduce the residual propylene oxide level to <300 mg/kg, a value set by the FDA. The treated polysaccharide is then packaged aseptically.

Prospects for improvement

Methods have been developed to increase process productivity and the final EPS concentration attained in the reactor. The methods used rely on either improving bulk mixing by employing improved impeller or reactor designs, or reducing the viscosity developed in the fermenter. For the former, a helical ribbon-screw impeller has been developed[50,51], and when used in combination with a Rushton turbine impeller, results in improved O_2 transfer by a factor of 2, as well as improved mixing characteristics caused by the liquid flow up the sides of the reactor and then down the centre. It is claimed that this system allows the production of 10% (w/w) xanthan at an overall productivity of 1 g/l/h. However, the power input required for commercial-scale production appears to make this system uneconomic. It should be emphasized that although EPS concentration and productivity is important, polysaccharides such as xanthan are sold primarily on the basis of their viscosifying power per unit weight. Consequently, improvements in EPS concentration and process productivity are only beneficial if the rheological properties of the EPS are maintained. As discussed above, commercial constraints necessitate the modification of existing fermentation capacity, and although new reactor systems have been examined at pilot-plant scale[29], to the authors' knowledge, these have not been used for commercial production.

Emulsion fermentation is an ingenious way of reducing the viscosity of EPS fermentations. The rationale for this system is that dispersion of the aqueous phase as discrete droplets within a water-insoluble hydrocarbon phase generates considerably less viscosity. For example, the viscosity of a 5.2% xanthan broth is reduced from 40,000 to 10,000 centipoise when it is in the form of an emulsion, and therefore xanthan concentrations between 7 and 10% can be produced routinely. Moreover, owing to the higher solubility of oxygen in the organic phase, there is a significant improvement in the oxygen mass transfer into the aqueous phase.

The hydrocarbon phase should have a boiling point $>100°C$, and paraffins $(>8°C)$, mineral or plant oils have been shown to be suitable[52]. The dispersed aqueous phase can be stabilized by the addition of nonionic emulsifiers with an HLB in the range 12–18, such as ethoxylated fatty acids and glycerols or sorbitol fatty acids and esters[52]. However, a careful balance has to be reached between obtaining a stable emulsion and yet facilitating adequate droplet coalescence and separation in order to have adequate control of the internal environment (for example, pH) of each discrete droplet. In addition, the emulsion has to be broken at the end of the fermentation, and good separation of the aqueous and hydrocarbon phase is required for most applications. However, there may be applications in which the EPS in the form of an emulsion is acceptable.

The main drawback to this system is that the effective aqueous volume of the reactor is reduced by the volume of the hydrocarbon phase. Thus an EPS concentration of 10% in the aqueous phase is equivalent to an overall reactor concentration of 5% when the aqueous-to-hydrocarbon phase is 1:1. Nevertheless, the higher EPS concentration in the aqueous phase may lead to a reduction in downstream processing costs (during the enzyme treatment, solvent precipitation or ultrafiltration), and these savings may be of sufficient magnitude to make this process economically attractive.

Structures and Properties

The importance of molecular conformation

Microbial polysaccharides are expensive on a weight basis relative to the average price of water-soluble polymers (Table 5). Therefore, successful commercial application depends on exploiting their unique properties, discussed below. There properties arise largely from the high M_r of the polysaccharides ($>10^6$), from the molecular conformations determined by the primary structures, and from association between molecules in solution. In multiphase systems, the interfacial properties of the polymers may also be important.

Table 5   The average price of water soluble polymers (£/kg) (1986)

Natural	Price
Alginate	4.2–5.7
Locust bean gum	5.7–11.4
Carageenan	4.2–7.1
Guar	0.8–2.0
Xanthan gum	5.7–10.0
Semi-synthetic	
Carboxymethylcellulose	0.8–2.8
Hydroxyethylcellulose	2.8–4.2
Synthetic	
Polyacrylates	1.4–4.6
Polyvinylalcohol	2.0–

The microbial polysaccharides usually possess a regular repeat unit consisting of between three and eight monosaccharides (Fig. 6). In most applications, the molecules exist in solution in an ordered conformation, which from solid-state data is believed to be a helix. This may be single (succinoglycan[53]), double (gellan[54-56]) or triple (scleroglucan[57,58]). Whether xanthan is a single or double helix has been a matter of dispute[59-64]. These helical conformations are stabilized by intramolecular hydrogen-bonds.

Intermolecular interactions are also important. They not only contribute to the stability of the multiple helices, but also influence solution behaviour, that is, solubility, viscosity and gel-formation. Thus a good interactive 'fit' between molecules will lead to insolubility (for

Biomaterials

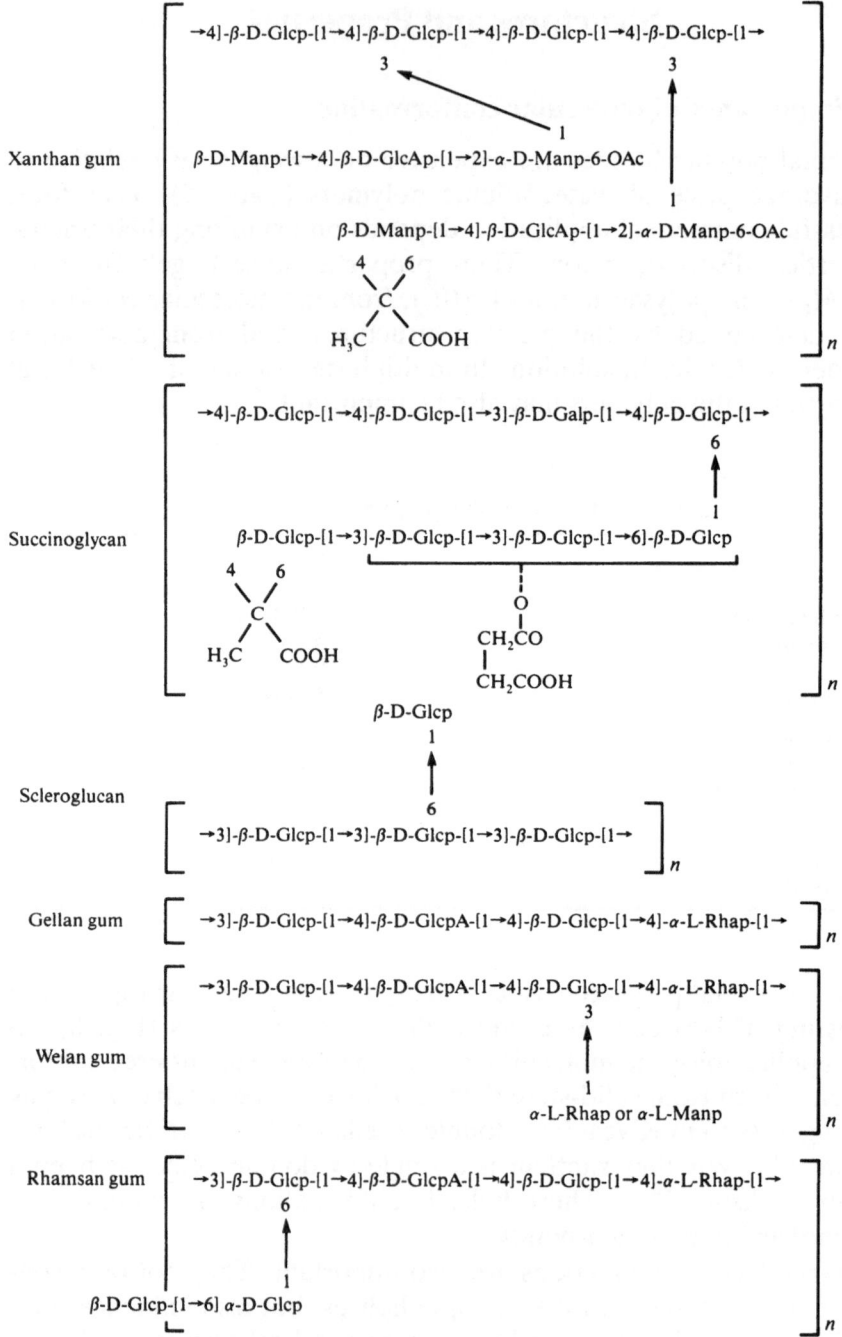

Fig. 6 Chemical structures of exopolysaccharides of potential industrial interest.

example, cellulose), whereas the introduction of the three-sugar-unit sidechain into the cellulose backbone gives the soluble xanthan molecule. These interactions may be 'finely tuned' by small changes in the structure of the side chains[64]. These changes may be brought about by choice of microbial strain, fermentation conditions, or by postfermentation chemical or enzymatic modification of the polysaccharide[65]. For example, deacetylation of native gellan gives a product which has improved crystallinity in the solid state and which forms much stiffer gels in solution[54,64]. The ability to adjust gel characteristics is important in food applications, whereas in enhanced oil recovery – a major potential use of xanthan – the presence of even traces of gel-like agglomerates in solution must be avoided (see later).

The helical conformation imparts rigidity to the molecule so that these giant polysaccharides are described as worm-like or semirigid rods. The persistence lengths of xanthan and gellan were found to be 35 nm and 72 nm, respectively[66-68]. These worm-like molecules sweep out large volumes in solution, in comparison to more flexible and compact molecules such as hydroxyethylcellulose, and these volumes overlap at relatively low concentrations. Consequently, the viscosities of solutions are relatively high at quite low concentrations (Fig. 7). Typically, microbial polysaccharides are used at concentrations from a few 100 p.p.m., for example, in enhanced oil-recovery operations, to a few 1000 p.p.m. in the majority of applications.

Rheology of solutions

This chapter is concerned mainly with microbial polysaccharides used either as viscosifiers or as gelling agents. Xanthan, succinoglycan, rhamsan, welan and scleroglucan are primarily viscosifiers, although they may be gelled by the addition of other components. Gellan is specifically a gelling agent. The following discussion relates to viscous, non-gelled solutions. Gels are discussed later.

When solutions of microbial polysaccharides are sheared, the molecules align in the shear field and the effective viscosity is reduced (Fig. 8). When the shear rate is decreased, the viscosity recovers immediately so that the reduction in viscosity is not a consequence of degradation (except when shear rates exceed 10^5/s). This strong pseudoplastic behaviour is an important property that distinguishes solutions of the microbial viscosifiers from solutions of other thickeners. Thus the solution viscosity must be defined as a function of shear

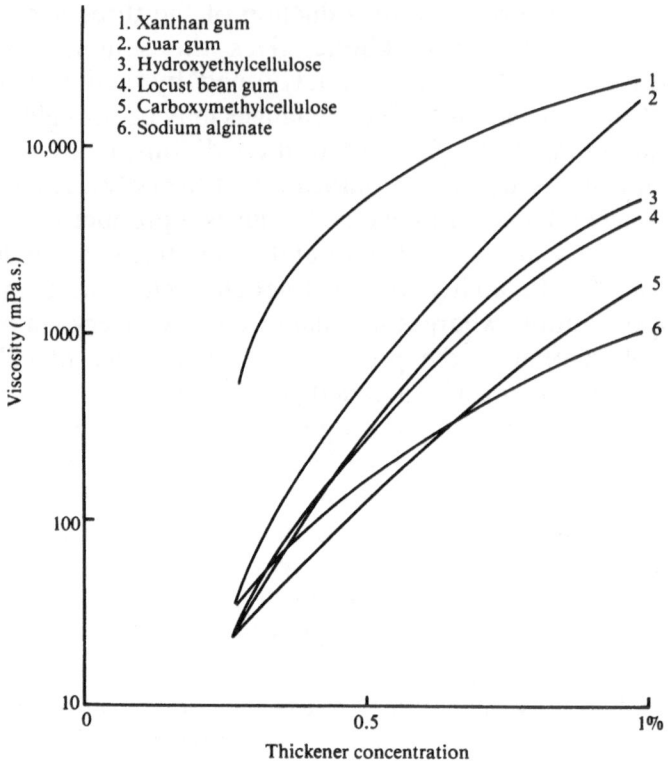

Fig. 7 Viscosity at varying concentrations of different thickeners.

rate. In practical terms, the pseudoplastic rheology means that high viscosity can be achieved at low shear rates, for example, to give good cling of a film or to inhibit particle sedimentation (Fig. 9), while maintaining good pumping or spraying characteristics of the viscosified system (conditions of high shear rate).

Dilute solutions of polysaccharides that have not been gelled will flow continuously when a very low stress is applied. The rate of flow may be very low in very viscous solutions, but there is no yield stress, that is, a stress that must be exceeded for flow to commence. However, some specification test methods that include an extrapolation to zero shear from measurements of stress at finite strain do give an 'apparent' yield. Although the word 'thixotropic' has been used loosely to describe solutions of polysaccharides, solutions actually show little true thixotropy, that is, delayed recovery of viscosity on

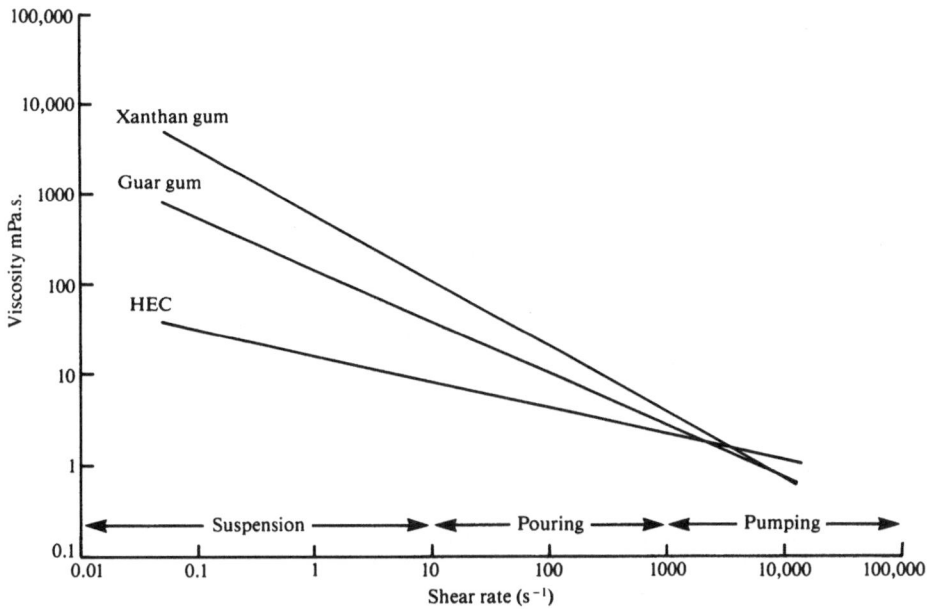

Fig. 8 Viscosity versus shear rate profile. Polymer concentration, $2.5 \, \mathrm{kg \, m^{-3}}$.

reducing shear rate. Recovery is immediate. These characteristics may be modified in concentrated solutions, or by the presence of dispersed solids or by cross-linking. Solutions of xanthan show little visco-elasticity, that is, they have a low tendency to 'stringiness'.

Temperature effects and stability

On heating solutions of the polysaccharides, a temperature is reached when the ordered conformation 'melts' to a disordered state. This critical temperature, T_m, depends on the detailed structure of the polysaccharide[69], the concentration and nature of ions in solution, pH, and nature and concentration of any miscible solvent that is present. At low ionic strengths, T_m increases linearly with the logarithm of ion activity[53,70,71]. High concentrations of salts or miscible solvents ($>0.5 \, \mathrm{mol \, dm^{-3}}$) will increase or decrease T_m, depending on the influence they have on hydrogen-bonding and the water structure[71]. In moderate to hard tap waters and in sea water, the values of T_m for commercial xanthan, succinoglycan, scleroglucan and welan are approximately 120°C, 70°C, 150°C and 150°C, respectively. The value

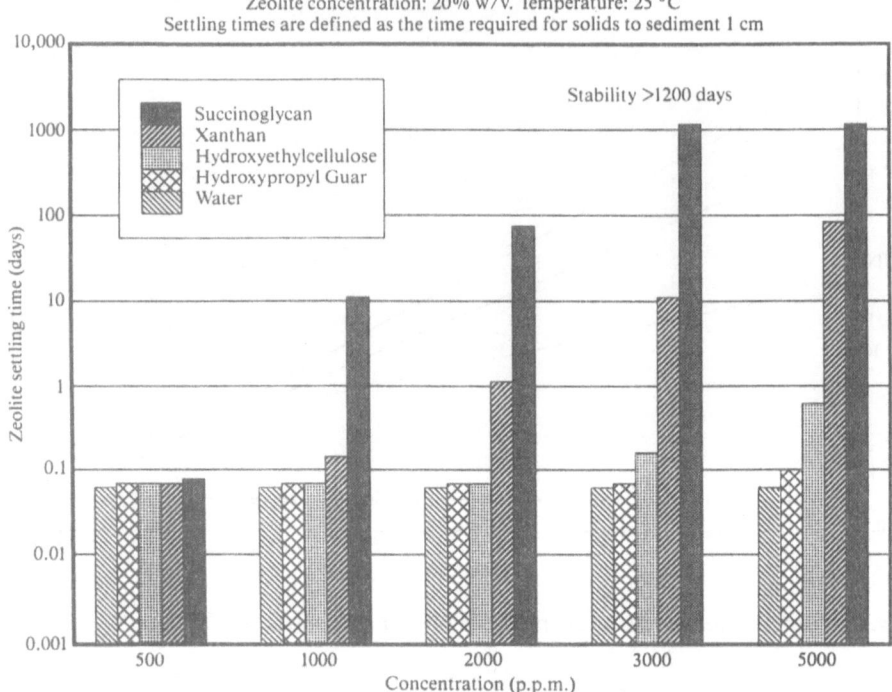

Fig. 9 Sedimentation of zeolite powder suspended in water with added viscosifiers.

of T_m for xanthan may be adjusted from ambient to over 200°C by the addition of appropriate salts[71,72].

Below T_m, the dependence of visosity on temperature is much weaker for solutions of microbial polysaccharides than for, say, hydroxyethyl-cellulose (Fig. 10). However, in their disordered conformations, that is above T_m, the solution behaviour of the polysaccharides is quite different. Above T_m, the more flexible disordered conformations confer a lower and less pseudoplastic viscosity to dilute solutions, while at concentrations exceeding a few per cent, the viscosity at moderate shear rates may actually be increased in comparison to the same solutions at temperatures below T_m. Indeed, for succinoglycan, the viscosity at normal use dosages essentially collapses above T_m (Fig. 10), a phenomenon of considerable value in some oilfield applications. On cooling solutions to below T_m, the return to ordered conformations leads to a variety of behaviours ranging from a return to the rheology observed before heating, to a partial restoration of viscosity, or gelling.

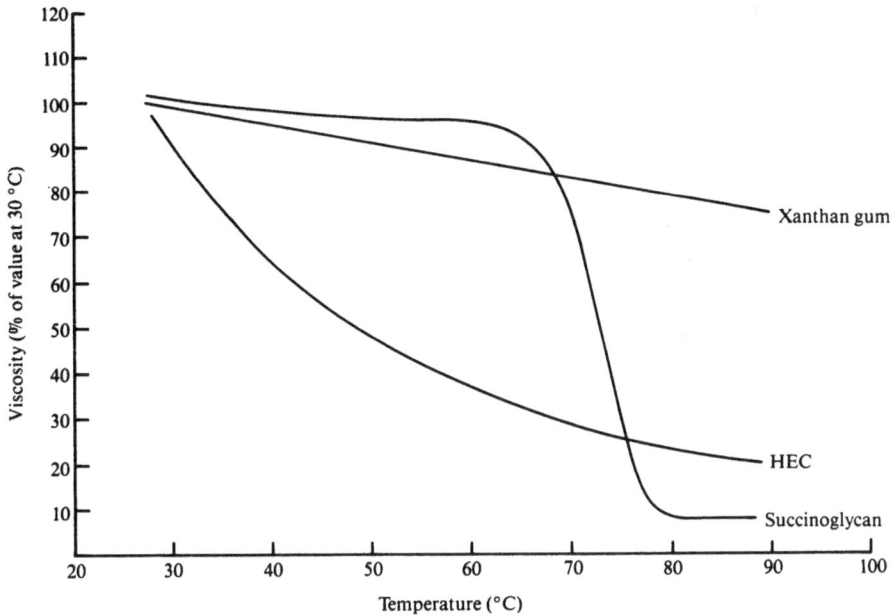

Fig. 10 The effect of temperature on viscosity.

Conformation also impacts dramatically on the resistance of the molecules to degradation by acid or alkali[71], free radicals[73], hydrolytic enzymes[74], and shear forces. Below T_m, the rate of chemical degradation of the microbial polysaccharides is approximately two orders of magnitude slower than for the disordered states, which behave much as hydroxyethylcellulose. Succinoglycan (Fig. 11) and rhamsan[75] are more resistant than xanthan to acid hydrolysis. The enhanced stability of the microbial polysaccharides in their ordered conformations probably arises from the restricted freedom of movement of the glycosidic bond in the backbones, thus raising the activation energy, and/or from the restricted access of enzymes, free radicals or protons to the backbone which is shielded from the side chains. Furthermore, it is possible that the integrity of multiple-helix states can be maintained until bonds in all of the participating molecules are broken in close proximity.

Adsorption, flocculation and dispersion

Microbial polysaccharides are used not only to modify the bulk rheological behaviour of products, such as cleaners, but also to give

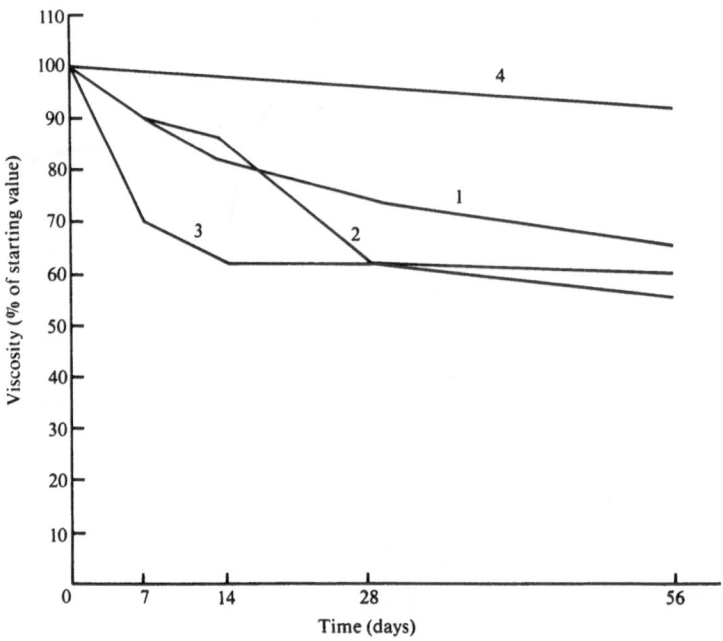

Fig. 11 Viscosity stability of an acid biopolymer solution. Temperature 20 °C. Xanthan gum in (1) 15% formic acid, pH 1.3; (2) citric acid 2%, pH 2; (3) phosphoric acid 5%, pH 1; and (4) succinoglycan in all acids.

'stability' to multiphase systems containing solids or droplets. The high viscosity of polysaccharide-containing systems greatly hinders gravitational settling of solids or creaming of emulsions, and reduces the rate of droplet coalescence in emulsions. When the particle or droplet size is small enough for the surface area to be significant in comparison to the volume, then interfacial effects become important. The microbial polysaccharides adsorb only weakly in comparison to some other water-soluble polymers such as some polyacrylamides. The addition of polysaccharides to disperse systems can lead to either flocculation or dispersion. Flocculation may be weak so that the particles or droplets remain suspended in a gel-like structure, or occasionally stronger, leading to aggregation and sedimentation of solids. These phenomena are strongly dependent on concentration.

Emulsan is a microbial polysaccharide which is particularly active at the oil/water interface where it is believed to form a 'skin' that inhibits droplet coalescence[76]. Alone it does not effectively lower interfacial tension and so it is used together with a conventional surfactant blend to aid emulsion formation. As little as 500 p.p.m. of emulsan can stabilize an emulsion of crude oil[77].

Gel formation

Gels, as distinct from viscous solutions that flow readily, are important for food applications and for some personal care products, but have few other uses so far. A potential application is the plugging of 'thief zones' in oil reservoirs. These are very permeable streaks in the reservoir that by-pass the less permeable oil-containing zones and so decrease the efficiency of water injection (see the section below on applications of xanthan gum).

Of the microbial polysaccharides, only xanthan has essentially worldwide approval as a food additive. Gellan has approval in Japan and restricted approval in the USA (see later). Xanthan forms cohesive, thermo-reversible gels when mixed with locust bean gum, provided that the gum concentration exceeds 0.3% (w/v)[8]. Gellan forms gels on cooling a warm solution to which cations such as calcium have been added. The mechanism of gellation is believed to involve end-to-end association of gellan molecules through double-helix formation[78]. Gel characteristics can be tailored by choice of concentration of polymer and ions[79-81]. Agar, gelatin and kappa-carrageenan are competitive products in foods, but need to be used at higher concentrations.

Polysaccharides form gels in the presence of sufficient transition metal cations under appropriate conditions of pH and redox, generally those conditions which promote the formation of the metal hydroxides. Polysaccharides which contain accessible *cis*-hydroxyl groups may also be gelled by borate anions. Even the low borate content of sea water is sufficient to impair the injectivity into reservoirs of some types of xanthan through the formation of micro-gels.

Specific properties of individual microbial polysaccharides and applications which exploit these characteristics are discussed below.

Xanthan Gum

Current status

Xanthan dominates the microbial polysaccharide market, estimated to be around 10,000 tonnes per annum worldwide. The food industry consumes approximately 60%; oil production, 10% and miscellaneous industrial applications/consumer products, 30%. Microbial polysaccharides are used in a wide variety of applications, but toothpaste, crop-protection formulations and textile/carpet printing probably account for about half of the consumption in this latter category. These are modest consumptions when compared with several other water-soluble polymers (for example, cellulosics, 140,000 tonnes per annum; polyvinyl alcohol, 50,000 tonnes per annum; both figures for Western Europe alone). Nevertheless, a trend towards the use of 'natural products' and replacement of solvent-based systems by water-based systems in several markets for environmental reasons, favour the growth of polysaccharide consumption.

Xanthan is more expensive than many other water-soluble polymers on the basis of active matter (Table 5). Nevertheless, the high activity at low concentrations and some unique technical properties enable xanthan to command a viable price in many applications.

The microbial polysaccharides are sold either as powders or as concentrated aqueous solutions containing about 8% (w/v) active matter. Although the 'liquid concentrates' occupy greater storage space and cost more to transport (per unit of active matter) than do powders, they can offer several advantages. They can be pumped and metered (a consequence of their high pseudoplasticity), the formation of lumps (so-called fisheyes) during solution make-up is eliminated and hydration is more rapid. Further, the viscosity achievable at a given concentration of active matter is generally higher for concentrates than for powders, a possible consequence of product degradation during the precipitation, drying and milling processes. In several applications it is preferred to make up a concentrated solution of biopolymer which is then added to the final formulation. Starting with powder, the maximum concentration of a solution that can be pumped is about 4% (w/v), half of that available in the 'liquid concentrates' produced from the fermentation broth directly by ultrafiltration. Despite this balance of advantages and disadvantages, it is the traditional powder product that is mainly used. Liquid product is often preferred for enhanced oil-recovery operations in order to achieve satisfactory injectivity.

Applications

FOOD

Xanthan gum is used in a wide variety of foods for several functions[82–86]. In salad dressings, xanthan stabilizes the emulsion as well as providing appealing flow and cling characteristics. In bakery products, xanthan is a foam stabilizer for doughs and a gelling agent for fillings where it inhibits the absorption of water from the filling by the pastry. In fruit juices, low concentrations of xanthan keep the pulp suspended. It is used in dairy products such as cottage cheese, yogourt or cheese dips, as a stabilizer, and in ice cream, sorbets and other frozen products to modify ice crystal structure and mouth feel[87]. In soups and sauces, xanthan has replaced traditional gums such as tragacanth as a thickener. Gels of xanthan with locust bean gum are said to give better texture and flavour release than starch gels in products such as pie fillings, jellies, aspics and patés. The consistent supply and quality of xanthan compared with plant gums are factors that favour the use of xanthan, in addition to any performance advantages. Often these accumulated advantages outweigh any additional cost of using xanthan rather than a cheaper plant gum. The synergistic rheology between xanthan gum and guar gum is frequently exploited to minimize costs[88].

OILFIELD APPLICATIONS

Xanthan is used frequently to viscosify water-based drilling fluids, either in combination with the traditional thickener, bentonite clay, or alone in clear muds. Good suspension of the cuttings is achieved with a low drop in pressure across the tubing.

Xanthan was extensively tested for enhanced oil-recovery operations during the late 1970s and 1980s. In this application the polymer is dissolved in the water injected into the reservoir to flush out the oil more rapidly. Often the viscosity of the oil is greater than that of water. Consequently the water tends to 'finger' through the oil to the production well, leaving much oil trapped behind. By increasing the viscosity of the water to match that of the oil, a more 'piston-like' displacement of the oil can be achieved. Strictly, it is the mobilities of the oil and water (parameters that include reservoir characteristics) that must be made comparable.

The injected, viscosified water must be highly shear-thinning so that the pressure-drop close to the wellbore is minimized (region of high shear), while the effective viscosity at a distance from the wellbore, where the oil displacement takes place, is maximized (region of low shear). Xanthan, scleroglucan, succinoglycan and polyacrylamides are candidates for this application. Polyacrylamide is more economic at low salt concentrations (0.5 to 1% sodium chloride, dependent on reservoir conditions), but is unacceptable in many injection waters (for example, sea water) or connate brines. Reservoirs with salty *in situ* water have been preflushed with sweet water, followed by a polyacrylamide flood, but with variable results, since some mixing of the *in situ* and injected waters invariably occurs. Long-term limitations of the use of fresh water, and the increasing requirement to re-inject produced-waters for environmental reasons, favour the future use of microbial polysaccharides. Injected solutions need to retain viscosity for a few years at reservoir temperatures which range from ambient to over 150°C (for example, the North Sea). Under simulated conditions in laboratory tests, and with optimized solution preparation conditions which include oxygen and free radical scavenging, it has been shown that xanthan could be used at temperatures up to 95°C (ref. 89).

Several field trials confirmed the technical feasibility of this approach, although several problems of the implementation were highlighted, including microbial and chemical degradation[90], high retention and poor reservoir characterization leading to inefficient use of the polymer. For this application, the liquid concentrate forms of the product are preferred, although broths and powders have been used in trials. Broths suffer from high transportation costs but are easily mixed and injected. The concept of well-head production has been proposed for projects that can justify such dedicated investment. There are difficulties in dissolving powders adequately, especially into the more saline waters, in order to make up solutions that can be injected into the fine pores of the rock without plugging them. Liquid concentrates would seem to strike the right balance of properties. The economics of this enhanced oil-recovery process are marginal at the time of writing, but many would consider that these processes will be the major consumer of xanthan in the next century.

The use of xanthan gels formed by crosslinking, usually with chromium ions, to plug very permeable streaks (thief zones) in reservoirs has also been tested in the field, with some success[91]. It is difficult to place the gel plug in the right position in a reservoir, and even when that is achieved, the result is sometimes of marginal benefit because of the complicated pattern of oil and water flow underground. The temperature limitation for the use of xanthan in this application is about 80°C.

OTHER INDUSTRIAL APPLICATIONS

A large proportion of usage in the general industrial sector is highly fragmented. A list of some of the more important applications is given in Table 6. Many more 'potential' applications have been proposed in reviews or patented, but many of these are not commercial realities. In most real applications the contributions of the polymer to the overall cost is relatively small, and the product is incorporated to achieve better performance for the consumer, even at higher cost. Such performance advantages can be exploited to good effect in the market place.

Table 6 General industrial applications of xanthan (excluding oilfield)

Application	Function	Specific references
Crop-protection formulations	Suspension	109
Suspension fertilizers (USA)	Suspension	
Carpet/textile printing	Cling, control penetration	110, 111, 112
Liquid feed supplements	Suspension	
Paints	Rheology, anti-syneresis	113
Liquid cleaners	Rheology, cling, suspension	
Ceramics	Suspension	114
Other minerals, zeolites	Suspension	
Gel explosives	Gelling	

Toxicology and safety

Short-term feeding trials with dogs and rats produced no toxic effects. Histopathological examination of rats and dogs subjected to long-term oral administration of xanthan at up to 1g/kg body weight indicated no adverse effects. Xanthan is not a sensitizing chemical nor is it a skin or eye irritant[92,93].

Xanthan is registered under the US Food and Drug Administration regulation 21 CFR 172.695 as a stabilizer and thickener. It is also listed in Annex 1 of the EEC Directive 80/597/CEE as a permitted thickening and gelling agent (food additive number E415).

The Newer Microbial Polysaccharides

Succinoglycan

While succinoglycan shares many of the characteristics of xanthan, it has some distinctive properties that make it particularly useful for some applications. First, the relatively low transition temperature of the molecule can be exploited in oilfield applications. Second, the rate of degradation by acid hydrolysis is lower. Third, it has demonstrated a lower tendency to flocculate clays and other fine solids. Fourth, there is no coproduction of cellulase during the fermentation process. However, some grades of xanthan have been processed to eliminate cellulase. Finally, succinoglycan is a more powerful viscosifier than xanthan.

The greatest application of succinoglycan is for workover and completion operations in the oilfield where use is made of the transition temperature[94]. In these operations, a viscous fluid is pumped down a well and comes into contact with the oil- or gas-producing zone. Some of the viscous fluid inevitably leaks into the formation and tends to impair the flow of oil or gas after the well is brought back into production. When using succinoglycan, the workover or completion fluid is used at a temperature below the transition temperature, T_m of the polymer. However, once the operation is completed, the temperature at the bottom of the well rises towards the reservoir temperature, which is above T_m. Any fluid which has leaked into the reservoir loses viscosity, and impairment of the formation is minimized. The value of T_m can be adjusted by dissolving salts into the brine[71].

In general industrial applications, succinoglycan is particularly suitable for promoting the suspension of fine particles (for example, zeolites, Fig. 9), for stabilizing emulsions and for acid cleaners (Fig. 11). Succinoglycan is not approved for food use.

Rhamsan

Although rhamsan, welan and gellan have related structures, solutions of these polymers show very different behaviours. Rhamsan is a better suspending agent than xanthan, partly because it is the more powerful viscosifier, but perhaps also because solutions exhibit a greater elastic modulus[95]. Current applications therefore include ceramic glazes, crop protection chemicals and emulsions. Rhamsan also is more stable in dilute acids than xanthan[75] and so it is used to thicken sanitary-ware cleaners, in order to improve cling. It is also used to suspend fertilizers and slurry explosives, since rhamsan is more tolerant than xanthan to high concentrations of some salts[96]. Solutions of rhamsan undergo a steep drop in viscosity above about 100 °C.

Welan

The rheological characteristics of solutions of welan resemble those of xanthan. An advantage claimed for welan is improved stability at temperatures over 100 °C (ref. 95). This is an increasingly important temperature range for drilling fluids, and current clay-based or alternative polymer-based fluids are inadequate. Potential competition to welan for this application comes from xanthan/formate fluids and from scleroglucan. A further potential application of welan is in mortars, grouts and concrete, as a replacement for methylcellulose[75]. The greater degree of pseudoplasticity that can be achieved is expected to improve workability and shape retention, and to prevent water separation during setting. Solutions of welan tend to form cohesive gels in the presence of high concentrations of salts, unlike solutions of xanthan, succinoglycan or rhamsan, which remain free flowing.

Gellan

The backbone of the gellan molecule resembles those of rhamsan and welan, but it is not substituted by sugar side groups (Fig. 5). The commercial product, Kelcogel, is made by the deacylation of native gellan which contains acetate and glycerate residues[95,97]. This treatment leads to stiffer gels so that the product is likened to carrageenan in some respects, but with improved acid stability[96]. Solutions of commercial gellan form brittle gels when cations, particularly divalent cations, are added, and like agar, gels of gellan show melting-setting point hysteresis. Gel hardness and modulus can be increased by increasing the concentrations of polymer and/or cations. Divalent cations also inhibit hydration so that a sequestering agent may be required to promote dissolution.

A clarified grade of gellan (Gelrite) is marketed as a substitute for agar in culture media for microbes and plant tissue. Improved clarity and lower use-levels (35–50%) compared to agar are claimed advantages. The most interesting market for this product will be food. Initial FDA approval for the use of gellan as an additive in low-solids jams and jellies, glazes, frostings and icings, has been granted. Full FDA approval in a wide range of foods is expected. Gellan has approval in Japan and is under consideration in the EEC. Mixtures of gellan with other hydrocolloids can be tailored to give a range of textures, and partial replacement of starch by gellan leads to better flavour release[97].

Scleroglucan

Scleroglucan is a nonionic polysaccharide made by the fermentation of fungi of the genus *Sclerotium*. The backbone consists of β-D-glucopyranosyl monomers linked (1–3) and approximately every third sugar is substituted with a further glucose monomer linked β(1–6). The molecules form triple helices. Probably as a consequence of this conformation, scleroglucan shows good resistance to degradation at elevated temperatures. When solutions of scleroglucan are cooled, a weak gel is formed[74]. It has been examined extensively as a candidate for enhanced oil recovery, but suffers some drawbacks. Firstly, it is difficult to dissolve from the powder, and the only liquid form available has been a dilute solution. Secondly, adsorption onto the rock is high.

A large amount of mycelial biomass is produced with the polymer during the fermentation, and cleaning up the polymer to a quality suitable for injection into reservoirs is difficult. Usually the broth is sheared to separate the mycelia and polymer, diluted and then filtered to remove the mycelia. Much energy is then required to remove the large amount of water added to dilute the broth. The good stability at elevated temperatures suggests that the polymer should be suitable for drilling fluids at high temperatures. Also, scleroglucan is nonionic, unlike the other microbial polysaccharide viscosifiers, and thus is compatible with cationic surfactants, found for example in bitumen emulsions. Scleroglucan has been available for evaluation for almost as long as xanthan, yet it does not appear to have been a commercial success, despite these potential markets. This is probably due to the difficulties of manufacturing a suitable form of the product at an acceptable cost.

Other viscosifying or gelling polysaccharides

A considerable number of microbial polysaccharides have been researched and a few produced in pilot or semi-commercial quantities. These include curdlan[98] which gels on heating, the microbial alginates developed by Tate and Lyle, several biopolymers researched by Kelco[8,99,115], and polysaccharides from methylotrophs. The commercial success rate has been low, despite considerable technical achievements. There are several reasons for this. Xanthan has become a well-established product, available from several manufacturers at a competitive price which, in part, arises from the scale of production. A new type of polysaccharide has to overcome several hurdles: high manufacturing costs arising from low production volumes and usually a poor productivity in comparison with xanthan; customers are reluctant to put themselves into the hands of a single supplier of a unique product; customers feel more comfortable with a product with a good track record; some products formulated with xanthan are registered, for example, crop-protection agents, and reregistration with a new polysaccharide would be expensive; customers formulating several products prefer to have just one additive rather than a range from which to choose, and xanthan is the obvious choice; and to seek food approval for a new product is extremely expensive and returns may not justify the costs.

Consequently, if a new polymer is to have any chance of success it must have a significant performance advantage over xanthan, or indeed any other nonmicrobial thickener. Those polymers discussed above do have some advantages for specific applications. Whether they can be a commercial success depends on whether the market for those specific applications is sufficiently attractive in terms of both price and volume.

Fibre- and film-forming polysaccharides

Many polysaccharides will form films or fibres, either by solvent precipitation or by drying. The physical characteristics can be modified by the addition of plasticizers, and water solubility eliminated by crosslinking. However, such films may swell and lose strength in water. The use of microbial polysaccharides for biodegradable packaging has been advocated, but the cost compared to alternative materials or to recycling conventional plastics is high.

Pullulan in particular has been proposed for packaging[10]. It is a linear polymer comprising maltotriose units in $\alpha(1-6)$ linkage, produced by the fungus *Aureobasidium pullulans*. Approximately 10 tonnes per annum is produced in Japan and it is converted into transparent films with low oxygen permeability. It is, however, fully water-soluble.

Bacterial cellulose is produced in the form of microfibrils by *Acetobacter* sp. and other bacteria under appropriate conditions of growth. Cellulose from *A. xylinum* is claimed to be of higher M_r than plant-derived cellulose, so that paper made from these fibres is stronger. Sony of Japan are using such paper to make the speaker cones for some headphones. Bacterial cellulose is also proposed as a thickener for niche cosmetic applications[101]. Certainly, it should have distinctive performance advantages in order to compete with other materials.

Microbial polysaccharides with useful surface-active properties

By adsorbing at interfaces, polymers can induce such phenomena as flocculation, dispersion, emulsion stabilisation or coalescence, enhanced flotation, and foam stabilization. All polymers, including xanthan, can influence these properties of multi-phase systems to a greater or lesser extent, but some microbial polysaccharides have been or are being developed specifically for these effects.

The emulsans are exocellular lipopolysaccharides produced by *Acinetobacter* sp.[102] and possess emulsion-stabilizing properties[76]. They also flocculate clays. They contain D-galactosamine and are *N*- and *O*-acylated with long-chain fatty acids. Just 500 p.p.m. in the aqueous phase can stabilize an emulsion for several months. The emulsans are used in conjunction with low-M_r surfactants which lower the interfacial tension and so promote emulsion formation. The role of the emulsan is to hinder droplet coalescence once the emulsion is formed. The product is currently formulated into cleaning fluids for oil tanks, and has been tested for stabilizing pumpable emulsions of heavy crude oils which, by themselves, can be handled only by heating in order to reduce viscosity. The effectiveness of the emulsan for this application seems to be strongly dependent on the nature of the crude oil.

The same research group that pioneered emulsans has also developed Biodispersans, microbial polysaccharides that act as dispersants for finely-divided minerals in water. They are excreted also by strains of *Acinetobacter calcoaceticus*[103]. Like many synthetic polymeric dispersants, Biodispersans will also flocculate minerals when used at lower concentrations. Another polysaccharide produced by *Zoogloea ramigera* is well known to promote flocculation and settling of the sludge in aerobic biotreatment systems[104]. Indeed, microbial polysaccharides play a key role in cell adhesion in a range of environments[105]. Chitosan, usually obtained by chemical treatment of insoluble chitin from the shells of crabs, is also proposed for use as a flocculant, since it has a slight cationic character. It can be produced from many fungi, for example, some *Mucor* sp. contain 27% chitosan. The authors are not aware of any commercial production of these products. Any new flocculant or dispersant would have to compete with established and effective, but petrochemical-based, products such as the polyacrylamides. They may be able to compete on the basis of environmental impact, since the synthetics are usually poorly biodegradable, and the presence of residual monomers in synthetics can be a cause for concern for the treatment of potable waters.

Microbial polysaccharides with other specific features

Many of the bacteria that produce succinoglycan co-produce cyclosophorans containing upwards of 17 glucose monomers in $\beta(1-2)$ linkage[106,107]. These molecules can form inclusion complexes with appropriate guests, in a similar fashion to the cyclodextrins. Whereas the cyclodextrins are produced commercially by the enzymatic treatment of starch, and their complexing properties have been extensively studied, the microbial production of cyclosophorans is not yet commercial and investigation into their properties has only recently begun.

Oligosaccharides play a vital role in recognition processes in animals, for example, in antigenic interactions, during growth and differentiation of cells, in malignancy, and for egg–sperm fusion. Oligosaccharides are also involved in the defence mechanisms and growth processes of plants[108].

Carbohydrates are very suitable 'words' for this chemical language, since many more isomers can be built from a limited number of 'letters' (monosaccharides) than is possible with amino acids or nucleotides. The synthetic pathways for these oligosaccharides are being cloned into bacteria, either to facilitate synthesis or to aid understanding.

Several polysaccharides, including scleroglucan, curdlan, mannans from yeast cell walls, and products of their hydrolysis, have shown anticancer activity[9]. Branched β-1,3-glucans are claimed to activate cellular immunity[1].

Conclusions

The commercial introduction of xanthan in the early 1960s provided a new rheological tool, distinctly different from the traditional thickening agents. The potential of xanthan was soon recognized in industries as diverse as food, explosives and oil recovery. The attractive growth of the xanthan market and the enormous potential use (pot of gold?) in enhanced oil-recovery applications provoked strong interest in both reducing manufacturing costs and in screening for alternative microbial polysaccharides. This was fertile territory for academic study too – a complex structure to be unravelled, fascinating rheology, and new process-engineering challenges. But the optimism of the past decades has hardly been realized. Manufacturing methods remain much as they were at the outset (an attempt to introduce continuous fermentation by Tate and Lyle/Hercules failed commercially), and xanthan remains by far the dominant player, despite several attempts to introduce new polymers. Why?

Xanthan is now a 'commodity' molecule, produced and sold on a large scale. The attractive food market provided xanthan with an important spring-board from which new applications were developed. Its price reflects both this scale of operation and a world over-capacity arising from the entry of several new players. These were attracted either by the market prospects or by the opportunity to fill fermenter capacity that had become under-used through improvements is antibiotic titres or through a temporary downturn in the citric-acid market. A new polymer is expensive to develop and commercialize, especially if food approval is required. Small-scale production implies high costs. Also *X. campestris* is an accommodating process organism compared to many others subsequently investigated. Thus a new polymer must find a very attractive market in order to compete against xanthan. Many potential markets are too small and fragmented to justify the investment, high technical service and market development that are required. The very success of xanthan has been a deterrent against the introduction of new microbial polysaccharides.

But xanthan does have its limitations – stability at the extremes of pH and temperature, compatability with cationic surfactants, for example – and new functionality is also required. Undoubtedly some new polymers will succeed. Gellan (if food approval is given), succino-glycan and rhamsan are candidates making good running. Perhaps the genetically engineered 'polytrimer' should also be included, if productivity can be improved, since this polymer should be compatible with cationics. But maybe the new pot of gold lies outside thickening and gelling.

Biodegradable plastics? Microbial polysaccharides may find niche opportunities for medical applications, but more general application, for example into packaging, is unlikely. Alternative, cheaper solutions to the environmental problems are on the horizon. These include the use of starch or cellulose fillers, and recycling, or simply improved disposal.

Perhaps the future for microbial polysaccharides lies in exploiting more fully the specificity of their interactions with animate and inanimate surfaces and other molecules, and the greater use of their information content. In the industrial context, processes such as mineral flotation, chiral separations, crystal growth modification, selective ion binding, and demand molecules that can interact specifically. In plants and animals the role of poly- and oligosaccharides in intercellular communication is under intense study. The addition of molecular modelling into the melting pot of the physical, chemical and biological sciences and engineering that lie behind the current products will surely open up new prospects for these fascinating and challenging molecules.

References

1. Lohman, D. (1990) Structural diversity and functional versatility of polysaccharides. In *Novel Biodegradable Microbial Polymers* (ed. Dawes, E.A.), pp. 333–348. Kluwer Dordrecht.

2. Eveleigh, D.E. (1973) Microbial monosaccharides and polysaccharides. In *Handbook of Microbiology* (eds Laskin, A.I. & Lechevalier, H.A.), pp. 89–96. CRC Press, Cleveland.

3. Sutherland, I.W. & Ellwood, D.C. (1979) Microbial exopolysaccharides—industrial polymers of current and future potential. In *Microbial Technology: Current State, Future Prospects* (eds Bull, A.T., Ellwood, D. C. & Ratledge, C.), pp. 107–150. Cambridge University Press, Cambridge.

4. Sutherland, I.W. (1982) Biosynthesis of microbial exopolysaccharides. *Adv. Microb. Physiol.* 23: 79–150.

5. Sutherland, I.W. (1990) Properties and potential of microbial exopolysaccharides. *Chimicaoggi* (Sept.), pp. 9–14.

Biomaterials

6. Kang, K.S. & Cottrell, I.W. (1979) Polysaccharides. In *Microbial Technology* (eds Peppler, H.J. & Perlman, D.), Vol. I, p. 197. Academic Press.

7. Baird, J.K., Sandford, P.A. & Cottrell, I.W. (1983) Industrial applications of some new polysaccharides. *Bio/Technology*, pp. 778–783.

8. Sandford, P.A. & Baird, J. (1983) Industrial utilization of polysaccharides. In *The Polysaccharides* (ed. Aspinall, G.O.), Vol. 2, pp. 411–490. Academic, London/New York.

9. Yalpani, M. & Sandford, P.A. (1987) Commercial polysaccharides: Recent trends and developments. *Industrial Polysaccharides. Progress in Biotechnology* (ed. Yalpani, M.), Vol. 3, pp. 311–335. Elsevier, Amsterdam.

10. Paul, F., Morin, A. & Monsan, P. (1986) Microbial polysaccharides with actual potential industrial applications. *Biotech. Adv.* **4**: 245–259.

11. Linton, J.D. (1990) The relationship between metabolite production and the growth efficiency of the producing organism. *FEMS Microbiol. Rev.* **75**: 1–18.

12. Rye A. R., Drozd, J. W., Jones, C. W. & Linton, J. D. (1988) Growth efficiency of *Xanthomonas campestris* in continuous culture. *J. Gen. Microbiol.* **134**: 1055–1061.

13. Linton, J.D., Watts, P.D., Austin, R.M., Haugh, D.E. & Niekus H.G.D. (1986) The energetics of microbial exopolysaccharide production from methanol by microorganisms possessing different pathways of C–1 assimilation. *J. Gen. Microbiol.* **132**: 779–788.

14. Linton, J.D., Evans, M.W. & Godley, A.R. (1987) Process for preparing a heteropolysaccharide, heteropolysaccharide obtained thereby, its use, and strain NCIB 11883. US Patent 4, 634, 667.

15. Linton, J.D., Evans, M., Jones, D. S. & Gouldney, D.N. (1987) Exocellular succinoglucan production by *Agrobacterium radiobacter* NCIB. 11883. *J. Gen. Microbiol.* **133**: 2961–2969.

16. Linton, J.D., Jones, D.S. & Woodard, S. (1987) Factors that control the rate of exopolysaccharide production by *Agrobacterium radiobacter. J. Gen. Microbiol.* **133**: 2979–2987.

17. Linton, J.D., Gouldney, D. & Woodard, S. (1988) Efficiency and stability of exopolysaccharide production from different carbon sources by *Erwinia herbicola. J. Gen. Microbiol.* **134**: 1913–1921.

18. Cornish, A., Linton, J.D. & Jones, C.W. (1987) The effect of growth conditions on the respiratory system of a succinoglucan-producing strain of *Agrobacterium radiobacter. J. Gen. Microbiol.* **133**: 2971–2978.

19. Jarman, T.R. & Pace, G.W. (1984) Energy requirements for microbial exopolysaccharide synthesis. *Arch. Microbiol.* **137**: 231–235.

20. Linton, J.D. & Niekus, H.G.D. (1987) The potential of one-carbon compounds as fermentation feedstocks. *Antonie van Leeuw.* **53**: 55–63.

21. Anderson, A.J., Hacking, A.J. & Dawes, E.A. (1987) Alternative pathway for the biosynthesis of alginate from fructose and glucose in *Pseudomonas mendocina* and *Azotobacter vinelandii. J. Gen. Microbiol.* **133**: 1045–1052.

22. Betlack, M.R., Campagne, M.A., Doherty, D.H., Hassler, R.A., Henderson, N.M., Vanderslice, R.W., Marelli, J.D. & Ward, M.B. (1987) Genetically engineered polymers: manipulation of xanthan biosynthesis. In *Industrial Polysaccharides. Progress in Biotechnology* (eds Yalpani, M.), Vol. 3, pp. 35–50. Elsevier, Amsterdam.

23. Vanderslice, R.W., Doherty, D.H., Capage, M.A., Betlack, M.R., Hassler, R.A., Henderson, N.M., Ryan-Graniero, J. & Tecklenburg, M. (1989) Genetic engineering of polysaccharide structure in *Xanthomonas campestris*. In *Biomedical and Biotechnological Advances in Industrial Polysaccharides* (eds Crescenzi, V., Dea, I.C.M., Paoletti, S., Stivala S. & Sutherland, I.W.), pp. 145–156. Gordon and Breach, New York.

256

24. Glazebrook, J. & Walker, G.C. (1989) A novel exopolysaccharide can function in place of the calcafluor-binding exopolysaccharide in the nodulation of Alfalfa by *Rhizobium meliloti*. *Cell*: 661–672.

25. Cornish, A., Greenwood, J.A. & Jones, C.W. (1988) Binding protein dependent glucose transport by *Agrobacterium radiobacter* grown in glucose limited continuous culture. *J. Gen. Microbiol.* **133**: 3099–3110.

26. Cornish, A., Greenwood, J.A. & Jones C.W. (1988) The relationship between glucose transport and the production of succinoglucan exopolysaccharide by *Agrobacterium radiobacter*. *J. Gen. Microbiol.* **134**: 3111–3122.

27. Mian, F.A., Jarman, T.R. & Rhigelato, R.C. (1978) Biosynthesis of polysaccharide by *Pseudomonas aeruginosa. J. Bact.* **134**: 418–422.

28. Rye, A.J. (1989) Growth energetics of *Xanthomonas campestris* in continuous culture. M. Phil. University of Leicester, UK.

29. Osterhuis, N.M.G. & Koerts, K. (1987) Method and reactor vessel for the fermentative preparation of polysaccharides, in particular xanthan. European Patent 0 249 288.

30. Rogovin, S.P. (1969) Continuous process for producing *Xanthomonas* heteropolysaccharide. US Patent 3, 485, 719.

31. Downs J. (1985) Process for preparing *Xanthomonas* heteropolysaccharide, heteropolysaccharides as prepared by the latter process and its use. European Patent 0 130 647.

32. Moraine, R.A. & Rogovin P. (1966) Kinetics of polysaccharide B-1459 fermentation. *Biotechnol. Bioengng.* **8**: 511–524.

33. Cripps, R.E., Ruffell, R.N. & Sturman, A.J. (1981) Fluid displacement with heteropolysaccharide solutions, and the microbial production of heteropolysaccharide. European Patent Application 0 040 445.

34. Kang, K.S. & McNeely, W.H. (1974) Polysaccharide and bacterial fermentation process for its preparation. Canadian Patent 945 091.

35. Hacking, A.J. (1986) In *Economic Aspects of Biotechnology. Cambridge Studies in Biotechnology* (eds Baddiley, J., Carey, N.H., Davidson, J.F., Higgins, I.J. & Potter, W.G.), Vol. 3, p. 111. Cambridge University Press, Cambridge.

36. Hitzman, D.O. (1974) Waterflood bacterial viscosifier. U.S. Patent 3, 801, 502.

37. Gozard, J.P., Melle, J. & Bron, A.C. (1988) Enzymatic treatment of polysaccharide biopolymers. US Patent 4, 775, 632.

38. Kang, K.S. & Burnett, D.B. (1976) Xanthan gum. UK Patent 1 488 645.

39. Colin, P. & Melle, V.G. (1971) Process for treating polysaccharides produced by fermentation. US Patent 3, 591, 578.

40. Colegrove, G.T. (1975) Clarification of Xanthan gum. UK Patent 1 443 507.

41. Rader, W.E. & Wong, J.C. (1981) Clarification of polysaccharide-containing fermentation products. European Patent Application 0 039 962 A1.

42. Nisbet, T.M., Linton, J.D. & Godley, A.R. (1986) Process for the preparation of polysaccharide aqueous solutions. European Patent Application 0 189 959 A2.

43. Drozd, J.W. & Rye A.J. (1988) Method for improving the filterability of a microbial broth. US Patent 4, 729, 958.

44. Smith, I.H. (1984) Precipitation of xanthan gum. European Patent Specification 0 068 706.

45. Hayashibara Seibuts (1975) A method for the production of pure pullulan. UK Patent Specification 1, 443, 918.

46. Ho, L. (1987) Concentration of aqueous pseudoplastic solutions by membrane ultrafiltration. European Patent Specification 0 069 523.

47. Van Lookeren Campagne, C.J. (1982) Treatment of polysaccharide solutions. European Patent Application 0 049 012.

48. Lee, H.L. (1981) Concentrated xanthan gum. US Patent 4, 299, 825.

49. Empey, R.A. & Pettit, D.J. (1973) Gum decontamination. UK Patent specification 1 409 706; European Patent Application 0 049 121 A1.

50. Merquiol, T. & Choplin, L. (1988) Mixing of viscoelastic fermentation broth with helical ribbon-screw (HRS) impeller. In *6th European Conference on Mixing*, pp. 465–472. Pavia, Italy, 24–26 May 1988.

51. Lee, S. S. X. Wang, D.I.C. (1989) A mechanistic study of highly viscous fermentations. Abstract, American Chemical Society, V 198, September, p.13.

52. Maury, L.G. (1982) Production of xanthan gum by emulsion fermentation. US Patent 4, 352, 882.

53. Gravanis, G., Milas, M., Rinaudo, M. & Clarke-Sturman, A. J. (1990) Conformational transition and polyelectrolyte behaviour of a succinoglycan polysaccharide. *Int. J. Biol. Macromol.* **8**: 195–200.

54. Upstill, C., Atkins, E.D.T. & Attwood, P.T. (1986) Helical conformations of gellan gum. *Int. J. Biol. Macromol.* **8**: 275–288.

55. Chandrasekaran, R., Millane, R.P. & Arnott C. (1988) The crystal structure of gellan. *Carbohyd. Res.* **175**: 1–15.

56. Chandrasekaran, R., Millane, R.P., Arnott, S., Atkins, E.D.T., Chen, C.S.H. & Sheppard, E.W. (1988) Conformation and shear stability of xanthan gum in solution. *Polym. Engng. Sci.* **20**: 512–518.

57. Yanaki, T., Kojima, T. & Norisuye, T. (1981) Triple helix of scleroglucan in dilute aqueous hydroxide. *Polym. J.* **13**: 1135–1143.

58. Yanaki, T. & Norisuye, T. (1983) Triple helix and random coil of scleroglucan in dilute solution. *Polym. J.* **15**: 389–396.

59. Moorhouse, R., Walkinshaw, M.D. & Arnott, S. (1977) Xanthan gum – molecular conformation and interactions. In *Extracellular Microbial Polysaccharides*, American Chemical Society Symp. Ser. No. 45 (eds Sandford, P.A. & Laskin, A.I.), pp. 90–102.

60. Isaac, D.H. (1985) Bacterial polysaccharides. In *Polysaccharides: Topics in Structure and Morphology* (ed. Atkins, E.D.T.). Macmillan, London.

61. Stokke, B.T., Elgsaeter, A. & Smidsrod, O. (1986) Electron microscopic study of single and double-stranded xanthan. *Int. J. Biol. Macromol.* **8**: 217–225.

62. Stokke, B.T., Smidsrod, O. & Elgsaeter, A. (1989) Electron microscopy of native xanthan and xanthan exposed to low ionic strength. *Biopolymers* **28**: 617–637.

63. Milas, M., Rinaudo, M., Tinland, B. & de Murcia, G. (1988) *Polym. Bull.* **19**: 567.

64. Atkins, E. (1990) Structure of microbial polysaccharides using X-ray diffraction. In *Novel Biodegradable Microbial Polymers* (ed. Dawes, E.A.), pp. 371–386, Kluwer.

65. Dea, I.C.M. (1987) The role of structural modification in controlling polysaccharide functionality. In *Industrial Polysaccharides. Progress in Biotechnology* (ed. Yalpani, M.), Vol. 3, pp. 207–216, Elsevier, Amsterdam.

66. Tinland, B., Maret, G. & Rinaudo, M. (1990) Reptation in semidilute solutions of wormlike polymers. *Macromolecules* **23**: 596–602.

67. Tinland, B. & Rinaudo, M. (1989) Dependence of the stiffness of the xanthan chain on the external salt concentration. *Macromolecules* **22**: 1863–1865.

68. Milas, M., Shi, X. and Rinaudo, M. (1990) On the physiochemical properties of gellan gum. *Biopolymers* **30**: 451–464.

69. Smith, I.H., Symes, K.C., Lawson, C.J. & Morris, E.R. (1981) Influence of the pyruvate content of xanthan on macromolecular association in solution. *Int. J. Biol. Macromol.* **3**: 129–134.

70. Holzworth, G. (1976) Conformation of the extracellular polysaccharide of

Xanthomonas campestris. Biochemistry **15**: 4333–4339.

71. Clarke-Sturman, A.M., Pedley, J.B. & Sturla, P.L. (1986) Influence of anions on the properties of microbial polysaccharides in solution. *Int. J. Biol. Macromol.* **8**: 355–360.

72. Clarke-Sturman, A.J. & Sturla, P.L. (1990) Aqueous polysaccharide compositions. US Patent 4, 900, 457.

73. Parsons, B.J., Phillips, G.O., Thomas, B., Wedlock, D.J. & Clarke-Sturman, A.J. (1985) Depolymerization of xanthan by iron-catalysed free radical reactions. *Int. J. Biol. Macromol.* **7**: 187–192.

74. Rinaudo, M. & Milas, M. (1987) On the properties of polysaccharides. Relation between chemical structure and physical properties. In *Industrial Polysaccharides. Progress in Biotechnology* (ed. Yalpani, M.), Vol. 3, pp. 217–223, Elsevier, Amsterdam.

75. Lockwood, B. (1989) Xanthan, rhamsan and welan gum: Properties and applications. Paper presented at the Chemspec Europe BACS Symposium.

76. Rosenburg, E., Zuckerberg, A., Rubinovitz, C. & Gutnick, D. L. (1979) Emulsifier of *Arthrobacter* RAG-1: Isolation and emulsifying properties. *Appl. Env. Microbiol.* **37**: 402–408. [Authors' note: renamed *Acinetobacter* RAG-1.]

77. Gutnick, D.L., Rosenberg, E. & Zosim, Z. (1983) Emulsans. US Patent 4, 395, 354.

78. Morris, V.J. (1990) Science, structure and applications of microbial polysaccharides. In *Gums and Stabilisers for the Food Industry* (eds Phillips, G.O. *et al.*), Vol. 5, pp. 315–328. IRL, Oxford.

79. Sanderson, G.R. & Clark, R.C. (1984) Gellan gum, a new gelling polysaccharide. In *Gums and Stabilisers for the Food Industry* (eds Phillips, G.O. *et al.*), Vol. 2, pp. 201–210. Pergamon.

80. Sanderson, G.R., Bell, V.L., Burgum, D.R., Clark, R.C. & Ortega, D. (1988) Gellan gum in combination with other hydrocolloids. In *Gums and Stabilisers for the Food Industry* (eds Phillips, G.O., *et al.*), Vol. 4, pp. 301–308. IRL, Oxford.

81. Sanderson, G.R., Bell, V.L., Clark, R.C. & Ortega, D. (1988) The texture of gellan gum gels. In *Gums and Stabilisers for the Food Industry* (eds Phillips, G.O. *et al.*), Vol. 4, pp. 219–229. IRL, Oxford.

82. Phillips, G.O., Wedlock, D.J. & Williams, P.A. (eds) *Gums and Stabilisers for the Food Industry*. Vol. 1, Pergamon (1982).

83. Phillips, G.O., Wedlock, D.J. & Williams, P.A. (eds) *Gums and Stabilisers for the Food Industry*. Vol. 2, Pergamon (1984).

84. Phillips, G.O., Wedlock, D.J. & Williams, P.A. (eds) *Gums and Stabilisers for the Food Industry*. Vol. 3, Elsevier (1986).

85. Phillips, G.O., Wedlock, D.J. & Williams, P.A. (eds) *Gums and Stabilisers for the Food Industry*. Vol. 4, IRL (1988).

86. Phillips, G.O., Wedlock, D.J. & Williams, P.A. (eds) *Gums and Stabilisers for the Food Industry*. Vol. 5, IRL (1990).

87. Szczesniak, A.S. & Farkas, E. (1962) Objective characterization of the mouthfeel of gum solutions. *J. Food Sci.* **27**: 381–385.

88. Collyer, S.G. (1984) The use of hydrocolloids in canned products. In *Gum and Stabilisers for the Food Industry* (eds Phillips, G.O. *et al.*), Vol. 2, pp. 349–356. Pergamon.

89. Ash, S.G., Clarke-Sturman, A.J., Calvert, R. & Nisbet, T.M. (1983) Chemical stability of biopolymer solutions. SPE Paper 12085.

90. Bragg, J.R., Gale, W.W., McElhannon, W.A., Davenport, O.W., Petrichuk, M.D. & Ashcraft, T.L. (1982) Loudon surfactant flood pilot test. SPE Paper 10862.

91. Moradi-Araghi, A., Beardmore, D.H. and Stahl, G.A. (1988) The application

of gels in enhanced oil recovery: theory, polymers and crosslinker systems. In *Water-soluble Polymers for Petroleum Recovery* (eds Stahl, G. A. & Schulz, D. N.), pp. 299–312. Plenum.

92. Booth, A.N., Hendrickson, A.P. & Deeds, F. (1963) Physiologic effect of three microbial polysaccharides on rats. *Toxicol. Appl. Pharmacol.* **5**: 478–484.

93. Woodard, G., Woodard, M.W., McNeely, W.H., Kovacs, P. & Cronin, M.T.I. (1973) Xanthan gum: safety evaluation. *Toxicol. Appl. Pharmacol.* **24**: 30–36.

94. Clarke-Sturman, A.J., den Ottelander, D. & Sturla, P.L. (1989) Succinoglycan. A new biopolymer for the oilfield. In *Oil-Field Chemistry: Enhanced Recovery and Production Stimulation*. American Chemical Society Ser. No. 396 (eds Borchardt, J. K. & Teh Fu Yen), pp. 157–168.

95. Moorhouse, R. (1987) Structure/property relationships of a family of microbial polysaccharides. In *Industrial Polysaccharides. Progress in Biotechnology* (ed. Yalpani, M.), Vol. 3, pp. 187–206. Elsevier, Amsterdam.

96. Sanderson, G.R. (1990) The functional properties and applications of microbial polysaccharides: A supplier's view. In *Gums and Stabilisers for the Food Industry*. (eds Phillips, G.O. *et al.*), Vol. 5, pp. 333–344. IRL, Oxford.

97. Moorhouse, R., Colegrove, G.T., Sandford, P.A., Baird, J.K. & Kang, K.S. (1981) PS–60: A new gel-forming polysaccharide. In *Solution Properties of Polysaccharides*. American Chemical Society Symp. Ser. No. 150 (ed. Brant, D.A.), pp. 111–124.

98. Harada, T. (1974) Production, properties and application of Curdlan. In *Extracellular Microbial Polysaccharides* (eds Sandford, P.A. & Laskin, A.), American Chemical Society, pp. 265–283.

99. Kang, K.S., Veeder, G.T. & Cottrell, I.W. (1983) Some novel bacterial polysaccharides of recent development. *Prog. Industrial Microbiol.* **18**: 231–253.

100. Jeanes, A. (1974) Dextrans and pullulans. In *Extracellular Microbial Polysaccharides* (eds Sandford, P.A. & Laskin, A.), American Chemical Society, pp. 284–298.

101. Anon (1990) Bacterial cellulose for cosmetic thickening. *Chemical Specialities*, May, p. 83.

102. Gutnick, D.L., Nestaas, E., Rosenburg, E. & Sar, N. (1989) Bioemulsifier production by *Acinetobacter calcoaceticus* strains, US Patent 4 883 757.

103. Rosenburg, E. & Ron, E.Z. (1990) Bacterial process for the production of dispersants. US Patent 4 921 793.

104. Ash, S.G. (1979) Adhesion of micro-organisms in fermentation processes. In *Adhesion of Micro-organisms to Surfaces* (eds Ellwood, D.C., Melling, J. & Rutter, P.), pp. 57–86, Academic.

105. Berkeley, R.C.W., Lynch, J.M., Melling, J., Rutter, P.R. & Vincent, B. (eds) (1980) *Microbial Adhesion to Surfaces*. Ellis Horwood.

106. Rizzo, R., Crescenzi, V., Gasparrini, F., Gargaro, G., Misist, D., Segre, A. L., Zevenhuizen, L.P.T.M. & Fokkens, R. H. (1988) Structural studies on cyclic beta-(1,2)-D-glucans produced by *Rhizobium trifolii* strain TA-1. In *Biomedical and Biotechnological Advances in Industrial Polysaccharides* (eds Crescenzi, V., Dea, I.C.M., Paoletti, S., Stivala, S.S. & Sutherland, I.W.), pp. 485–493, Gordon and Breach.

107. Zevenhuisen, L.P.T.M. (1990) Recent developments in rhizobium polysaccharides. In *Novel Biodegradable Microbial Polymers* (ed. Dawes, E.A.), pp. 387–402, Kluwer.

108. Cutler, H.C. (ed.) (1988) *Biologically Active Natural Products. Potential Use in Agriculture*. American Chemical Society Symp. Ser. No. 380.

109. Colegrove, G.T. (1983) Agricultural applications of microbial polysac-

charides. *Ind. Eng. Chem. Prod. Res. Dev.* **22**: 456–560.

110. Fitzgerald, E.E. (1983) Xanthan gum in textile printing applications. *Pap. Natl. Tech. Conf. AATCC*, pp. 306–313.

111. Dawson, T.L. (1981) Foam dyeing and printing of carpet. *J. Soc. Dyers Colourists* **97**: 262–274.

112. Racciato, J.S. (1976) Better gum for carpet printing. *Am. Dyestuff Reporter* **51**, November: 69.

113. Evans, N. (1986) Use of xanthan gum in water-based paints. *Paint Resin* **56**: 27–30.

114. Glennon, R.J. & van Winkle, T.L. (1976) *Proc. Am. Ceramic Soc.*, 81–83.

115. Sandford, P.A., Cottrell, I.W. & Pettitt, D.J. (1984) Microbial polysaccharides: New products and their commercial applications. *Pure Appl. Chem.* **56**: 879–892.

5   Microbial cellulose

D. Byrom

ICI Bioproducts and Fine Chemicals, PO Box 1, Billingham,
Cleveland TS23 1LB, UK.

Introduction

Cellulose is probably the most abundant biologically produced polymer and has been important to man for many centuries in the wood, textile and paper industries. It is a major constituent of plant, and some algal and fungal cell walls. It is less widely known that cellulose is also produced by several different types of micro-organisms. Cellulose is a homopolymer of glucose in $\beta(1-4)$ linkage, but is generally found in nature in close association with xylans, other polysaccharides and polymers such as lignin. By contrast, the cellulose synthesized by bacteria is free from contamination with other materials. For this reason bacteria have been studied to elucidate the route of cellulose biosynthesis. The properties deriving from the microfibrillar nature of cellulose produced by microorganisms has led to commercial interest in the material.

Cellulose-Producing Organisms

Cellulose is synthesized by a limited range of bacteria. *Acetobacter* species, Gram-negative aerobic organisms, are the most abundant producers and have been studied most extensively. Production by this species was noted as early as 1886 by Brown in a study of an acetic acid 'ferment'[1].

A number of *Acetobacter* species have been reported to produce cellulose, including *A. xylinum*, *A. aceti*, *A. acetigenum* and *A. pasteurianum*, although the true species status of some of these is open to question.

Organisms of this genus have been used in industrial processes to make microbial cellulose on a large scale. Much of this chapter will describe work carried out on *Acetobacter* species, but it is worth first recording the known instances of cellulose synthesis in other organisms.

Rhizobium produces microfibrils of cellulose that attach the organism to the root-hair surface, although this is not an obligatory step for infection[2]. *Agrobacterium tumefaciens* secrete cellulose often when in association with plant cells. The cellulose appears to aid the attachment of the bacteria to the cells and to entrap further organisms[3,4].

Deinema and Zevenhuizen[5] identified several different organisms from activated sludge that were capable of synthesizing cellulose. These included *Pseudomonas*, *Achromobacter*, *Alcaligenes*, *Aerobacter* and *Azotobacter*. The identification of cellulose was based on alkali insolubility, susceptibility to cellulase, X-ray diffraction and infrared spectral analysis. Another organism commonly found in sludge, *Zooglea ramigera*, also produces cellulose.

Anaerobic organisms have also been reported to synthesize cellulose in significant amounts. *Sarcina ventriculi* is an anaerobic coccoid organism that forms packets of cells bound together by a matrix of fibrous material. The material was identified as cellulose from its hydrolysis products, solubility characteristics and staining reactions[6]. Considerable amounts of cellulose were accumulated by the organism – up to 19% of the cell dry weight[7].

Synthesis and Characteristics of Microbial Cellulose

Acetobacter is aerobic and in static culture elaborates a network of interlocking cellulose fibres (Fig. 1) at the air–liquid interface, forming a floating pellicle. The organism can synthesize cellulose from a variety of carbon sources including hexoses (glucose, fructose) and their corresponding acids, for example, gluconate, compounds such as pyruvate, glycerol and dihydroxyacetone, and intermediates of the tricarboxylic acid cycle such as succinate. Little cellulose is produced from pentoses[8–10].

In resting cell suspensions approximately 1.5×10^8 glucose residues per hour can be polymerized[11]. The weighted average degree of polymerization (DPw) has been measured using viscometry and ranges from 2190 to 3470 (refs 12,13). In contrast to plant cells where the DPw of cell wall cellulose is constant during its synthesis, the DPw of *Acetobacter* cellulose increases linearly and is dependent on the organism's generation time[14].

Production will occur in the absence of an exogenous nitrogen supply[15] and in the presence of inhibitors of protein synthesis[16], indicating that *de novo* protein synthesis is not necessary for the process. However, cells producing cellulose must be metabolically capable in oxidative processes[17–19].

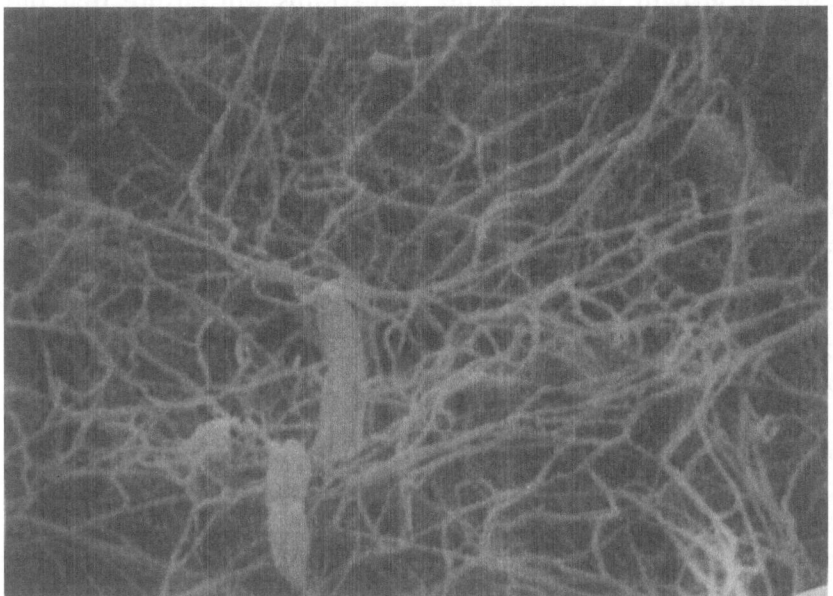

Fig. 1 Network of cellulose fibres from *Acetobacter*.

As mentioned, in static culture *Acetobacter* produce a pellicle of cellulose at the air–liquid interface. It is believed that this functions to bring cells to the environment that has the highest oxygen tension. Indeed, ATP estimations show that viable cells are located at the air–pellicle surface[20]. Alternatively, Williams and Cannon[21] suggested that the cellulose protects cells from ultraviolet light, enhances the ability of the organism to compete in colonization of a natural substrate and conserves water on or in association with that substrate.

Repeated subculture in agitated conditions rapidly results in the isolation of spontaneous cellulose-negative (cel⁻) mutants[22]. It has been suggested that cellulose synthesis is shear-sensitive and that this, coupled with a high spontaneous mutation rate to cel⁻, is responsible for the phenomenon. It is my view that the speed of onset of the effect is likely to be due to a combination of factors. Any cells in the population that are spontaneously cel⁻ in shaken culture may have a growth advantage over cel⁺ organisms, since the former do not need to divert large amounts of carbon and energy to cellulose synthesis. On subculture, many cel⁻ cells, but only a few cel⁺ cells, will be transferred, as the latter are aggregated in flocs of cellulose fibres. The difference in growth kinetics of the two strains will ensure that the trend towards the cel⁻ phenotype accelerates with each subculture. In controlled experiments, where the effects of aggregation on cel⁺ cells were eliminated, evidence was obtained that the spontaneous mutation rate of *A. xylinum* was not excessively high – less than 10^{-5} mutations per cell per generation – suggesting that previously reported high rates may have been due to effects similar to those described above[23]. The instability of cellulose production in agitated conditions was one of the major problems to be solved in the commercialization of microbial cellulose.

Microbial cellulose is produced in both surface and stirred culture as a ribbon of highly crystalline cellulose I extruded from the cell. The width of the ribbon lies between that of fibrils of molecular size and that of conventional cellulose fibres of plant origin. The ribbon is of the order of 40 to 100 nm in width and of indeterminate length. It is very difficult to measure or estimate the length of individual ribbons because they are produced as a mass of interlocking fibres in culture (Fig. 1). Individual ribbons are composed of bundles of microfibrils that associate by hydrogen-bonding[24]. Each microfibril has a diameter of 3 to 4 nm and is in turn composed of three subelementary fibrils each of about 1.78 nm twisted into a left-handed helix (Fig. 2). The crystallite size for the cellulose in the dry state is 7.5 nm and in the wet state 6.5 nm (ref. 25).

The subelementary fibrils are extruded from a row of pores in the bacterial cell wall parallel to the longitudinal axis of the cell[26,27]. The pores are highly localized and occupy only about 0.5% of the outer membrane. Cellulose formation is associated with rows of particles (assumed to be multienzyme complexes) below the pores. The mechanism by which the glucan chains of the microfibril are assembled into the ribbon is unknown, but the processes of polymerization and crystallization are coupled, the latter limiting the rate of polymerization. Haigler and Benziman[25] discuss these events in more detail.

It is possible to alter the morphology of the normal ribbon at two levels by inclusion of specific chemicals in the growth medium of the organism.

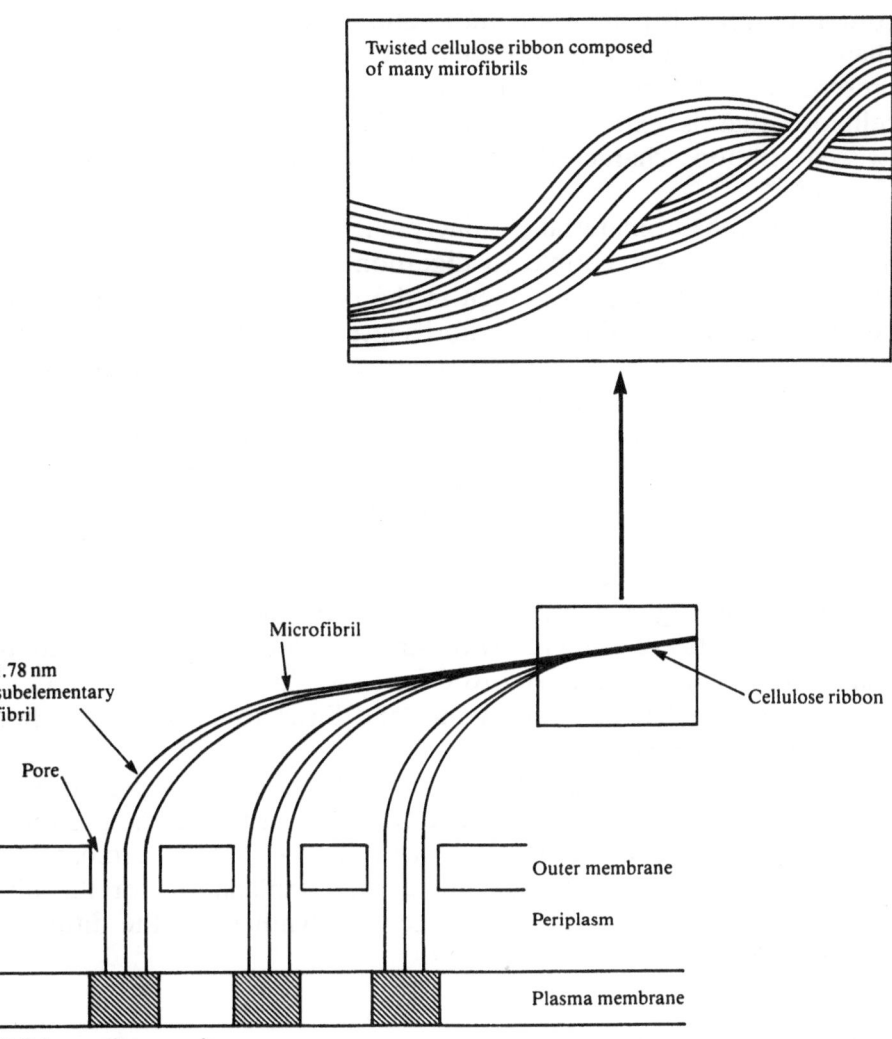

Fig. 2 Formation of cellulose ribbon.

Calcofluor white ST

Calcofluor disrupts the normal assembly of ribbons by interfering with the association of subelementary fibrils to give microfibrils. The Calcofuor molecule is planar in conformation and binds along the glucan chains of the cellulose molecule by hydrogen-bonding. In the presence of Calcofluor, bacteria produce a broad band of amorphous cellulose that is extruded perpendicular to the cell surface. The twisted ribbon characteristic of normal cellulose synthesis is absent[28,29].

The broad-band material is noncrystalline and, in contrast to normal cellulose, is susceptible to digestion by cellulase. Normal synthesis of the glucan chain is seen in the presence of Calcofluor, and because the compound binds to fibrils after they exit the cell, crystallization and bundle formation must be an extracellular event although closely associated with the cell.

Addition of Calcofluor to cells actively synthesizing cellulose causes immediate cessation of normal ribbon formation and production of broad-band material. Ribbon formation resumes when the cells are transferred to medium without Calcofluor. The rate of cellulose synthesis is stimulated 200–400% in the presence of Calcofluor, suggesting a linkage between synthesis and crystallization[30].

Similar effects are observed with other direct cellulose dyes such as Congo Red, Trypan Blue, Benzo Orange R, Amidine Red, Amidine Black, Diamine Violet, Chrysamine and Sirus Red and with other fluorescent brighteners[25].

The formation of cellulose I was not observed in the presence of some direct dyes by Kai *et al.*[31]. The cellulose formed in the presence of direct Red 80 and Congo Red was cellulose II and cellulose IV, respectively. The authors did not have an explanation for the difference in the effect of the dyes or for the synthesis of the different cellulose forms in these experiments.

Carboxymethyl cellulose

Carboxymethyl cellulose (CMC) interferes with ribbon assembly at a higher level than Calcofluor. Cells incubated with glucose and 0.1% CMC produce ribbons in which the bundles of microfibrils fail to associate into a coherent ribbon[32]. The ribbon assumes the appearance of a rope that has been untwisted to expose the component strands. The differing effects of Calcofluor and CMC suggest a hierarchical mechanism of ribbon assembly. Cellulose synthesis rate was also stimulated by CMC[33], although to a lower extent than by Calcofluor.

270

Other effectors

Purified pea xyloglucan will also prevent the association of microfibrils into bundles when added to cultures of *Acetobacter*[34].

Biochemistry

It was established fairly rapidly in the study of biosynthesis in *Acetobacter* that the immediate precursor of cellulose was uridine diphosphate-glucose (UDPG) (see ref. 10 for a discussion of this topic). All the enzymes involved in the synthesis of UDPG are enzymes of intermediary metabolism. The final enzyme in the pathway, that which catalyses the addition of a glucose residue to the growing polymer chain, is cellulose synthase or UDPG: 1,4-β-D-glucan-4-β-D-glucosyltransferase, EC 2.4.1.12. This enzyme, its cellular location, regulation and genetics, have been seen as the key to understanding the mechanism of cellulose synthesis in bacteria.

Cellulose synthase has been difficult to purify and obtain in a highly active state in cell-free preparations. Membrane fractions from *Acetobacter* were demonstrated by several authors to be capable of *in vitro* synthesis of 1,4-β-D-glucan (cellulose) from UDPG[35–37], but the rates of cellulose formation were very low compared to those measured *in vivo*[10]. It was then discovered that factors other than substrate and enzyme are important in stimulation of the rate in cell-free systems. Aloni *et al.*[38] prepared total membrane fractions in the presence of polyethylene glycol-4000 (PEG-4000) and found 3- to 10-fold the activity of that previously reported. They could further stimulate the activity by addition of GTP. GTP stimulation was seen only if the membranes were made when PEG-4000 was present. An unidentified protein factor was absent from membranes prepared without PEG. The factor was found to dissociate easily from membranes prepared by the usual techniques. The rate of cellulose synthesis in the presence of PEG, GTP and the protein was 200-fold that seen previously.

Digitonin can be used to solubilize cellulose synthase, allowing the enzyme to be used in conventional biochemical studies without the complication of having to have membrane material present[39]. The digitonin-solubilized enzyme responds to similar regulatory factors as the membrane-bound version, being activated by GTP but only in the presence of the protein factor. Calcium promotes an enzyme/protein factor association in membranes but not in the solubilized system.

The protein factor was identified by Ross and co-workers[40] as an enzyme that converts GTP to a cyclic diguanylic acid (Fig. 3). The cyclic nucleotide is the activator of cellulose synthesis and is itself broken down by a second enzyme, which is also membrane-bound. The second enzyme is inactivated by calcium ions. This basic model for control of cellulose synthesis was elaborated later to include a third enzyme. The current state of understanding[41] is as follows.

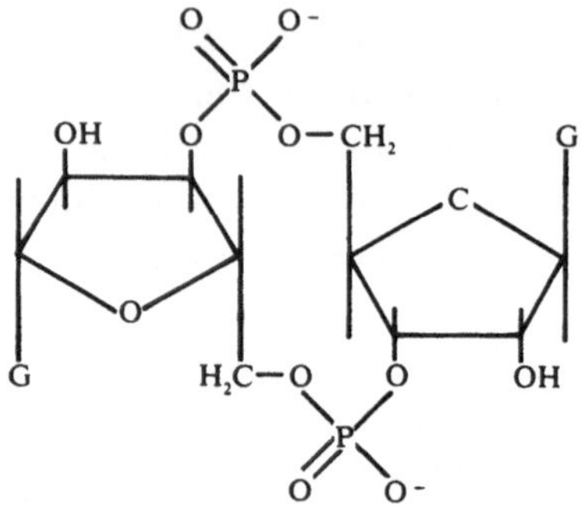

Fig. 3 Structure of 3′,5′ -cyclic diguanylic acid.

The protein factor, diguanylate synthase, converts two molecules of GTP into a 5′ triphosphate dimer which condenses internally to form the cyclic diguanylic acid. This compound can associate with cellulose synthase to cause activation. Deactivation is caused by two membrane phosphodiesterases, one which cleaves the activator to give the open dimer again, and the other which further degrades the dimer to two molecules of 5′ GMP. The phosphodiesterases have been designated PDE-A and PDE-B. PDE-A is inactivated by calcium ions which can therefore prevent degradation of the cyclic form of the diguanide and thus provide a mechanism for the control of cellulose synthesis (Fig. 4).

Further confirmation of the synthesis of cellulose in these cell-free systems was obtained by Lin *et al.*[42] who demonstrated the presence of fluorescent brightener-binding fibrils in the *in vitro* reaction mixtures. The fibrils bound gold-labelled cellobiohydrolase and were degraded by the enzyme if incubation was extended for more than a few minutes. The material gave an electron diffraction pattern consistent

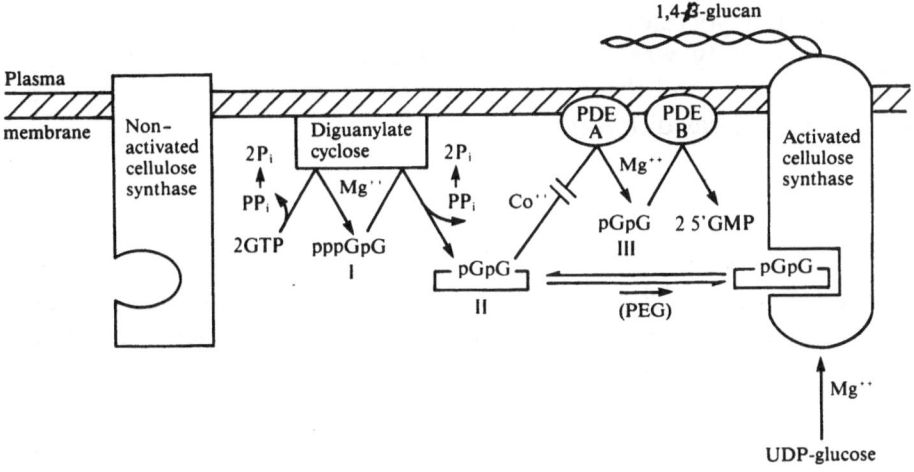

Fig. 4 Proposed model for regulation of cellulose synthesis. See text for explanation.

with that of cellulose, and the results indicated that cellulose I had been formed. The fibril diameter was about 1.7 nm, similar to that found *in vivo* when optical brighteners are used to disrupt crystallization.

Although the *in vitro* systems described so far had been extremely useful in elucidating the control of cellulose synthesis, they had not allowed the identification and isolation of the cellulose synthase protein. An enzyme-entrapment technique was used to purify two polypeptides from membrane preparations[43]. Their mobility on denaturing gels suggested relative molecular masses of 83K and 93K. The 93K peptide was sensitive to trypsin digestion, whereas the 83K band was insensitive and still displayed synthase activity, the implication being that it was a strong candidate for the cellulose synthase. There was evidence from gel-filtration chromatography that the native protein was a tetramer or octamer.

Cellulose synthase was located in the cytoplasmic membrane by sucrose density centrifugation[44]. Earlier studies using ultrastructural evidence from electron microscopy had found a linear row of 12-nm particles on the inner face of the outer membrane. It was concluded that the enzyme appeared to be in the outer membrane. Bureau and Brown[44] identified the membrane fractions using markers characteristic for cytoplasmic and outer membrane, namely a range of dehydrogenases and NADH oxidase for the cytoplasmic membrane[31] and

2-keto-3-deoxyoctulosonic acid for the outer membrane. Cellulose synthase activity was clearly associated with the cytoplasmic membrane fraction, and the product from the assay was confirmed to be cellulose by X-ray diffraction.

Genetics

There is little formal genetics known for the genus *Acetobacter*. However, several authors have isolated plasmids from members of the genus and gene transfer systems have been devised.

A. xylinum ATCC 10245 contains several plasmids of 16 kilobases (kb), 44 kb and 64 kb. Large plasmids of 200–300 kb are also seen[45]. The size distribution of the plasmids alters dramatically in cells that are mutated to a cel⁻ phenotype, suggesting that the plasmids are involved in cellulose synthesis[46].

The 16-, 44- and 64-kb plasmids could be mobilized by a broad host range helper plasmid. The 44-kb plasmid was mobilized into a cel⁻ mutant but failed to complement it, indicating that the wild-type allele of the mutated gene was probably not located on that plasmid or that the mutation could not be complemented in *trans*.

Complex interactions between the plasmids and between the plasmids and the chromosome were suggested by hybridization studies using cloned plasmid fragments as probes[47].

Genetic evidence has been used to confirm the biosynthetic pathway as postulated from the biochemical data. Attempts have been made to isolate mutants in the structural genes for the cellulose synthase from *A. xylinum* ATCC 10245[23], but mutants exerting pleiotropic effects on cellulose synthesis were instead obtained. In later work, cel⁻ mutants of *A. xylinum* ATCC 23768 were complemented by a 2.8-kb cloned fragment of DNA from the wild-type organism. Biochemical analysis showed that the mutants were deficient in the enzyme UDPG pyrophosphorylase. Confirmation of the activity encoded by the 2.8-kb fragment was obtained by complementation of a mutant of *Escherichia coli* lacking that enzyme. This was the first example of the cloning of a gene with a known function in cellulose synthesis. The gene was named *celA*[48].

Other attempts to isolate cel⁻ mutants have also yielded mixed results. *Acetobacter* mutants unable to make pellicles have been isolated[37,49]. The isolates from one group still made some cellulose and had normal cellulose synthase activity. The cellulose they produced was abnormal in that it was cellulose II rather than cellulose I. Alterations in the lipopolysaccharide component of the cell wall were also seen.

Fukaya *et al.*[50] have constructed shuttle vectors to enable DNA fragments to be moved in and out of *Acetobacter* with more ease.

Six transposon insertions in the chromosome of *Agrobacterium tumefaciens*, all of which resulted in a cel⁻ phenotype, have been mapped. One additional insertion caused overproduction of cellulose. No data on the biosynthetic step(s) affected by the mutations were reported[51]. The fact that several genes involved in cellulose synthesis have been labelled in this way should make them prime targets for cloning work in the future.

An alternative approach to the cloning of the cellulose synthase gene was taken by Saxena *et al.*[52]. Their technique did not rely on the production of mutants. Oligonucleotide probes were constructed corresponding to the N-terminal amino-acid sequence of the purified 83K catalytically active subunit of the enzyme. Hybridization of the probes with size-fractionated, cloned *A. xylinum* ATCC 53582 DNA detected a 9.5-kb fragment that contained an open reading frame (ORF) encoding an 80K protein. The nucleotide sequence of the beginning of the ORF corresponded to that encoding the first 20 amino acids of the purified 83K protein. Two potential glycosylation sites were identified in the sequence, supporting earlier observations that the gene product may be a glycoprotein[43]. There was no evidence of expression of the gene in the *E. coli* host.

The oligonucleotide probes hybridized to DNA extracted from other strains of *A. xylinum*, indicating a substantial degree of homology between the cellulose synthase genes from these organisms. Additional evidence that the 83K protein is the UDPG-binding subunit of cellulose synthase was obtained from photoaffinity probe-binding experiments[53].

Commercial Applications and Industrial Production

Microbial cellulose has been available for many years as a dessert called Nata de Cocoa[54]. Nata is sold mainly in the Philippines, although ironically I have bought and eaten this delicacy in the eastern United States not far from an industrial laboratory working on microbial cellulose for other applications.

Despite its use as Nata, microbial cellulose has been regarded as a laboratory curiosity for many years and a product in search of a market. However, progress has recently been made towards commercial applications with the invention of the stirred tank process for production of the material. This production route is the only one likely to provide cellulose in sufficient quantities and at a low enough price to allow commercial development. Before describing the process, however, it is informative to review the applications that have been envisaged for microbial cellulose since it first attracted the attention of the industrial community.

Applications

Many patents have been published describing varied uses of microbial cellulose. The applications cited in food and food processing have been as dispersing aids[55], thickeners[56], melt preventer in ice cream[57], in improvement of mouth feel in processed foods[58] and as a fibrous food when dispersed with a stabilizer[59–61]. Microbial cellulose has also been used to make composites by inclusion of particles to form sheets[62], addition of reinforcing fibres alone or in combination with inorganics[63], addition of glass or polyamide fibres[64,65], addition to polystyrene to form plastic films for food packaging[66] and paper formation in combination with a filler such as clay[67]. A slurry of the fibrils has been patented as an aid to flocculation of conventionally produced cellulose fibres in paper making[68]. The first patented use for microbial cellulose was as a membrane for filtration or other uses[69]. Membranes have also been made by stretching a pellicle on a former and allowing it to dry in air[70]. Other groups have used the material to make reverse osmosis membranes[71,72]. Miscellaneous uses as an insulating material[73], as a fine crystalline product when ground for use in paints, pharmaceuticals and cosmetics[74] and a viscosity improver in food, cosmetics and paints[75] have also been recorded.

Brown and colleagues have filed patents on many aspects of cellulose production. These have included increasing the hydrophilicity of fibres such as polyesters, nylon and cotton by growing cellulose on them[76]; production of formed cellulose articles on or in structures permeable to oxygen so that either hollow or solid articles made of cellulose can be manufactured[72]; cultivation of microbes in a magnetic field to produce amorphous cellulose[78]; use of cellulase in the preculture stage of a fermentation to improve yield[79]; use of microbial cellulose to increase the strength of, for example, documents[80]; use of cellulose in manufactured articles where it has been produced in the presence of agents that interfere with crystallization[81]; modification of the structure of cellulose by addition of agents such as carboxymethyl cellulose to the fermentation to give cellulose with greater clarity, resilience, absorptive properties and rewettability[82]; production of cellulose by strains of organism that form antiparallel ribbons of cellulose with greater strength than usual[83]; and production of cellulose by a particular strain of *Acetobacter*[84].

Patents have been published describing the use of the pellicle product in wound dressings[85,86] and artificial skin[87], and as an oral plaster when liquid-loaded with drugs[88]. The drape and absorptive properties of the pellicle are claimed to be useful in these applications. The *Acetobacter* cells are removed from the pellicle by caustic treatment and washing with water.

Microbial cellulose produced in stirred culture has been patented for use as a coating to make high-quality paper[89], as a wet-laid sheet with high tensile strength[90] and for the production of nonwoven paper or fabric[91].

Despite all these patents, I am aware of only one commercial product made from microbial cellulose (with the exception of Nata). Ajinomoto have used the caustic-treated pellicle to form the speaker diaphragm in the headphones of personal stereos. When pressed and dried the material becomes exceptionally stiff and is particularly suitable for this use[92].

Production

Many of the applications given above are for pellicle-derived material. Production of pellicles is slow and is therefore expensive. The ability to intensify the production of cellulose in reliable stirred fermentations had to be achieved if the product was to be commercially exploited. An added bonus was the greater ease of handling associated with the nonpellicle material for removal of cells and fabrication into sheets and other articles.

The main barrier to using in stirred culture the strains previously studied is their instability – the capacity to produce cellulose is rapidly lost. Loss of a plasmid(s) involved in cellulose biosynthesis is a possible explanation for this.

PELLICLES

Several systems have been developed to attempt to either produce pellicles in a more useful/efficient way or to synthesize cellulose in agitated conditions – the forerunners of the true stirred tank processes described below. These systems are briefly reviewed here.

In one of the processes, fresh medium is added on top of a growing pellicle in a continuous or intermittent fashion[93]. Cellulose has also been produced in flowing medium in order to align the fibrils along the flow path. The fibrils may be formed into thread or yarn[94]. A second process using flowing medium involved another organism, *Sphaerotilus natans*. The organism was grown on a pitted plate over which the medium flowed and the cellulose and cells were periodically scraped off[95].

AGITATED CULTURE

The first attempts at production of cellulose in agitated systems relied on low-shear regimes. A process using a gently stirred reactor aimed at minimizing shear has been patented. Biomass production was minimized by limiting the nitrogen or phosphorus supply, and the carbon substrates for cellulose synthesis were plant-derived carbohydrates such as starch, sucrose or glucose[96]. An alternative approach used a bubble tower fermenter to produce cellulose in low-shear conditions[97–99].

Cellulose has been separated from cells and medium components by alkali treatment and heat. In one case, heating was achieved with ultrasound[100].

Large-scale stirred tank fermentation has been developed by two groups, one at Cetus/Weyerhaeuser and the other at ICI.

Weyerhaeuser have operated a process developed on their behalf by Cetus[101]. The approach required the selection of a strain that was stable for cellulose production in stirred culture. The organism reverted to a cel⁻ phenotype at a frequency of less than 0.5% over 42 to 45 generations. Once stable operation in stirred culture had been obtained, higher-yielding mutants were isolated. The mutants had a reduced capacity to produce gluconic acid, which is excreted in large quantities by the parent strain when grown on glucose. Gluconate production was reduced by about 50-fold in strain 1306-11 and was not detectable in strain 1306-21. Cellulose yield from these mutants was reported to be substantially higher than that from the parent organism, although absolute figures are difficult to quote owing to the way the experiments are described[101].

Fermentation is operated in a fed batch mode and the cellulose is separated from cells by hot caustic treatment followed by washing with water. The product is available as an 18% solids suspension, although the company are developing a drying process capable of producing a powder form of the material. Production has been demonstrated in commercial scale fermenters up to 200,000 litres. Current yields are greater than 0.2 grammes of cellulose per gramme of glucose used.

Weyerhaeuser are assessing the use of the product in applications as diverse as metal recovery, paper coating, hydraulic fracture fluid in oil and gas recovery and as a viscosity modifier in the food industry. The company has patented uses of the product as a paper-coating material, nonwoven fabric and high-tensile-strength wet-laid sheet (see above). The material is estimated to cost £5 to £10 per kilogramme when manufactured at scale (5000 to 10,000 tonnes per annum).

The process developed by ICI[102] is also a stirred tank fermentation operated in fed batch mode, but it is run in glucose limitation. An initial charge of glucose is used by the organism and then glucose is pumped into the fermenter at a rate that maintains the supernatant concentration below 0.5 grammes per litre. Gluconate production is minimized in this way. With this process, yields of 0.18 grammes of cellulose per gramme of glucose have been achieved without modification of the production organism. The fermentation can be carried out as a draw-and-fill process, which is a means of maximizing plant productivity for an essentially batch process. At the end of a fermentation batch, the fermenter is emptied leaving approximately 10% of the contents as an inoculum for the next fermentation. Fresh sterile medium is added to fill the fermenter and the culture is allowed to grow. At the end of the cellulose production phase the culture is harvested in a similar way to that described previously, and the whole process is repeated. The procedure can be repeated several times. Harvesting produces a cake of 25% solids material (Fig. 5).

The cost projections for large-scale production are similar to those quoted by Weyerhaeuser. A higher-yielding strain of *Acetobacter* has been isolated from natural sources for use in this process[103].

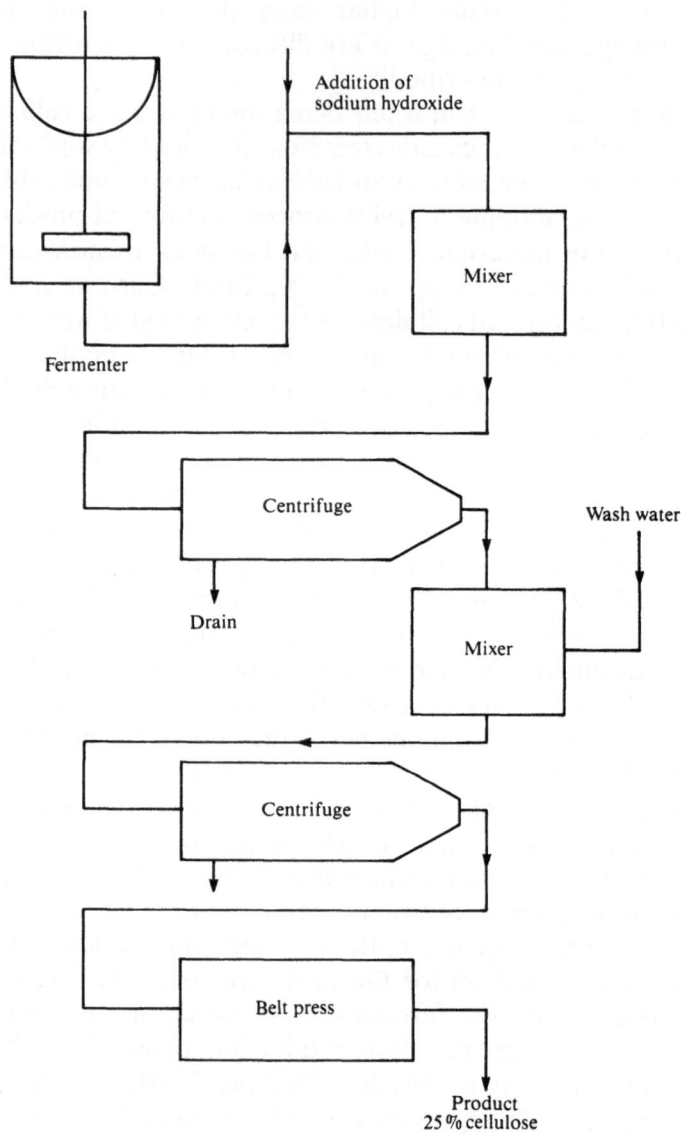

Fig. 5 Outline process for production of microbial cellulose.

Concluding Remarks

Research on microbial cellulose over many years has enabled our understanding of the process to move from the descriptive, where the phenomenon of cellulose production was documented, to a point where the physiology and biochemistry have been elucidated and the genetic and molecular organization of biosynthesis have begun to be unravelled. Commercial interest in the material was hampered by lack of a suitable production process. The advent of stirred tank processes provides an opportunity for the industrial evaluation of the utility of the material, and will determine whether the product is to find a place in the fermentation industry or remain an interesting academic curiosity.

References

1. Brown, A.J. (1886) *J. Chem. Soc. Lond.* **49**: 432–439.
2. Napoli, C., Dasso, F. & Hubbel, D. (1975) *Appl. Microbiol.* **30**: 123–131.
3. Matthysee, A.G., Holmes, K.V. & Gurlitz, R.H.G. (1981) *J. Bacteriol.* **145**: 583–595.
4. Delmer, D. (1983) *Adv. Carbohydr. Chem. Biochem.* **41**: 105–153.
5. Deinema, M.H. & Zevenhuizen, L.P.T.M. (1971) *Arch. Microbiol.* **78**: 42–57.
6. Canale-Parola, E., Borasky, R. & Wolfe, R.S. (1964) *J. Bacteriol.* **81**: 311–318.
7. Canole-Parola, E. & Wolfe, R.S. (1964) *Biochim. Biophys. Acta* **82**: 403–405.
8. Schramm, M., Gromet, Z. & Hestrin, S. (1957) *Biochem. J.* **67**: 669–679.
9. Benziman, M. & Burger-Rachamimov, H. (1962) *J. Bacteriol.* **84**: 625–630.
10. Aloni, Y. & Benziman, M. (1982) In *Cellulose and Other Natural Polymer Systems* (ed. Brown, R.M. Jr), pp. 341–361. Plenum, New York.
11. Weinhouse, H. & Benziman, M. (1974) *Biochem. J.* **138**: 537–542.
12. Marx-Figini, M. & Pion, B.G. (1974) *Biochim. Biophys. Acta* **338**: 382–393.
13. Marx-Figini, M. & Pion, B.G. (1976) *Makromol. Chem.* **177**: 1013–1020.
14. Marx-Figini, M. (1982) In *Cellulose and Other Natural Polymer Systems* (ed. Brown, R.M. Jr), pp. 243–271. Plenum, New York.
15. Hestrin, S. & Schramm, M. (1954) *Biochem. J.* **58**: 345–352.
16. Webb, T.E. & Colvin, J.R. (1967) *Can. J. Biochem.* **45**: 465–476.
17. Gromet, Z., Schramm, M. & Hestrin, S. (1957) *Biochem. J.* **67**: 679–689.
18. Benziman, M. & Mazover, A. (1973) *J. Biol. Chem.* **248**: 1603–1608.
19. Swissa, M. & Benziman, M. (1976) *Biochem. J.* **153**: 173–179.
20. Legge, R.L. (1990) *Biotech. Adv.* **8**: 303–319.
21. Williams, W. S. & Cannon, R.E. (1989) *Appl. Envir. Microbiol.* **55**: 2448–2452.
22. Schramm, M. & Hestrin, S. (1954) *J. Gen. Microbiol.* **11**: 123–129.
23. Valla, S. & Kjosbakken, J. (1982) *J. Gen. Microbiol.* **128**: 1401–1408.
24. Gardner, K.H. & Blackwell, J. (1974) *Biochim. Biophys. Acta* **343**: 232–237.
25. Haigler, C.H. & Benziman, M. (1982) In *Cellulose and Other Natural Polymer Systems* (ed. Brown, R.M. Jr), pp. 273–297. Plenum, New York.
26. Brown, R.M. Jr, Willison, J.H.M. & Richardson, C.L. (1976) *Proc. Natl. Acad. Sci. U.S.A.* **73**: 4565–4569.

27. Zaar, K. (1979) *J. Cell Biol.* **80**: 773–777.
28. Haigler, C.H., Brown, R.M. Jr & Benziman, M. (1980) *Science* **210**: 903–906.
29. Brown, R.M. Jr, Haigler, C.H. & Cooper, K. (1982) *Science* **218**: 1141–1142.
30. Benziman, M., Haigler, C.H., Brown, R.M. Jr & Cooper, K. (1980) *Proc. Natl. Acad. Sci. U.S.A.* **77**: 6678–6682.
31. Kai, A., Kido, H. & Ishida, N. (1990) *Chem. Lett.* **6**: 949–952.
32. Haigler, C.H., White, A.R., Brown, R.M. Jr & Cooper, K. (1982) *J. Cell Biol.* **94**: 64–69.
33. Ben-Hayim, G. & Ohad, I. (1965) *J. Cell Biol.* **25**: 191–207.
34. Hayashi, T., Marsden, M.P.F. & Delmer, D.P. (1987) *Plant Physiol.* **83**: 384–389.
35. Glaser, L. (1958) *J. Biol. Chem.* **232**: 627–636.
36. Colvin, J.R. (1980) *Planta* **149**: 97–107.
37. Swissa, M., Aloni, Y., Weinhouse, H. & Benziman, M. (1980) *J. Bacteriol.* **143**: 1142–1150.
38. Aloni, Y., Delmer, D.P. & Benziman, M. (1982) *Proc. Natl. Acad. Sci. U.S.A.* **79**: 6448–6452.
39. Aloni, Y., Cohen, R., Benziman, M. & Delmer, D.P. (1983) *J. Biol. Chem.* **258**(7): 4419–4423.
40. Ross, P., Aloni, Y., Weinhouse, C., Michaeli, D., Weinberger-Ohana, P., Meyre, R. & Benziman, M. (1985) *FEBS Lett.* **186**(2): 191–196.
41. Ross, P., Weinhouse, H., Aloni, Y., Michaeli, D., Weinburger-Ohana, P., Mayer, R., Braun, S., de Vroom, E., van der Marel, G.A., van Boom, J.H. & Benziman, M. (1987) *Nature* **325**: 279–281.
42. Lin, F.C., Brown, R.M. Jr, Cooper, J.B. & Delmer, D.P. (1985) *Science* **230**: 822–825.
43. Lin, F.C. & Brown, R.M. Jr (1989) In *Cellulose and Wood: Chemistry and Applications* (ed. Schuerch, C.), pp. 473–492. Wiley, New York.
44. Bureau, T.E. & Brown, R.M. Jr (1987) *Proc. Natl. Acad. Sci. U.S.A.* **84**: 6985–6989.
45. Valla, V., Coucheron, D.H. & Kjosbakken, J. (1983) *Arch Microbiol.* **134**: 9–11.
46. Valla, S., Coucheron, D.H. & Kjosbakken, J. (1986) *J. Bacteriol.* **165**(1): 336–339.
47. Valla, S., Coucheron, D.H. & Kjosbakken, J. (1987) *Mol. Gen. Genet.* **208**: 76–83.
48. Valla, S., Coucheron, D.H., Fjaervik, E., Kjosbakken, J., Weinhouse, H., Ross, P., Amikam, D. & Benziman, M. (1989) *Mol. Gen. Genet.* **217**: 26–30.
49. Saxena, I.M. & Brown, R.M. Jr (1989) In *Cellulose and Wood: Chemistry and Technology* (ed. Schuerch, C.), pp. 537–557. Wiley, New York.
50. Fukaya, M., Okumura, H., Masai, H., Uozumi, T. & Beppu, T. (1985) *Agric. Biol. Chem.* **49**: 2091–2097.
51. Robertson, J.L., Holliday, T. & Mattysse, A.G. (1988) *J. Bacteriol.* **170**(3): 1408–1411.
52. Saxena, I.M., Lin, F.C. & Brown, R.M. Jr (1990) *Plant Mol. Biol.* **15**: 673–683.
53. Lin, F.C., Brown, R.M. Jr, Drake, R.R. Jr & Haley, B.E. (1990) *J. Biol. Chem.* **265**: 4782–4784.
54. Lapuz, M.M., Gallardo, E.M. & Pallo, M.A. (1967) *Philip. J. Sci.* **96**(2): 91.
55. Japanese Patent JP 63199201, 12 February 1987.
56. Japanese Patent JP 61212295, 20 September 1986.
57. Japanese Patent JP 62022554, 30 January 1987.
58. Japanese Patent JP 62294047, 21 December 1987.

59. Japanese Patent JP 62083854, 17 April 1987.
60. Japanese Patent JP 2053454, 22 February 1990.
61. Japanese Patent JP 1085052, 30 March 1989.
62. Japanese Patent JP 1156600, 11 December 1989.
63. Japanese Patent JP 1014400, 18 January 1989.
64. Japanese Patent JP 1014399, 18 January 1989.
65. Japanese Patent JP 1013789, 18 January 1989.
66. Japanese Patent JP 63309529, 16 December 1987.
67. Japanese Patent JP 63295793, 2 December 1988.
68. World Patent WO 8908148, 8 September 1989.
69. East German Patents DD 88307 and 92136, 1972.
70. Japanese Patent JP 63294794, 12 December 1988.
71. Japanese Patent JP 1199604, 11 August 1988.
72. Japanese Patent JP 1193335, 3 August 1989.
73. East German Patent DD 132512, 4 October 1978.
74. Japanese Patent JP 61221201, 1 October 1986.
75. Japanese Patent JP 61113601, 31 May 1986.
76. US Patent US 4378431, 29 March 1983.
77. US Patent US 684844, 21 December 1984.
78. US Patent US 719505, 3 April 1985.
79. European Patent EP 258038, 2 March 1988.
80. US Patent US 199906, 28 May 1988.
81. US Patent US 199606, 31 May 1988.
82. European Patent EP 346507, 20 December 1989.
83. European Patent EP 347481, 27 December 1989.
84. US Patent US 4954439, 4 September 1990.
85. European Patent EP 206830, 30 December 1986.
86. British Patent GB 2131701, 27 June 1984.
87. European Patent EP 85904787, 30 September 1985.
88. Japanese Patent JP 1050815, 27 February 1989.
89. European Patent EP 289993, 9 November 1988.
90. US Patent US 4863565, 5 September 1989.
91. World Patent WO 8901074, 9 February 1989.
92. European Patent EP 200409, 5 November 1986.
93. Japanese Patent JP 62265990, 11 November 1987.
94. US Patent US 4745058, 17 May 1988.
95. US Patent US 4320198, 16 March 1982.
96. French Patent FR 877387, 26 May 1987.
97. Japanese Patent JP 62087099, 21 April 1987.
98. Japanese Patent JP 1243997, 28 September 1989.
99. US Patent US 905481, 8 September 1986.
100. Japanese Patent JP 63068093, 26 March 1988.
101. US Patent US 900086, 28 August 1986.
102. European Patent EP 279506, 24 August 1988.
103. European Patent EP 323717, 12 July 1989.

6 Hyaluronic acid

David A. Swann & Jing-wen Kuo

Medchem Products Inc., 232 West Cummings Park, Woburn, Mass 01801, USA

Introduction

This review will focus in a selective way on the structure and function of hyaluronic acid (HA)*, and the development of HA-containing products, and, we hope, shed some light on the factors that have influenced the development of these biomaterials.

Since the discovery of HA by Meyer[1] in 1934, this substance has been studied extensively by many investigators and there is a wealth of information[2-5] regarding its chemical structure, physical and biological properties, synthesis by cells and tissues, catabolism in animals and humans, and more recently, function as a drug or device to treat or aid in the treatment of clinical disorders.

Hyaluronic Acid Structure

Hyaluronic acid is defined as a linear polysaccharide composed of two alternatively linked sugars, D-glucuronic acid (A), and N-acetylglucosamine (B), and the polysaccharide chain has the empirical formula $[A-B]_n$. This structure was established by enzymatic and chemical studies[6] on umbilical-cord HA and was confirmed by chemical synthesis[7]. Hyaluronic acids isolated and purified from all sources, from the capsules of bacteria to the connective tissues of vertebrates, appear to have this same basic structure. The presence of minor amounts of other constituents has been reported[8], but it seems probable that these constituents are contaminants or analytical artefacts[9].

The only differences between the properties of HA samples isolated from different sources are the relative molecular mass (M_r), the organization of the HA polysaccharide chains and the degree of contamination with other tissue constituents that either copurify or interact with HA.

M_r values for isolated and purified HA samples reported in the literature vary from an average low value of 70,000 for bovine vitreous HA[10] to a value of 12×10^7 for an HA preparation isolated from bovine synovial fluid[11]. There is reason to think, however, that some of the high values reported for the M_r of HA reflect the occurrence of

*Throughout this chapter HA is used as an abbreviation for hyaluronic acid, sodium hyaluronate and hyaluronan.

aggregates or complexes involving interactions between HA polysac-charide chains[12–14] or with other matrix constituents that may be present as contaminants. The M_r values will also depend on the purification methods employed, in that these may allow the cleavage of the HA polysaccharide chains by either chemical, enzymatic or mechanical means. In addition, the conditions such as the ionic strength of the solvent, the pH, and the counterion employed in the M_r measurement procedure will also affect the M_r values obtained. The different methods used to determine the average M_r will also produce different values because they are based on different assumptions. The principal methods used to determine the M_r of the HA samples are: viscosity[15], sedimentation/diffusion[10], light-scattering[16], gel-permeation chro-matography[17], and radiological methods to determine the number of end groups[18]. A further consideration is the relationship of the HA sample analyzed to the HA in the tissue source and how representative it is. All HAs studied so far have contained a distribution of molecules with a wide range of M_r values[15,18,19].

A characteristic property of HA, which is important in its use as a biomaterial, is the high viscosity in aqueous HA solutions at relatively low solute concentrations[20]. These solutions also exhibit elastic properties[21] with the result that HA products such as those used for ophthalmic applications are generally called viscoelastics. The vis-coelastic properties of HA solutions are determined primarily by the concentration and the M_r of the HA polysaccharide chains.

Other factors, such as the conformation of the HA chains and the occurrence of intra-chain hydrogen bonds[22], the presence of water molecules that form bridges between adjacent carboxyl and *N*-acetyl groups[23], limit rotation at the glycosidic linkages, and the ability of HA to form networks[24], are also important in determining the viscoelastic properties of HA solutions.

The contribution of the conformation and three-dimensional struc-ture of the HA chains to the viscoelastic properties of HA solutions can be estimated by performing viscosity measurements at pH values of 7 and 11. Under the appropriate alkaline conditions the viscosity of HA solutions is drastically decreased without cleavage of the glycosidic bonds and without decreasing the M_r[25]. The decrease in viscosity at high pH was reversed when the pH was lowered to 7. The low viscosity at high pH was presumably due to the disruption of hydrogen-bonds with a resulting collapse of the HA molecules, so that they then occupy a much smaller hydrodynamic volume.

Under physiological conditions of ionic strength and pH, HA chains in solution are thought to have a random-coil configuration that possesses a certain degree of stiffness[15,16,20]. It is clear, however, from a number of X-ray studies, that HA chains can assume different helical[26–29], as well as double-helical, configurations[30,31]. Optical measurements[32] have also provided evidence that a double-stranded structure for HA can occur.

Biological Properties

Hyaluronic acid occurs in tissues in different forms and in different concentrations. In tissues, such as the rooster comb dermis and umbilical cord where it is present in the extracellular matrix in abundant quantities, the majority of the HA, associated with varying amounts of proteins, can be readily extracted by low-ionic-strength aqueous solutions[33]. In other tissues, such as cartilage, there are relatively small amounts of HA in the extracellular matrix and this HA occurs as a component in a complex organized extracellular matrix, where it interacts specifically with proteoglycans and glycoprotein constituents[34,35]. In synovial fluid, HA occurs in solution and is associated with plasma proteins[36] and joint tissue-specific glycoproteins[37]. This mixture of macromolecules contributes to the viscoelastic and joint tissue-lubricating properties of synovial fluid[38,39]. In other connective tissue, extracellular matrix HA appears to be associated with other constituents with which it interacts in a specific manner[40–43].

Initial studies on the biological properties of HA focused on the physical characteristics of the molecule, namely its polyanionic nature and high M_r. As a consequence of its physical structure, HA is thought to contribute to the osmotic pressure exerted by tissues[44], to exert exclusion effects[45], and to influence the transport[46] of other solute molecules. Its ability to act as a lubricant for joint soft tissues also appears to be an M_r dependent property[39]. In addition to these functions, other studies have shown that HA has a wide variety of biological properties that appear to be related to the chemical structure of HA molecules, as well as to its polymeric characteristics. The synthesis of HA and its accumulation in extracellular matrices is associated with the migration of cells as part of developmental[47] and wound-healing[48] sequences of events, as well as tumorigenesis[49]. In

289

addition, HA has been shown to influence the migration[50] and adhesion of neutrophils[51], cell aggregation[52] and proliferation[53], as well as leukocyte functions[54].

Hyaluronic Metabolism

Early studies showed that HA synthesis occurred by the alternate and sequential addition of the glucuronic acid and *N*-acetylglucosamine units from nucleotide intermediates[55]. The HA synthetase is located at the plasma membrane[56], and the addition of sugar units occurs at the reducing end of the chain followed by extrusion of the HA polysaccharide chain to the exterior of the cell[57]. There is thought to be a close association between HA synthesis, which appears to be a requirement for the completion of mitosis, and the sequence of events in the cell cycle[58]. There are also relationships between HA synthesis and cell density in culture[59], and HA synthesis is increased in proliferating cultures[60].

Catabolism of HA occurs intracellularly following uptake by cells with HA receptors[61]. Mammalian hyaluronidases are endo-β hexosaminidases[62] located in lysosomes[63]. The principal route of catabolism following diffusion of HA from the tissue site appears to be transport and degradation by cells in the lymphatics and liver[64]. The upper M_r limit for the renal excretion of HA is about 25,000[65].

Perhaps one of the most important properties of HA relating to its current and potential use as a biomaterial is that it does not normally initiate an antibody-producing response following injection into animals and humans[66,67]. This is to be expected because HAs with identical chemical structure have such a wide distribution in many biological systems. Recent studies, however, have detected naturally occurring antibodies that cross-react with HA[68], and with the appropriate HA antigen mixture, antibodies to HA can be produced in mice[69].

Development of Medical Applications of Hyaluronic Acid

The initial use of HA as a biomaterial was as a product prepared from bovine vitreous and used as a vitreous substitute in retina-reattachment surgery[70]. An experimental product containing HA and collagen prepared from human umbilical cords[71] was then used for the same purpose. Later studies in animals and humans employed purified preparations of human umbilical-cord HA[72]. This product was also used in early studies to treat lameness in horses[73] and osteoarthritis in humans[74].

Although these early studies were important in that they explored the use of HA for medical applications and laid a foundation for further studies, the present success of HA as a biomaterial in ophthalmology is based on a use that became apparent only after the introduction of plastic intraocular lenses as implants following cataract extraction. The use of HA as a viscoelastic in these procedures was invented by Miller[75]. This use triggered a new era in the development of HA biomaterials and was largely responsible for the present commercial interest in the development of HA products for a variety of medical applications.

The majority of the HA used for ophthalmic applications is manufactured from rooster combs. The use of this tissue source was based on studies that showed rooster comb HA had a high M_r,[76] and could be isolated in a pure form, free of proteins and in high yield by methods that were suitable as manufacturing procedures[33,77].

Manufacturing Issues

As already outlined, the structure of HA, its relationship to other tissue constituents and to agents that cause degradation, are well understood. It is possible, therefore, to devise manufacturing procedures for HA to achieve the desired physical properties, chemical purity and biocompatibility. Most, if not all, of the procedures used to extract HA from tissues[33,78,79], the purification and fractionation by sedimentation[11,15,33,76], treatment with chloroform[80,81], differential precipitation with detergents[82], ion-exchange[83] and gel-permeation[84] chromatography and filtration[85] methods, have been well decribed in the literature.

The methods used to test the purity of HA preparations are based on chemical and enzymatic methods[6,33,76,77] to determine the presence of HA and the absence of other macromolecular constituents, such as proteins, nucleic acids or other polysaccharides. Physical methods principally employing viscosity and light-scattering measurements[15] are employed to determine the M_r of the HA and the mechanical properties of the product. Biological methods, such as intra-vitreal implantation in owl monkeys[86], are then used to determine that the product is biocompatible and free of contaminants which may cause inflammation.

More recently there has been a great deal of interest in manufacturing HA by bacterial fermentation processes for both medical, as well as cosmetic, applications. One of the ophthalmic viscoelastic products (Viscoat®) currently approved for use in the United States contains a low M_r HA that is manufactured by a fermentation procedure.

Patent and Regulatory Issues

Perhaps the simplest assessment of the commercial interest in HA as a biomaterial is the number of patents that have been awarded. In a recent search for HA and HA-related patents awarded worldwide, a total of 488 were recovered. Over the years 1952–59, 5 patents were awarded compared to 6 in the 1960s, 35 in the 1970s, 63 between 1980 and 1985, and 377 between 1986 and 1990.

The majority of these patents (56%) were assigned to Japanese companies and reflect their interest in manufacturing HA principally by fermentation methods for cosmetic applications, but also for medical uses. The United States accounted for about 20% of the patents, Sweden about 5% and France, Italy and the United Kingdom about 3% each.

The areas of technology covered by these patents also demonstrate a wide range of manufacturing methodologies, product forms, and applications. A selection of patents in areas other than cosmetics are outlined in Table 1.

Of particular interest to the US market for HA products is the patent by Balazs[90] which was allowed by the US Patent Office because it constituted a 'special fraction of HA' with certain noninflammatory characteristics defined by a test in owl monkeys. Despite the prior sale of the patented product and inaccuracies in the claims with regard to the ORD characteristics of this putative special fraction, a Boston court delivered a finding indicating that all HA fractions having the noninflammatory characteristics of 'the special fraction' would infringe the patent. Essentially, this may mean that the patent applies, with narrow exception, to all properly purified, high-M_r, naturally

Table 1 A selection of patents on hyaluronic acid*

	Reference numbers
Manufacturing	
Extraction and purification processes	87–98
Bacterial fermentation	99–120
Heat sterilization	121
Low M_r HA	122–126
Formulations	127–130
Analytical procedures	131–134
Medical applications	
Ophthalmic	135–144
Wound healing	145–147
Adhesion prevention	148–150
Surgical dressing	151
Topical dressing	152–154
Collagen matrices	155–158
Skin substitute	159, 160
Bone graft substitute	161
Artificial cartilage	162
Pain relief	163
Tissue lubricant	164, 165
Crosslinked implants	148, 166–171
Coating of biomaterials	172–181
Drug delivery	182–198
Treatment of skin infections	199, 200
Nasal treatments	201, 202
Stimulation of phagocytosis	203, 204
Immunogenicity modifying agent	205–207
Neoplasm inhibitor	208
Sperm separation medium	209
Cell culture medium	210
Epidermal cell growth inhibitor	211
Intra-articular injection	212
Miscellaneous	
Beverage	213, 214
Hair treatments	215–219
Antistatic agent	220
Thermal printing	220–224

*Patents concerned with cosmetic applications of HA are not included.

occurring HA, not just the 'special fraction' defined in the patent. There is still outstanding litigation regarding this patent and it will be interesting to see how it is finally resolved. Whatever the outcome of these ongoing proceedings, this patent has caused a great deal of controversy.

Another factor that has been, and still is, a major issue in the development of HA products is the extensive regulatory process that must be completed before commercial sale. The regulatory sequence of events which led to the approval of viscoelastic HA products for use in ophthalmology was also quite an unusual one.

Initially, in the early 1970s, the Food and Drug Administration (FDA) considered that HA products were drugs, and safety and efficacy studies with MedChems Products' HA products, HYVISC®, for ophthalmic and equine orthopaedic applications were conducted accordingly. While these studies were in progress, Pharmacia submitted a 510K application to the FDA for the approval of Healon® for ophthalmic indications. This application considered that HA was a 'substantially equivalent' substance to collagen, which was approved as a device before 1976. The FDA approved this application and Healon became available for sale in 1979. Thereafter the FDA reinvestigated the classification of HA viscoelastics for use in ophthalmology and reclassified them as class III devices which are subject to pre-market approval before commercial sale. At the present time, HA products for human, ophthalmic, orthopaedic and wound-healing applications are still considered as class III devices in the United States, whereas HYVISC® and other products used for the intra-articular treatment of equine osteoarthritis are classified as drugs.

Current and Future HA Products

The first generation of HA products currently being used in ophthalmology, orthopaedics, and some that are being developed for wound-healing applications, have employed aqueous solutions of HA. With these products the mechanism of action and efficacy are dependent on the viscoelastic properties of the HA. In the case of the ophthalmic viscoelastic products, these products aid in the surgical procedures by acting as space-filling agents, maintaining a separation between tissues. These products coat tissue surfaces and act as a lubricant, thus protecting the tissues from trauma. The HA products used to treat synovitis associated with equine osteoarthritis have a similar mechanism of action, in that they improve joint function by acting as a lubricant for the joint tissues. In addition to the viscoelastic lubricating effect, there appears to be other beneficial effects derived from the intra-articular injection of HA to treat joint dysfunction. In the case of

HYVISC®, an initial beneficial effect is observed in the period following the intraarticular injection and this effect persists beyond the time that the implanted HA remains in the joint[225].

This long-term effect has also been observed in a recent clinical study of ORTHOVISC® used to treat reducing displacement of the disc in temporomandibular joint dysfunction[226]. A single injection was found to be efficacious, by both objective and subjective criteria, for periods up to 6 months. This persisting efficacy is thought to be due to the ability of HA to modify the behaviour of both resident cells in joint tissues as well as transient cells, such as phagocytes and lymphocytes.

The presently developed HA solution products have been used in clinical medicine only for relatively specialized indications and for a relatively short period of time. It is expected, therefore, that there will be a continued growth in the use of these viscoelastic products. The estimated annual potential markets for these products in the US are listed in Table 2.

Table 2 Potential markets for HA products in the United States ($MM)

Orthopaedic–synovial fluid substitute	
Temporomandibular joint	$50–100
Arthroscopy	$100–200
Arthritis	$800–1,500
Hip replacement	$50
Diagnostic	$500–1,000
Sports injuries/trauma	$50–100
Ophthalmology	$250
Wound management	$200–500

Once these initial viscoelastic products have become well established, it seems likely that a second generation of HA products will be developed which will utilize different forms, such as films, coatings and crosslinked implants[227]. In addition, the fact that HA is a naturally occurring biocompatible polymer makes it an ideal candidate for use as a drug-delivery vehicle for many medical indications. As more information becomes available about the mechanisms by which HA is able to modify the behaviour of cells and also the relationship that appears to exist between HA synthesis and the cell cycle, it seems likely that other specific applications for HA therapy will be devised.

References

1. Meyer, K. & Palmer, J. (1934) The polysaccharide of the vitreous humour. *J. Biol. Chem.* **107**: 629–634.

2. Everd, D. & Whelan, J. (eds) (1989) *The Biology of Hyaluronan*, Ciba Foundation Symposium, Vol. 143. Wiley, Chichester, UK.

3. Laurent, T.C. (1987) Biochemistry of hyaluronan. *Acta Otolaryngol. (Stockh.) suppl.* **442**: 7–24.

4. Swann, D.A. (1980) Chemistry and biology of the vitreous body. *Int. Rev. Exp. Pathol.* **2**: 1–57.

5. Comper, W.D. & Laurent, T.C. (1978) Physiological function of connective tissue polysaccharides. *Physiol. Rev.* **58**: 255–315.

6. Meyer, K. (1958) Chemical structure of hyaluronic acid. *Fed. Proc.* **17**: 1075–1077.

7. Flowers, H.M. & Jeanloz, R.W. (1964) The synthesis of the repeating unit of hyaluronic acid. *Biochemistry* **3**: 123–125.

8. Varma, R., Varma, R.S., Allen, W. & Wardi A.H. (1978) Neutral sugars from the hyaluronate-peptide of vitreous humour. *Carbohydr. Res.* **32**: 386–395.

9. Katzman, R.L. (1974) Absence of arabinase in bovine brain hyaluronic acid as analyzed by gas-liquid chromatography. *Biochim. Biophys. Acta* **372**: 53.

10. Varga, L. (1955) Studies on hyaluronic acid prepared from the vitreous body. *J. Biol. Chem.* **217**: 651–658.

11. Silpananta, P., Dunstone, J.R. & Ogston, A.G. (1968) Fractionation of a hyaluronic acid preparation in a density gardient. Some properties of the hyaluronic acid. *Biochem. J.* **109**: 43–50.

12. Welsh, E.J., Rees, D.A., Morris, E.R. & Madden, J.K. (1980) Competitive inhibition evidence for specific intermolecular interactions in hyaluronate solutions. *J. Mol. Biol.* **138**: 375–382.

13. Silver, F.H. & Swann, D.A. (1982) Laser light scattering measurements on vitreous and rooster comb hyaluronic acids. *Int. J. Macromol.* **4**: 425–429.

14. Turner, R.E., Lin, P.Y. & Cowman, M.K. (1988) Self-association of hyaluronate segments in acqueous NaCl solution. *Arch. Biochem. Biophys.* **265**: 484–495.

15. Laurent, T.C., Ryan, M. & Pietruszkiewicz, A. (1960) Fractionation of hyaluronic acid: The polydispersity of hyaluronic acid from the bovine vitreous body. *Biochem. Biophys. Acta* **42**: 476–485.

16. Laurent, T.C. & Gergely, J. (1955) Light scattering studies on hyaluronic acid. *J. Biol. Chem.* **212**: 325–333.

17. Laurent, U.B.G. & Granath, K.A. (1983) The molecular weight of hyaluronate in the aqueous humour and vitreous body of rabbit and cattle eyes. *Expt. Eye Res.* **36**: 481–492.

18. Swann, D.A., Silver, F.H., Sotman, S.L. & Hermann, H. (1982) Measurements of reducing end groups on bovine vitreous humor hyaluronic acid by reaction with ^{14}[C] cyanide. *Biochem. J.* **207**: 409–414.

19. Swann, D.A. (1969) Studies on the structure of hyaluronic acid. Characterization of the product formed when hyaluronic acid is treated with ascorbic acid. *Biochem. J.* **114**: 819–825.

20. Ogston, A.G. & Stanier, J.E. (1951) The dimensions of the particle of hyaluronic acid complex in synovial fluid. *Biochem. J.* **49**: 585–590.

21. Ogston, A.G. & Stanier, J.E. (1953) Composition and properties of hyaluronic acid complex of ox synovial fluid. *Discuss. Faraday Soc.* **13**: 275–280.

22. Atkins, E.D.T., Meader, D. & Scott, J.E. (1980) Model for hyaluronic acid incorporating four intramolecular hydrogen bonds. *Int. J. Biol. Macromol.* **2**: 318–319.

23. Heatley, F. & Scott, J.E. (1988) A water molecule participates in the secondary

structure of hyaluronan. *Biochem. J.* **254**: 489–493.

24. Scott, J.E., Cummings, C., Brass, A. & Chen, Y. (1991) Secondary and tertiary structures of hyaluronan in aqueous solution investigated by rotary shadowing-electron microscopy and computer simulation. *Biochem. J.* **274**: 699–705.

25. Swann, D.A. (1970) On the state of hyaluronic acid in a connective tissue matrix. In *The Chemistry and Biology of the Intercellular Matrix* (ed. Balazs, E.A.), Vol. II, p. 743. Academic, New York.

26. Atkins, E.D.T. & Sheehan, J.K. (1973) Hyaluronates: relation between molecular conformations. *Science* **179**: 562–564.

27. Sheehan, J., Atkins, E. & Nieduszynski, I. (1975) Studies on the connective tissue polysaccharides. Two-dimensional packing schemes for three-fold hyaluronate chains. *J. Mol. Biol.* **91**: 153–163.

28. Winter, W.T., Smith, P.J.C. & Arnott, S. (1975) Hyaluronic acid: structure of a fully extended 3-fold helical sodium salt and comparison with the less extended 4-fold helical forms. *J. Mol. Biol.* **99**: 219–235.

29. Winter, W.T. & Arnott, S. (1977) Hyaluronic acid: The role of divalent cations in conformation and packing. *J. Mol. Biol.* **177**: 761–784.

30. Sheehan, J.K., Gardner, K.H. & Atkins, E.D.T. (1977) Hyaluronic acid: A double helical structure in the presence of potassium at low pH and found also with the cations ammonium, rubidium and caesium. *J. Mol. Biol.* **117**: 113–135.

31. Arnott, S., Mitra, A.K. & Raghunathan, S. (1983) Hyaluronic acid double helix. *J. Mol. Biol.* **169**: 861–872.

32. Staskus, P.W. & Johnson, W.C. Jr (1988) Double-stranded structure for hyaluronic acid in ethanol-aqueous solution as revealed by circular dichroism of oligomers. *Biochemistry* **27**: 1528–1534.

33. Swann, D.A. (1968) Studies on hyaluronic acid. I. The preparation and properties of rooster comb hyaluronic acid. *Biochim. Biophys. Acta* **156**: 17–30.

34. Hardingham, T. & Muir, H. (1972) The specific interaction of hyaluronic acid with cartilage proteoglycans. *Biochim. Biophys. Acta* **279**: 401–405.

35. Tengblad, A. (1981) A comparative study of the binding of cartilage link protein and the hyaluronate-binding region of the cartilage proteoglycan to hyaluronate-substituted sepharose gel. *Biochem. J.* **199**: 297–305.

36. Silpanata, P., Dunstone, J.R. & Ogston, A.G. (1969) Protein associated with hyaluronic acid in ox synovial fluid. *Austr. J. Biol. Sci.* **22**, 1031–1037.

37. Swann, D.A. (1978) Macromolecules of synovial fluid. In *The Joints and Synovial Fluid* (ed. Sokoloff.), Vol. 1, pp. 407–435. Academic, New York.

38. Swann, D.A., Slayter, H.S. & Silver, F.H. (1981) The molecular structure of lubricating glycoprotein—I, the boundary lubricant for articular cartilage. *J. Biol. Chem.* **256**, 5921-5925.

39. Swann, D.A., Radin, E.L., Nazimiec, M., Weisser, P.A., Curran, N. & Lewinnek, G. (1974) Role of hyaluronic acid in joint lubrication. *Ann. of Rheum. Dis.* **33**: 318–326.

40. Delpech, B. & Halavent, C. (1981) Characterization and purification from human brain of a hyaluronic acid-binding glycoprotein, hyaluronectin. *J. Neurochem.* **36**: 855–859.

41. Yamada, K.M., Kennedy, D.W., Kimata, K. & Pratt, R.M. (1980) Characterization of fibronectin interactions with glycosaminoglycans and identification of active proteolytic fragments. *J. Biol. Chem.* **255**: 6055–6063.

42. Isemura, M., Yosizawa, Z., Koide, T. & Ono, T. (1982) Interaction of fibronectin and its proteolytic fragments with hyaluronic acid. *J. Biochem. (Tokyo)* **91**: 731–4.

43. D'Souza, M. & Datta, K. (1986) A novel glycoprotein that binds to hyaluronic

acid. *Biochem. Int.* **13**: 79–88.

44. Laurent, T.C. & Ogston, A.G. (1963) The interaction between polysaccharides and other macromolecules and the osmotic pressure of mixtures of serum, albumin and hyaluronic acid. *Biochem. J.* **89**: 249–253.

45. Ogston, A.G. & Phelps, C.F. (1960) The partition of solutes between buffer solutions and solutions containing hyaluronic acid. *Biochem. J.* **78**: 827–833.

46. Preston, B.N., Laurent, T.C., Comper, W.D. & Checkley, G.J. (1980) Rapid polymer transport in concentrated solutions through the formation of ordered structures. *Nature* **287**: 499–503.

47. Toole, B.P. & Gross, J. (1971) The extracellular matrix of the regenerating newt limb: Synthesis and removal of hyaluronate prior to differentiation. *Dev. Biol.* **25**: 57–77.

48. Abatangelo, G., Martelli, M. & Vecchia, P. (1983) Healing of hyaluronic acid-enriched wounds: Histological observations. *J. Surg. Res.* **35**: 410–416.

49. Toole, B.P., Biswas, C. & Gross, J. (1979) Hyaluronate and invasiveness of the rabbit V2 carcinoma. *Proc. Natl. Acad. Sci. U.S.A.* **76**:6299–6303.

50. Hakansson, L. & Venge, P. (1987) The molecular basis of the hyaluronic acid-mediated stimulation of granulocyte function. *J. Immun.* **138**: 4347–52.

51. Forrester, J.V. & Lackie, J.M. (1981) Effect of hyaluronic acid on neutrophil adhesion. *J. Cell. Sci.* **50**: 329–344.

52. Wasteson, A., Westermark, B., Lindahl, U. & Ponten, J. Aggregation of feline lymphoma cells by hyaluronic acid. *Int. J. Cancer* **12**: 169–178.

53. Goldberg, R.L. & Toole, B.P. (1987) Hyaluronate inhibition of cell proliferation. *Arth. Rheum.* **30**: 769–778.

54. Hakansson, L., Hallgren, R., Venge, P., Artursson, G. & Vedung, S. (1980) Hyaluronic acid stimulates neutrophil function in vitro and in vivo. A review of experimental results and a presentation of a preliminary clinical trial. *Scand. J. Infect. Dis. (Suppl.)* **24**: 54–57.

55. Roden, L. (1980) Structure and metabolism of connective tissue proteoglycans. In *The Biochemistry of Glycoproteins and Proteoglycans* (ed. Lennarz, W.J.), pp. 267–371. Plenum, NewYork.

56. Mian, N. (1986) Character of a high Mr plasma membrane bound protein and assessment of its role as a constituent of Hyaluronate synthetase complex. *Biochem. J.* **237**: 343–357.

57. Prehm, P. (1983) Synthesis of hyaluronate in differentiated teratocarcinoma cells. Mechanism of chain growth. *Biochem. J.* **21**: 191–198.

58. Brecht, M., Mayer, U., Schlosser, E. & Prehm, P. (1986) Increased hyaluronate synthesis is required for fibroblast detachment and mitosis. *Biochem. J.* **239**: 445–450.

59. Cohn, R.H., Cassiman, J.J. & Bernfield, M.R. (1976) Relationship of transformation, cell density and growth control to the cellular distribution of newly synthesized GAG. *J. Cell Biol.* **71**: 280–293.

60. Tomida, M., Koyama, H. & Ono, R. (1975) Induction of hyaluronic acid synthetase activity in rat fibroblasts by media change of confluent cultures. *J. Cell Physiol.* **86**: 121–130.

61. Laurent, T.C., Fraser, J.R.E., Pertoft, H. & Smedsrod, B. (1986) Binding of hyaluronate and chondroitin sulphate to liver endothelial cells. *Biochem. J.* **234**: 653–658.

62. Gibian, H. (1968) Hyaluronidases. In *The Amino Sugars* (ed. Jeanloz, R. & Balazs, E.A.), Vol. IIB, pp. 181–200. Academic, New York.

63. Aronson, N.N. Jr & Davidson, E.A. (1967) Lysosomal hyaluronidase from rat liver. 1. Preparation. *J. Biol. Chem.* **242**: 437–440.

298

64. Fraser, J.R.E., Appelgren, L.E. & Laurent, T.C. (1983) Tissue uptake of circulating hyaluronic acid. A whole body radiographic study. *Cell Tissue Res.* **233**: 285–293.

65. Laurent, T.C., Lilja, K., Brunnberg, L. *et al.* (1987) Urinary excretion of hyaluronan in man. *Scand. J. Chem. Invest.* **47**: 793–799.

66. Richter, W. (1974) Non-immunogenicity of purified hyaluronic acid preparations tested by passive cutaneous anaphylaxis. *Arch. Allergy* **47**: 211–17.

67. Richter, A.W., Ryde, E.M. & Zetterstrom, E.O. (1979) Non-immunogenicity of a purified sodium hyaluronate preparation in man. *Int. Arch. Allergy Appl. Immun.* **59**: 45–48.

68. Underhill, C.B. (1982) Naturally-occurring antibodies which bind hyaluronate. *Biochem. Biophys. Res. Commun.* **108**: 129–30.

69. Fillet, H.M., McCarty, M. & Blake, M. (1986) Induction of antibodies to hyaluronic acid by immunization of rabbits with encapsulated streptococci. *J. exp. Med.* **164**: 762–776.

70. Hruby, K. (1961) Hyaluronsaure als Glaskorparersatz bei Metzhautablosung. *Klin. Mbl. Augenheilk* **138**: 484–496.

71. Balazs, E.A. & Sweeney, D. (1966) The replacement of the vitreous body in the monkey by reconstituted vitreous and by hyaluronic acid. *Mod. Probl. Ophthalmol.* **4**: 230.

72. Balazs, E.A., Freeman, M.I., Kloti, R. *et al.* (1972) Hyaluronic acid and replacement of vitreous and aqueous humor. *Mod. Probl. Ophthalmol.* **10**: 3–21.

73. Rydell, N. & Balazs, E.A. (1971) Effect of intra-articular injection of hyaluronic acid on the clinical symptoms of osteoarthritis and on granulation tissue formation. *Clin. Orthop.* **80**: 25–32.

74. Peyron, J.G. & Balazs, E.A. (1974) Preliminary clinical assessment of Na-hyaluronate injection into human arthritic joints. *Path. Biol.* **22**: 731–736.

75. Miller, D., O'Connor, P. & Williams, J. (1977) Use of Na-hyaluronate during intraocular lens implantation in rabbits. *Ophthal. Surg.* **8**: 58–61.

76. Laurent, T.C. (1955) A comparative study of physico-chemical properties of hyaluronic acid prepared according to different methods and from different tissues. *Arkiv. Kemi.* **11**: 487–496.

77. Swann, D.A. (1968) Studies on hyaluronic acid. II. The protein components of rooster comb hyaluronic acid. *Biochim. Biophys. Acta* **160**: 96–105.

78. Boas, N. (1949) Isolation of hyaluronic acid from the cock's comb. *J. Biol. Chem.* **181**: 573–575.

79. Jeanloz, R.W. & Forchielli, E. (1950) Studies on hyaluronic acid and related substances. 1. Preparation of hyaluronic acid and derivatives from human umbilical cord. *J. Biol. Chem.* **186**: 495–511.

80. Sevag, M.G. (1934) Eine neue physikalische Enteiweibungsmethode zur darstellung biologisch wirksamer Substanzen. *Biochem. Z.* **273**: 419–429.

81. Blix, G. & Snellman, O. (1945) On chondroitin sulphuric acid and hyaluronic acid. *Ark. Kemi Mineral. Geol.* **19A, No. 32**: 1–19.

82. Scott, J.E. (1960) Aliphatic ammonium salts in the assay of acidic polysaccharides from tissues. *Meth. Biochem. Anal.* **8**: 145–197.

83. Berman, E.R. (1963) Distribution and localization of various molecular species of hyaluronic acid in the bovine vitreous body. *Exp. Eye Res.* **2**: 1–11.

84. Caygill, J.C. (1971) Evidence for the aggregation of hyaluronate protein fractionated by chromatographic onagarose. *Biochim. Biophys. Acta* **244**: 421–426.

85. Ogston, A.G. & Stanier, J.E. (1950) On the state of hyaluronic acid in synovial fluid. *Biochem. J.* **46**: 364–376.

86. Constable, I.J. & Swann, D.A. (1972) Evaluation of the biological vitreous

Biomaterials

substitutes. The inflammation response in normal and altered. *Arch. Ophthalmol.* **88**: 544–548.

87. Hadidian, Z. & Pirie, N.W. (1952) High-viscosity hyaluronic acid. Patent US 2583096.

88. Hadidian, Z. & Pirie, N.W. (1952) High-viscosity hyaluronic acid. Patent US 2585546.

90. Balazs, E.A. (1979) Ultrapure hyaluronic acid and its use. Patent US 4141973.

91. Green Cross Corp. (1981) Production of hyaluronic acids from connective tissues. Patent JP 58084801.

92. Takahashi, T., Komiyama, Y., Oogushi, T. *et al.* (1986) Isolation of hyaluronic acid from cockscomb for pharmaceutical and cosmetic uses. Patent JP 61171703.

93. Cullis-Hill, D. (1986) Hyaluronic acid. Patent WO 8606728.

94. Kubota, H. (1979) Pyrogens and fungi removing agents. Patent JP 54067024.

95. Seki, M., Ogawa, Z., Inoue, Y. *et al.* (1988) Purification of hyaluronic acid. Patent JP 63270701.

96. Kono, S., Nishimura, H, Ishii, H. *et al.* (1989) Alumina for purification of hyaluronic acid. Patent JP 01313503.

97. Kitagawa, H., Chiba, S., Saegusa, H. *et al.* (1990) Purification of hyaluronic acid. Patent JP 02103204.

98. Yashizawa, Y., Yamada, K. & Fukui, F. (1990) Ultrafiltration membrane in purification of high-molecular-weight hyaluronic acid. Patent JP 02047101100.

99. Bracke, J.W. & Thacker, K. (1984) Hyaluronic acid from bacterial culture. Patent WO 8403302.

100. Brown, K.K. & Cooper, H. (1985) Use of ultrapure hyaluronic acid to improve animal joint function. Patent EP 143393.

101. Brown, K.K., Ruiz, L.L.C. & Van de Rijn, I. (1985) Ultrapure hyaluronic acid. Patent EP 144019.

102. Nimrod, A., Greenman, B., Kanner, D. *et al.* (1986) High molecular weight sodium hyaluronate by fermentation. Patent WO 8604355.

103. Akasaka, H. (1986) Bacterial production of hyaluronic acid. Patent JP 61219394.

104. Kurato, Y., Shibata, S., Kawahara, S. *et al.* (1987) Hyaluronic acid production by *Streptococcus*. Patent JP 62051999.

105. Morita, Y., Ushiyama, M. & Fujii, M. (1986) Manufacture of hyaluronic acid by fermentation method. Patent JP 61063294.

106. Miyazaki, T., Ochi, H., Shintani, K. *et al.* (1987) Fermentative production of hyaluronic acid. Patent JP 62215397.

107. Miyazaki, T., Ochi, H., Shintani, K. *et al.* (1987) Microbial production of hyaluronic acid. Patent JP 62289198.

108. Morikawa, K., Hamai, A. & Horie, K. (1987) Manufacture of hyaluronic acid for ophthalmological use. Patent JP 62257901.

109. Murata, H., Himoni, M. & Ishii, S. (1988) Manufacture of pyrogen-free hyaluronic acid from *Streptococcus* culture. Patent JP 63012293.

110. Miyamori, T., Numazawa, R., Sakimae, A. *et al.* (1988) High molecular-weight hyaluronic acid manufactured by micro-organisms cultured in the presence of saccharides containing hydro-oyaromatic substituents. Patent EP 266578.

111. Chiba, S., Kitagawa, H. & Myoshi, T. (1988) Fermentative manufacture of high-molecular weight hyaluronic acid. Patent JP 63129991.

112. Chiba, S., Kitagawa, H., Hashimoto, M. *et al.* (1988) Enhancement of hyaluronic acid manufacture with *Streptococcus* equi ATCC 9527. Patent JP 63141594.

113. Kono, S., Ishii, H. Kogure, M. (1988) Chitosan for removal of bacteria from

hyaluronic acid fermentation mixtures. Patent JP 63276493.

114. Hosoya, H., Kimura, M. & Endo, H. (1989) Hyaluronic acid and its manufacture with hyaluronidase-deficient *Streptococcus* mutant. Patent JP 01067196.

115. D'Hinterland, L.D., Normier, G., Durand, J. *et al.* (1989) Manufacture of high molecular weight hyaluronic acid by *Streptococcus pyogenes*. Patent FR 2617849.

116. Miyamoto, T., Numazawa, R. & Sakimae, A. (1989) Fermentative manufacture of hyaluronic acid. Patent JP 01225491.

117. Swann, D.A., Sullivan, B.P, Jamieson, G. *et al.* (1990) Hyaluronic acid and its manufacture with *Streptococcus zooepidemicus*. Patent US 4897349.

118. Sato, I., Shindo, T. & Kurokawa, Y. (1990) Hyaluronic acid fermentation, and effects of nitrogen sources. Patent JP 02058502.

119. Hashimoto, M., Saegusa, H., Chiba, S. *et al.* (1990) Sodium hyaluronate manufacture with *Streptococcus* equi. Patent US 4946780.

120. Weigel, P.H. & Papaconstantinou, J. (1991) Cloning of DNA encoding hyaluronate synthase. Patent WO 9103559.

121. Brown, K.K., Greene, N.D., Trump, S.L. *et al.* (1987) Sterilizable high-molecular weight hyaluronic acid for injection. Patent EP 228698.

122. Della Valle, F. & Aurelio, R. (1985) Hyaluronic acid fractions having therapeutic activity. Patent BE 900810.

123. Hildesheim, J. (1987) Non-inflammatory heat-resistant low mol. wt. hyaluronic acid, a process for extracting it from crude sources, and compositions containing it. Patent EP 239335.

124. Shindo, T., Hatakeyama, M. & Fujii, M. (1989) Hyaluronic acid and salts with low viscosity and nonenzymic process for their preparation. Patent JP 01266102.

125. Sugitani, H., Sugitani, T., Nozawa, T. *et al.* (1990) Manufacture of hyaluronic acid alkali salts having low degree of polymerization with *Streptococcus* or *Pasturella*. Patent JP 02234193.

126. Akasaka, H. & Yamaguchi, T. (1991) Production of low-molecular weight hyaluronic acid by shear. Patent WO 9104279.

127. Kurokawa, Y., Shindo, T. & Fujii, M. (1989) Manufacture of powdered sodium hyaluronate. Patent JP 01313502.

128. Egawa, K., Mori, S. & Yoshizawa, M. (1990) Mucopolysaccharide prosthetic fibers for surgical and dental use. Patent JP 02014019.

129. Kurokawa, Y., Hiraki, J. & Fujii, M. (1990) Manufacture of powdered sodium hyaluronate. Patent JP 02142801.

130. Shinohara, M. (1990) Manufacture of hyaluronic acid beads. Patent JP 02167301.

131. Chichuibu, K. (1988) Methods of assaying high molecular weight hyaluronic acid and reagent kits for the assay. Patent EP 283779.

132. Brandt, R.K. & Hedloef, M.E. (1988) Specific binding inhibition assay for determining hyaluronic acid. Patent EP 271461.

133. Corning Glass Works (1989) Proteoglycans and anti-keratan sulfate antibody for immunoassay of hyaluronic acid in body fluids. Patent JP 01029767.

134. Ghosh, P. & Kongtawelert, P. (1990) Inhibition method and kit for the detection and quantification of hyaluronan (hyaluronic acid) using polycationic material-coated solid support. Patent WO 9007121.

135. Pape, L.G. (1982) Ophthalmological procedures. Patent US 4328803.

136. Chang, A.S., Boyd, J.E., Koch, H.O. *et al.* (1985) Chondroitin sulfate/sodium hyaluronate compositions. Patent EP 136782.

137. Shimizu, H. Green Cross Corp. (1985) Eye lotions containing hyaluronates. Patent JP 60084225.

138. Benanti, G. (1988) Anti-inflammatory formulations containing lysine salt of 4-diphenylacetic acid. Patent EP 261565.

139. Iwao, J., Iso, T., Uemura, O. *et al.* (1989) Manufacture of eye lotions containing sodium hyaluronate. Patent WO 8900044.

140. Nakada, K., Ogasawara, M. & Ichikawa, M. (1989) Manufacture of intraocular lens. Patent JP 01032859.

141. Iwata, S., Miyauchi, S. & Sakamoto, T. (1989) Hyaluronic acid (salts) for treatment of disorders in corneal epithelium. Patent JP 01238530.

142. Kita, K. (1989) Therapeutic film containing hyaluronic acid and fibronectin for treatment of corneal damage. Patent JP 01279836.

143. DeVore, D.P., Swann, D.A. & Sullivan, B.P. (1990) Sodium hyaluronate composition for eye surgery. Patent US 4920104.

144. Morita, T., Iso, T. & Kawashima, Y. (1990) Ophthalmic solutions containing hyaluronic acid (salts), benzalkonium chloride, and chelating agents. Patent JP 02164829.

145. Drenk, F. (1990) Topical compositions comprising fibroblast growth factor and hyaluronic acid for wound-healing promotion. Patent DE 3900198.

146. Takacsi Nagy, G., Takacsi, N.G., Tethey, I. *et al.* (1990) Hyaluronic acid metal complexes for epithelization acceleration. Patent WO 9010020.

147. Pierschbacher, M.D., Polarek, J.W., Petrica, M.P. *et al.* (1990) Polypeptide-polymer conjugates for wound healing. Patent WO 9006767.

148. De Belder, A.N. & Maelson, T. (1986) Gel for preventing adhesion between body tissues. Patent WO 8600912.

149. Goldberg, E.P. & Yaacobi, Y. (1989) Hydrophilic polymers for preventing tissue adhesions during surgery. Patent WO 8911857.

150. Lindblad, G. & Buckley, P. (1990) Dextran-hyaluronic acid mixture for prevention of adhesions between body tissues. Patent WO 9010031.

151. Green Cross Corp. (1983) Surgical dressings containing hyaluronic acid. Patent JP 58057319.

152. Shinohara, M., Kanda, T. & Igrashi, H. (1986) Bandages containing hyaluronic acid and amino acids and their derivatives. Patent JP 61187866.

153. Ishimura, F., Miyata, K., Katayose, H. *et al.* (1988) Films containing hyaluronic acid for manufacturing bandages. Patent JP 63159452.

154. Sugitani, H., Sugitani, T., Nozawa, T. *et al.* (1990) Manufacture of bandage from hyaluronic acid salts for wound healing. Patent JP 02268765.

155. Silver, F.H., Berg, R.A., Birk, D.E. *et al.* (1985) Biodegradable matrix. Patent WO 8504413.

156. Wallace, D.G., Reihanian, H., Pharriss, B.B. *et al.* (1988) Injectable implant composition containing a particulate biomaterial especially crosslinked collagen, and a lubricant which has improved intrudability and injectability. Patent EP 251695.

157. Silver, F.H., Berg, R.A., Doillon, C.J. *et al.* (1989) Biodegradable collagen compositions for treatment of skin wounds. Patent EP 314109.

158. Tsunenaga, M., Tominaga, N., Nishiyama, T. *et al.* (1989) Hyaluronic acid-containing aqueous solution of aqueous dispersion of collagen for tissue substitution or repair. Patent EP 338813.

159. Sakurai, K. & Takuyasu, K. (1989) Manufacture of skin substitute with poly(vinyl alcohol) and hyaluronic acid. Patent JP 01274767.

160. Shimada, M. Takigami, S., Uchida, T. *et al.* (1988) Manufacture of skin protectors from protein membranes on which hyaluronic acid is immobilized. Patent JP 63300770.

161. Breeze, J.H. (1989) Biodegradable, osteogenic, bone graft substitute. Patent GB 2215209.

162. Sakurai, K., Miyazaki, M., Tokuyasu, K. *et al.* (1989) Manufacture of basic materials for artificial cartilage. Patent JP 01268559.

163. Schultz, R.H., Wollen, T.H., Greene, N.D. *et al.* (1987) Hyaluronates for the treatment of pain associated with trauma. Patent EP 243867.

164. Yoneda, T. (1988) Aqueous lubricant compositions for gynaecological vaginal examinations. Patent JP 63218617.

165. Hills, B.A. (1989) Lubricants for joints and body tissue containing hyaluronates and phospholipids. Patent WO 8901777.

166. Balazs, E.A. & Leshchiner, A. (1985) Water-insoluble hyaluronic acid preparation. Patent DE 3434082.

167. Balazs, E.A. & Leshchiner, A. (1985) Water-insoluble hyaluronic acid. Patent FR 2556728.

168. Sakurai, K., Ueno, Y. & Okuyama, T. (1985) Crosslinked hyaluronic acid and its use. Patent EP 161887.

169. Maelson, T. & Lindqvist, B.L. (1986) Gel of crosslinked hyaluronic acid for use as a vitreous humor substitute. Patent WO 8600079.

170. Balazs, E.A. & Leshchiner, A. (1986) Crosslinked gels of hyaluronic acid and products containing these gels for cosmetics and pharmaceuticals. Patent US 4582865.

171. Balazs, E.A., Leshchiner, A. *et al.* (1989) Method for preparing hylan and novel hylan product. Patent EP 320164.

172. Balazs, E.A., Wedlock, D.J. & Phillips, G.O. (1984) Polymeric articles modified with hyaluronate. Patent US 4487865.

173. Balazs, E.A. & Leshchiner, A. (1985) Hyaluronate-coated polymeric articles. Patent US 4500676.

174. Halpern, G., Campbell, C., Beavers, E.M. *et al.* (1985) Hydrophilic coatings for plastics. Patent DE 3529758.

175. Sakurai, K. & Ueno, Y. (1986) Crosslinked glycosaminoglycan composites as artificial organs. Patent JP 61154567.

176. Kodama, A., Sakai, T., Tsuda, K. *et al.* (1986) Antithrombotic materials from polymers coated with mucopolysaccharides, collagens and crosslinking aldehyde compounds. Patent JP 61191364.

177. Maelson, T., Ahrgren, L. & De Belder, A.N. (1986) Shaped article. Patent EP 193510.

178. Maelson, T. (1987) Process for the manufacture of crosslinked carboxy group-containing polysaccharides. Patent SE 452469.

179. Della Valle, F. & Romeo, A. (1988) Crosslinked esters of hyaluronic acid, their preparation, and their pharmaceutical, cosmetic, medical and surgical uses. Patent EP 265116.

180. Halpern, G. & Gould, J.U. (1988) Prosthetic plastic article containing a top coat comprising an albumin and polysaccharide mixture. Patent US 4722867.

181. Leshchiner, A., Larsen, N.E., Balazs, E.A. *et al.* (1988) Cross-linked hyaluronate gels for percutaneous embolization. Patent EO 291177.

182. Aktiebolaget, L. (1956) Preparations with protracted corticotropic (ACTH) effect. Patent UK 761061.

183. Della Valle, F., Romeo, A. & Lorenzi, S. (1986) Hyaluronic acid containing medicaments for topical use. Patent EP 197718.

184. Balazs, E.A., Larsen, N.E. & Leshchiner, A. (1987) Drug delivery systems based on hyaluronan, derivatives thereof and their salts and method of producing same. Patent EP 224987.

185. Della Valle, F. & Romeo, A. (1987) Hyaluronic acid esters and their medical and cosmetic uses and formulations. Patent EP 216453.

186. Keller, N., Olejnk, O. & Abelson, M.B. (1987) Sustained release

compositions containing polysaccharides especially hyaluronic acid for injection into the eye. Patent EP 244178.

187. Cullis-Hill, D. & Ghosh, P. (1988) Polysaccharide sulfate metal complexes for anti-inflammatory compositions. Patent WO 8807060.

188. Ellwood, D.C. (1988) Hyaluronic acid derivatives as sustained-release pharmaceuticals. Patent EP 296740.

189. Hamilton, R.G., Fox, E.M., Acharya, R.A. *et al.* (1989) Water-insoluble biocompatible derivatives of hyaluronic acid, their manufacture and use. Patent WO 8902445.

190. Mohler, M.A. & Nguyen Tue, H. (1989) Topical pharmaceuticals containing sparingly water-soluble enzymes for the prevention of fibrin deposition or adhesion formation in wounds. Patent EP 297860.

191. Machida, M. & Arakawa, M. (1989) Manufacture of slow-release pharmaceutical microgranules. Patent JP 01156912.

192. Myazaki, T., Terasawa, K., Inoe, K. *et al.* (1989) Sustained-release subcutaneous or intramuscular preparations containing hyaluronic acid (salts). Patent JP 01287041.

193. Kawamori, R., Nomura, M., Kubota, S. *et al.* (1989) Sustained release insulin ophthalmic solutions. Patent JP 01294633.

194. Dalla Valle, F. & Romeo, A. (1989) Preparation of crosslinked carboxy polysaccharides as biodegradable plastic materials for cosmetics and pharmaceuticals. Patent EP 341745.

195. Hoshi, K., Yanagawa, A., Mizushima, Y. *et al.* (1989) Superoxide dismutase-containing topical drug preparations. Patent JP 01319427.

196. Kitano, S. & Kitazato, K. (1990) Sustained-release preparations for physiologically active peptides. Patent JP 02000213.

197. Koide, M. & Konishi, A. (1990) Biocompatible materials for preparation of prosthetics and microcapsules. Patent JP 02001287.

198. Maelson, T. & Lindqvist, B. (1990) Phosphorus-crosslinked hyaluronate gels, their use and method for producing them. Patent WO 9009401.

199. Nimrod, A. & Greenman, B. (1987) Heavy metal salts of hyaluronic acid useful as antimicrobial agents. Patent WO 8705517.

200. Maddren, L.T. (1990) Topical treatment of herpesvirus hominis. Patent WO 9014095.

201. Miyoshi, T. & Kitagawa, H. (1990) Sustained-release nasal sprays containing hyaluronic acid. Patent JP 02032013.

202. Nakamura, M. & Shudo, K. (1990) Nasal drops containing sorbitol, pyrrolidonecarboxylic acid, and hyaluronic acid for treatment of snoring. Patent JP 20253728.

203. Saito, S. (1981) Synergistic composition for stimulating phagocytic activity, comprising a solution containing muramidase and a hyaluronic acid salt. Patent FR 2466990.

204. Wenge, P.S.W., Haakansson, L.D. & Haellgren, H. (1988) Method of enhancing the host defense with hyaluronic acid. Patent US 4725585.

205. Uhlin, A.G. (1980) Immunization agent, and use thereof for producing antisera. Patent EP 14995.

206. Ueno, H., Motoyuki, C. & Okanari, E. (1990) Hyaluronic acid-modified superoxide dismutase with improved in vivo half-life. Patent JP 02231078.

207. Sakurai, K. & Miyazaki, T. (1990) Modification of non-human superoxide dismutase with glycosaminoglycans. Patent JP 02273176.

208. Sakurai, K., Horie, K. & Sakamoto, T. (1986) Mucopolysaccharides as neoplasm inhibitors. Patent JP 61000017.

209. Bergman, P., Steen, Y. & Ingelman, B.G.A. (1987) Process for separating motile sperm cells in vitro, fertilization method and penetration medium containing hyaluronate for sperm cells. Patent WO 8702382.

210. Skinsnes, O.K. & Matsuo, E. (1976) Mycobacteria culture medium and method for *in vitro* cultivation of leprosy mycobacteria. Patent US 3983003.

211. Horie, K., Matsubara, T. & Tokuyasu, K. (1989) Hyaluronates as inhibitors of epidermal cell growth in blood vessels. Patent JP 01290631.

212. Lindblad, G.T.O. (1986) Hyaluronic acid preparation for treating inflammation of skeletal joints. Patent WO 8605984.

213. Yamagami, Y., Oguro, Y. & Muto, T. (1986) Tonic pharmaceuticals containing chondroitin sulfuric acid and hyaluronic acid. Patent JP 61047418.

214. Hiramori, T. & Nishimura, M. (1987) Beverages containing hyaluronic acid, its derivatives, and salts obtained from the Dead Sea. Patent JP 62224268.

215. Pola Chemical Industries, Inc. (1984) Hair conditioners containing hyaluronic acid. Patent JP 59110612.

216. Tanaka, N. Yoshida, S. (1986) Hair tonics containing hyaluronic acid-protein complexes and placental extract. Patent JP 61043104.

217. Furuya, T. (1987) Hair wave-setting preparations containing mucopolysaccharides. Patent JP 62042914.

218. Solomon, H., Catsimpoolas, N., Klein, J. *et al.* (1988) Treatment of alopecia by compositions containing omental lipids and sodium hyaluronate. Patent WO 8804931.

219. Scott, I.R. (1988) Skin treatment composition and hair growth stimulant comprising hyaluronic acid fragments. Patent EP 295092.

220. Yagi, H. (1990) Electrically conductive compositions. Patent JP 02014229.

221. Saito, M. (1987) Water-thinned ink composition. Patent JP 62004764.

222. Arisawa, K., Takagishi, I. & Kawabata, K. (1988) Drying-resistant water-based ink compositions for writing apparatus. Patent JP 63210181.

223. Oka, T. (1989) Anticlogging water-thinned writing ink compositions. Patent JP 01174575.

224. Arai, N., Nojima, M. & Kanda, N. (1990) Thermal printing material having hyaluronic acid. Patent JP 02025378.

225. Brown, T.J., Laurent, U.B. & Fraser, J.R. (1991) Turnover of hyaluronan in synovial fluids: elimination of labelled hyaluronan from the knee joint of the rabbit.

226. Bertolami, C.N., Gay, T., Swann, D.A. *et al.* (Manuscript in preparation.)

227. Kuo, J.-W., Swann, D.A. & Prestwich, G.D. (1991) Chemical modification of hyaluronic acid by carbodiimides. *Bioconjugate Chem.* **2**: 232–241.

7 Alginates

Ian W. Sutherland

Institute of Cell and Molecular Biology,
Edinburgh University, Edinburgh EH9 3JH, Scotland, UK

Algal Alginates

Occurrence and extraction

The main sources of alginates and the current source of all commercial alginate material are species of the brown algae (Phaeophyceae). These are found in coastal waters in cold and temperate areas of the world. Although some of these resources are potentially very large indeed, their commercial value cannot always be realized because of the remoteness of the location and the difficulties attending harvesting and shipment. It is however possible that some geographical areas with such resources may be developed in the future, as the alginate industry is expanding by about 10% per annum. Another problem is encountered with some algal species from which the alginate is less readily extracted and processed. Estimation of current annual production of alginate world-wide is extremely difficult, but is probably upwards of 25,000 tonnes. Successful attempts to cultivate seaweeds for alginate production have been made in the Republic of China, but the contribution of this source to total world production cannot be accurately determined and is probably not currently significant. It is probably unlikely that extensive production of alginate from such cultivated material will be seen in the foreseeable future.

The major species used for alginate production are *Ascophylla*, *Laminaria* and *Macrocystus*. On the US Pacific coast, the giant kelp *Macrocystus pyrifera* is the major source of alginate, while on the Canadian Atlantic coast *Ascophyllum nodosum* is used. *M. pyrifera* grows extensively on rocks in water ranging from 8 to 25 metres in depth. As the kelp beds are frequently 1 mile in length, harvesting can be achieved mechanically 2–4 times annually. In the alginate-producing countries of western Europe (France, Spain, Scotland and Norway), *A. nodosum*, *Laminaria hyperborea* and *L. digitata* provide commercial sources of alginates. *L. hyperborea* is a sublittoral algal species growing on rocks at depths of 1.2–18 metres. It and the related *L. digitata* are perennial plants which are damaged if harvested too regularly. The *L. digitata* grows in a narrow band on either side of the low tide mark. Because of the site of growth and the smaller size of the algae, harvesting of both species of *Laminaria* is obviously more difficult than is the case for *Macrocystus*. In all these brown algae, the alginate is the major polysaccharide present and it may comprise up to 40% of the dry weight (Table 1). It is found in the intracellular matrix where, in the native state, the polysaccharide exists as a mixed salt of various cations found in sea-water – Mg^{2+}, Ca^{2+}, Sr^{2+}, Ba^{2+}, Na^+ and so on. Because of the selectivity of binding of cations, the proportions of the different cations are not the same as are found in sea-water. Thus, the native alginate is mainly present as an insoluble gel, cross-linked by Ca^{2+}.

Table 1   Alginate content of commercially significant marine algae

Species	Alginate as % dry weight
Ascophyllum nodosum	20–30
Ecklonia maxima	30–38
Laminaria digitata	15–40
(fronds)	(15–26)
(stipes)	(27–33)
Laminaria hyperborea	14–24
(fronds)	(9–19)
(stipes)	(19–23)
Macrocystus pyrifera	13–14

The alginate obtained from the different algal species varies in composition, and further differences are observed between the different parts of the algae. Thus the stipe of *Laminaria* owes much of its mechanical rigidity to the presence of alginate of high guluronic acid content, while the blades are more flexible and contain alginate of higher mannuronic acid content. Alginate prepared from two species of the order Ectocarpales, *Stilophora rhizoides* and *Leathesia difformis*, is very high in polyguluronic acid content. This is consistent with the brittle texture of these algal species[1]. The polyguluronic acid content of 67% in the alginate from *L. difformis* is among the highest reported for mature vegetative algal tissue.

As *M. pyrifera* is mechanically harvested, it is processed when wet. The other algal species are normally collected and dried before transport to the processing plant. Alginate is then extracted from dried, milled algal material from *Laminaria* after preliminary treatment with dilute mineral acid to remove or degrade associated neutral homopolysaccharides, including laminarin and fucoidin. Simultaneously, the alkaline earth cations are exchanged for H^+. The alginate is thereafter converted from the insoluble protonated form to the soluble sodium salt by careful addition of sodium carbonate. To avoid any possible β-elimination reactions, the pH must be kept below a value of 10. After extraction, the alginate can be further purified and then converted to either a salt or acid form or to an esterified derivative. Other algal species can be similarly treated, although the higher content of phenolic compounds and neutral polysaccharides found in some types, may present some problems.

Propylene glycol alginate can be prepared by interaction between alginate and propylene oxide under pressure. Although other esters can be prepared, none are used commercially.

Composition and structure

The alginate product from the marine algae is a linear polyuronate containing variable amounts of the uronic acids D-mannuronic acid (M) and L-guluronic acid (G). All algal alginates are composed solely of these two uronic acids. The two monosaccharides are found as random sequences in the linear polymer. The intact polysaccharide may thus contain sequences that are entirely or almost entirely composed of D-mannuronosyl residues, other sequences that are entirely or almost entirely L-guluronic acid and others that contain a random mix of the two monomers (Fig. 1). It is therefore a copolymer composed of irregular sequences of the two component monosaccharide residues. The extent and composition of these sequences, together with the relative molecular mass (M_r), determine the properties of the

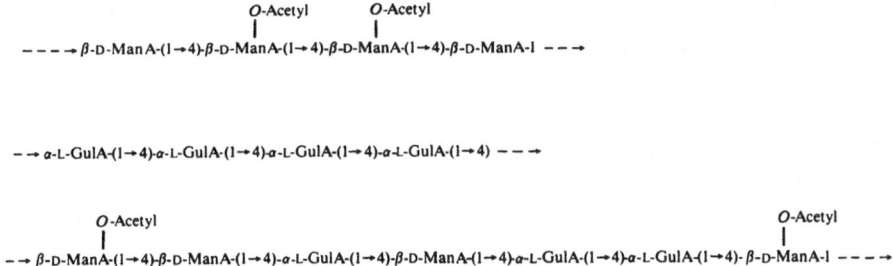

Fig. 1 The block structure of a typical alginate from *Azotobacter vinelandii*. Algal alginates resemble this closely except that they have no acetyl groups.

alginates. For example, in the natural state, the stipe of *Laminaria* contains alginate rich in L-guluronosyl residues. This confers a high degree of rigidity, whereas the fronds are much more flexible and contain much higher amounts of D-mannuronic acid. The uronic acid composition of the material from seaweeds also varies considerably throughout the year. Preparations of alginate from *L. digitata* have been shown to contain long homopolymeric sequences of each type which are then joined by sequences of mixed residues. Alginate from other sources may possess more random arrangement of the component uronic acids.

Analysis of alginates is more difficult than for many other polysaccharides, as acid hydrolysis can lead to considerable destruction of the component uronic acids. Initially, laborious hydrolytic and fractionation methods were used for gravimetric determination of the proportions of the homopolymeric and mixed sequences making up the polymer structure. This approach was later superseded by circular dichroism analysis which matches the linear spectra of the alginate samples to model spectra for well-characterized homopolymeric blocks[2]. The use of ^{1}H- and ^{13}C-NMR has contributed greatly to knowledge of alginate structures, although it may first be necessary to depolymerize the samples slightly to reduce their viscosity. Through such analyses, it has proved possible to determine the frequency of the monomers together with the frequencies of the four possible diad structures F_{GG}, F_{MG}, F_{MM} and F_{GM}. It is now also possible to estimate with considerable accuracy the eight possible triad frequencies and from these results to calculate the average block length. The composition and the block structure vary for different alginate preparations and will depend on the source and the time of harvesting of the algal material. Some typical examples obtained by Skjaek-Braek and colleagues[3] are shown in Table 2 for various alginate preparations from four different algal species.

Table 2 Composition of some algal alginates[3]

Algal species	F_{MM}	F_{MG}	F_{GG}	F_{GGG}	F_{MGM}
Laminaria digitata	0.43	0.15	0.27	0.22	0.1
Laminaria hyperborea	0.2	0.1	0.6	0.56	0.06
Ascophyllum nodosum	0.32	0.25	0.19	0.13	0.17
Macrocystis pyrifera	0.42	0.15	0.28	0.24	0.11

The viscosity of alginate solutions depends mainly on the M_r of the material under study. As samples are normally polydisperse, knowledge of the distribution of M_r can be of importance in determining their suitability for industrial applications. Light scattering can be used for determination of average M_r but is less valuable for establishing the M_r distribution. Alternatively, gel permeation chromatography can provide some data, although there is a lack of polysaccharide standards other than dextran and pullulan for calibration. More recently, low-angle laser light scattering has been used to obtain M_r values for several algal alginates[4], yielding values of the order of 210K for the samples under test. Values of 80.3K and 290K have been obtained for alginate samples from *A. vinelandii* and *Pseudomonas aeruginosa* respectively.

Algal and bacterial alginates, like all biopolymers, are biodegradable. Enzymes degrading alginates have been isolated from many sources including marine and terrestrial bacterial isolates and marine gastropods (Table 3). All the enzymes characterized so far are lyases (eliminases) cleaving the polysaccharides by β-elimination. Although the enzymes show specificity towards homopolymeric structures, they yield the same product from each of the uronic acids originally present and are thus of relatively little value in structural determination of alginates.

Table 3 Enzymes degrading alginates

	Source and specificity of alginate lyases	
Source	Specificity	Products*
Gram-negative bacteria		
Azotobacter vinelandii	Polymannuronic acid	
A. vinelandii phage	Polymannuronic acid	
Beneckea pelagia	Polyguluronic acid	
Klebsiella pneumoniae	Polyguluronic acid	
Pseudomonas aeruginosa	Polymannuronic acid	2, 2+
P. maltophilia	Polymannuronic acid	
Marine bacterium A3	Exolyase, polymannuronic acid	1
Marine bacterium	Endolyase, polymannuronic acid	
Marine bacterium	Polymannuronic acid	3, 4, 4+
Gram-positive bacteria		
Bacillus circulans	Endolyase, polymannuronic acid	
Eukaryotes		
Dendryphiella salina	Not known	

* Degree of polymerization.

Ion binding and gel formation by alginates

Alginates can selectively bind alkaline earth metal ions. The selectivity increases with the guluronic acid content of the alginate sample[5]. Polymannuronic acid is essentially nonselective. Ba^{2+}, Sr^{2+} and Ca^{2+} ions are more strongly bound by alginate than other cations. This property is utilized in a number of industrial applications.

The calcium ions are selectively bound between sequences of poly-guluronosyl residues, the ions being held between diaxially linked L-guluronic acid residues which are in the $^1C_4$ chair conformation (Fig. 2). Because of this conformation in which the Ca^{2+} ions are packed into the interstices between polyguluronic acid chains associated pair-wise, the name 'egg-box' sequences has been applied by Rees and colleagues and provides the currently accepted model for the arrangement of contiguous sequences of guluronosyl residues found in inter-strand junction zones. A strong dependence exists between the length of the polyguluronic acid block and the ability to form a junction zone. Little change is observed in the circular dichroism spectra on removal of water, indicating that the 'egg-box' structure remains undisturbed. In contrast, very large changes are seen when calcium polygalacturonic acid mixtures are similarly treated[6]. As can be seen in Fig. 3, Mg^{2+}

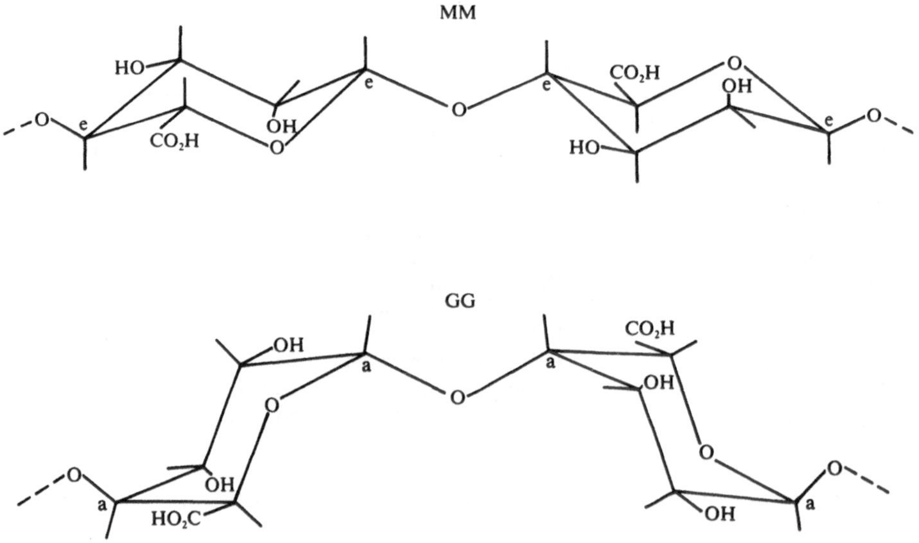

Fig. 2   Structural geometry of MM and GG dimers.

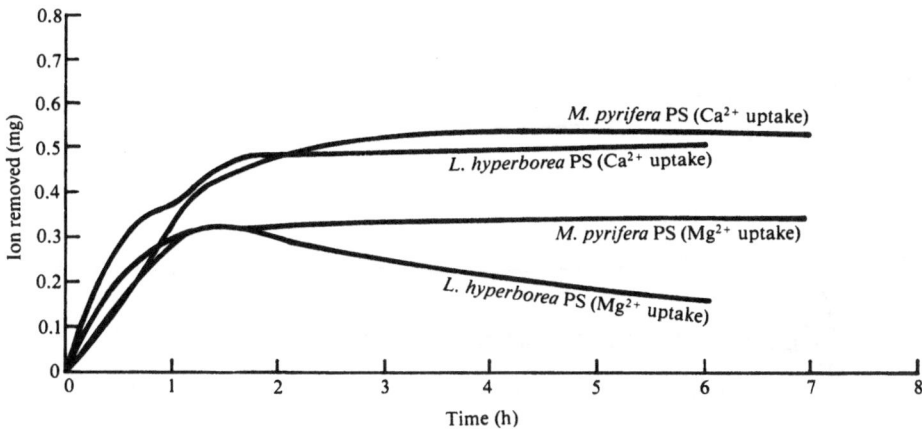

Fig. 3 Ca²⁺ and Mg²⁺ uptake by algal alginates *L. hyperborea* and *M. pyrifera*. Ca²⁺ or Mg²⁺ available for adsorption. Unpublished results of Geddie (1991). PS, polysaccharide.

ions are more weakly bound to alginate than are Ca²⁺ ions. They do not induce gelation, while monovalent cations also show interactions with the polyanionic materials but also fail to form gels.

The amount of divalent cations needed for incipient gel formation in solutions of sodium alginate decreases as the affinity for the cation increases. Ion-binding and gel formation are related processes. Thus, Mg²⁺ is poorly bound and does not induce gelation. The modulus of rigidity of a gel formed after dialysis against a cation parallels the affinity for that cation, the gel strength being strongly increased with increasing guluronosyl content of the alginate. Interestingly, the highly acetylated bacterial alginates differ from the algal material both in the extent and the selectivity of cation binding. However, if the polysaccharides are deacetylated, selectivity similar to that of the algal material is revealed (Fig. 4). Chemical acetylation of algal alginates reduces ion binding and the selectivity of the process.

Alginate biosynthesis

The biosynthetic mechanism for alginate has been extensively studied in bacterial systems but relatively little information has been obtained from algae. The polysaccharides from both eukaryotic and prokaryotic

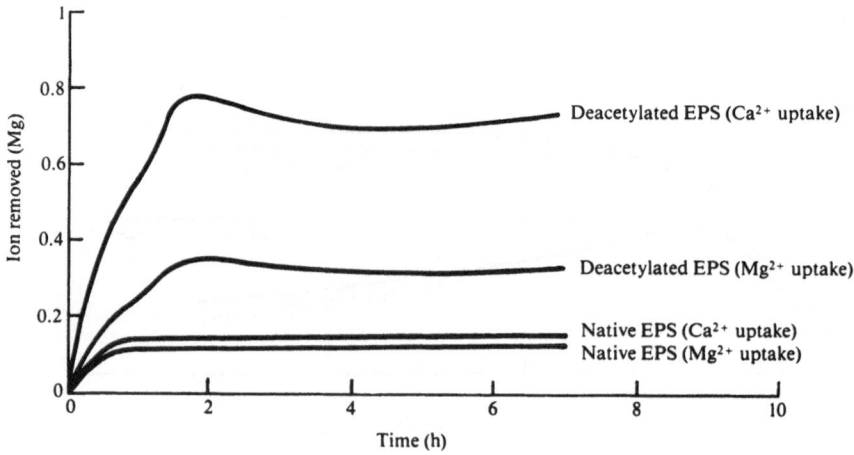

Fig. 4 Ca^{2+} and Mg^{2+} uptake by *P. aeruginosa* EPS native and deacetylated form of EPS. Unpublished results of Geddie (1991). EPS, exopolysaccharide.

types of organism show similarities in their mechanism of synthesis. Monosaccharides are activated by phosphorylation and conversion to nucleoside diphosphate sugars. GDP-D-mannose is converted to GDP-dehydrogenase by the enzyme GDP-mannose dehydrogenase and provides an activated source of the D-mannuronosyl residues. These are polymerised to yield a homopolymer of mannuronic acid. In bacteria, but not in algae, *O*-acetyl groups derived from acetyl CoA are then attached to some of the mannuronic acid residues by an acetylase enzyme. Subsequently, a polymannuronic acid C5-epimerase converts some of the mannuronosyl residues to guluronic acid (Fig. 5). This activity requires the presence of Ca^{2+} in an exchange of H_5 from the aqueous solvent. The concentration of the divalent cation influences the product of the reaction. This has been demonstrated for algal systems by analysing the frequency of diads (adjacent residues of different types) after epimerization. Results can be seen in Table 4. In the laboratory, the overall guluronic acid content of some algal alginates has been increased using a bacterial epimerase by 60–70%. Mutants of *Pseudomonas* producing polymannuronic acid have been isolated and complemented with a gene controlling epimerase production cloned from the wild-type (heteropolymer-producing) strain[7]. Clearly, the presence of the epimerase is not essential for polymerization of the bacterial polysaccharide.

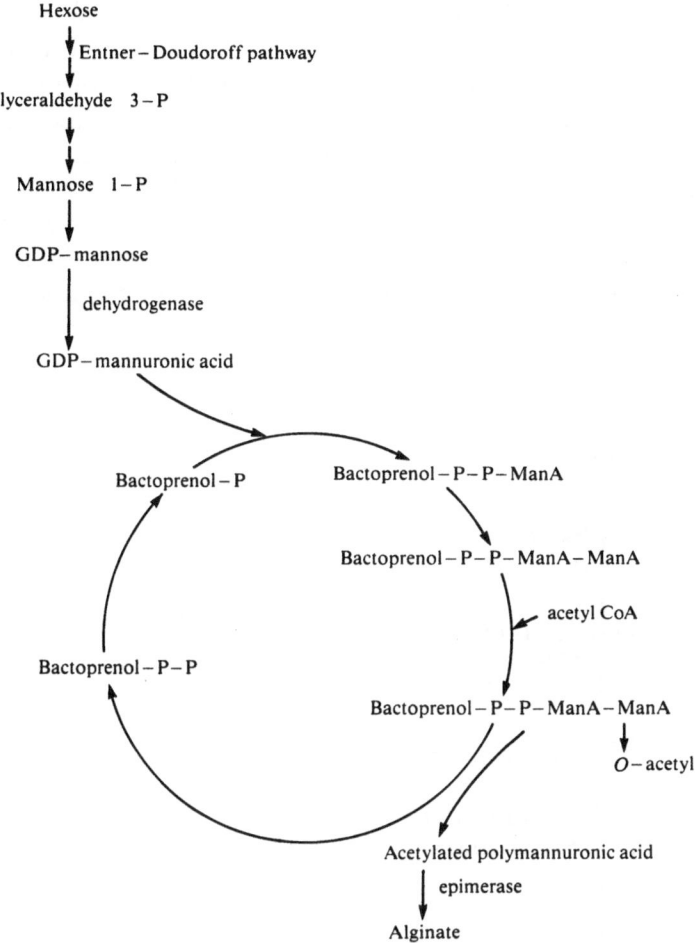

Fig. 5 The probable mechanism of bacterial alginate synthesis.

Table 4 Effect of calcium ion concentration on alginate epimerization. (Use of *A. vinelandii* epimerase on (1) algal alginate and (2) polymannuronic acid)

Ca^{2+}	Mannuronic acid (%)	Guluronic acid (%)	F_{MM}	F_{MG}	F_{GG}
(1) 0.85 mM	62	38	0.51	0.11	0.27
3.4 mM	69	31	0.46	0.23	0.08
(2) 0.68 mM	43	57	0.4	0.17	0.26

Results of Larsen *et al.*[13].

Biomaterials

The alginate composition resulting from the activity of an epimerase in algal material has been demonstrated by analysing the uronic acids in alginate produced during the light and dark periods of growth. In the light period, the ratio of mannuronic acid to guluronic acid was 2.43 whereas in the dark period it was 1.99. Enzyme activity varied with the algal species examined. It was observed that the frequency of introduction of guluronic acid diads was favoured by low concentrations of Ca^{2+}. It seems probable that in algae as in bacteria, a simple homopolymer is first exported from the cells then modified by an extracellular enzyme (Fig. 6).

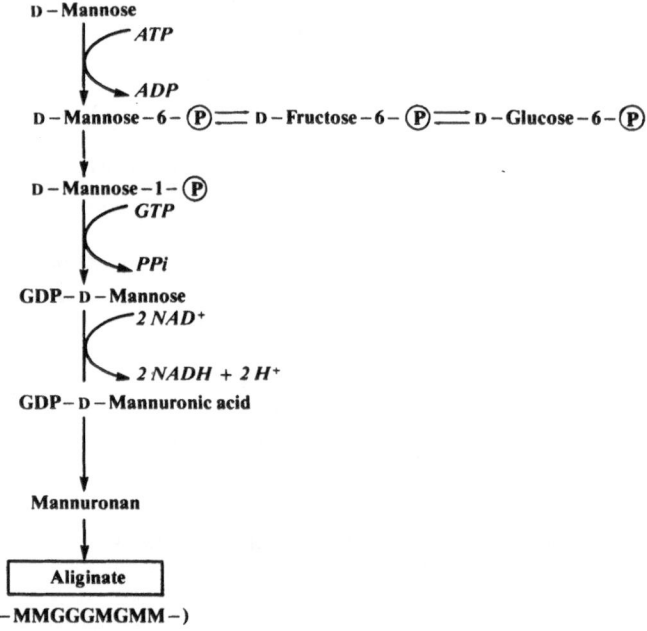

Fig. 6 Possible mechanism of algal alginate biosynthesis.

The activity of the epimerase in the algae is influenced by various factors including the divalent ions present. In algae grown in media with varying amounts of phosphate, the guluronic acid content increased with increasing phosphate. This observation applied to old and new tissue, indicating that the epimerase system was active in non-elongation tissue as well as in the new actively growing tissue[8]. The activity of the enzyme under these conditions caused a decrease in contiguous mannuronosyl sequences and an associated increase in the proportion of blocks of polyguluronic acid.

From studies on cell wall formation in *Fucus* zygotes, Quatrano[9] has suggested that fertilization might result in the activation of an alginate epimerase essential for the synthesis of polymer of high guluronic acid content found deposited into the extracellular wall of the algal cells soon after fertilization. Indeed, the wall material initially appears to be made of Ca^{2+} and alginate interacting to yield a calcium alginate matrix or gel needed for the subsequent deposition of fibrous wall components. Epimerization of a pre-formed polymer in this manner is not unique to alginate. It closely resembles part of the even more complex biosynthetic process observed for heparin. In dermatan synthesis, analyses of the polymer obtained after incubation of chondroitin in ^3H_2O-containing medium showed that tritium first accumulated in L-iduronosyl residues[10]. Tritium was later found in D-glucuronosyl residues. When dermatan was used as substrate, the reverse was true. Thus, epimerization of the uronosyl residues in the polymer involves an abstraction of the C5 hydrogen of the target sugar residue. Inversion then occurs and is followed by reinsertion of a hydrogen from the aqueous medium. The similarities and the complexities of the reaction mechanism of polysaccharide lyases and epimerases and the role of the conformation adopted by the carbohydrate have been evaluated by Feingold and Bentley[11].

Bacterial alginate synthesis has received considerable study in *Azotobacter vinelandii* and *Pseudomonas aeruginosa*, the latter organism having proved amenable to genetic studies while the enzyme polymannuronic acid 5-epimerase has been isolated from culture supernates of the former. The extracellular enzyme has been partially purified by ammonium-sulphate precipitation, ion-exchange chromatography and affinity chromatography on alginate linked to Sephadex. It was also subsequently immobilized by coupling to epoxy-activated polyacrylamide beads[12]. The epimerase from *A. vinelandii* can act on nonacetylated blocks of mannuronic acid larger than two contiguous units. The polysaccharide substrate must have a minimum M_r of 10^6. The enzyme requires the presence of Ca^{2+} both to stabilize the protein and for activity; in *in vitro* experiments, maximum activity is observed when substrate and Ca^{2+} are present in equimolar concentrations. The cation concentration also affects the pattern of introduction of guluronic acid residues by the epimerase. At low Ca^{2+} concentrations, epimerization favours introduction adjacent to other guluronosyl residues. At higher cation concentrations, the mode of enzyme action is altered to a more random process. This can be seen from the results of Larsen *et al.*[13] in Table 4. The overall content of guluronic acid in algal alginates has been increased from about 40% to 62–71% through treatment with the bacterial enzyme. The epimerase is unable to modify mannuronosyl residues carrying *O*-acetyl groups (Fig. 7).

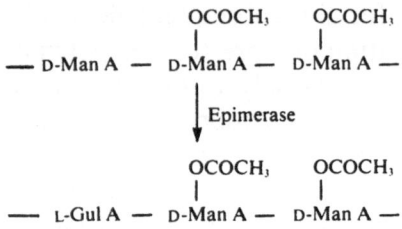

Fig. 7 The influence of *O*-acetyl groups on the action of the epimerase. The *O*-acetylated mannuronic acid residues are not subject to epimerization, being apparently 'protected' from the action of the enzyme.

Genetic studies using *P. aeruginosa* initially faced several problems. The alginate-synthesizing strains isolated from patients with cystic fibrosis were unstable and reverted to non-exopolysaccharide-producing types. Many of the enzymes postulated to function in the early stages of alginate synthesis were barely detectable owing to low levels of activity. Eventually the isolation of a stable polysaccharide-producing strain by Darzins and Chakrabarty[14] was followed by production of a number of alg⁻ mutants which were used for the cloning of *alg* genes. Subcloning and complementation analysis revealed that most of the alg⁻ mutants could be allocated to seven complementation groups. Some of the genes identified and the activities of their products are shown in Fig. 8.

Microbial Alginates

Alginates produced from bacterial cultures have been proposed as alternative sources to algal alginates, but the bacterial biopolymers also

(a)

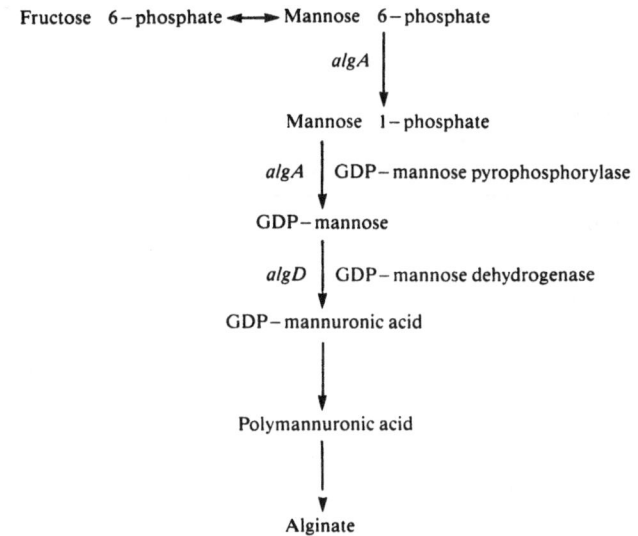

Phosphomannoseisomerase

Fructose 6-phosphate ←→ Mannose 6-phosphate

algA

Mannose 1-phosphate

algA | GDP-mannose pyrophosphorylase

GDP-mannose

algD | GDP-mannose dehydrogenase

GDP-mannuronic acid

Polymannuronic acid

Alginate

(b)

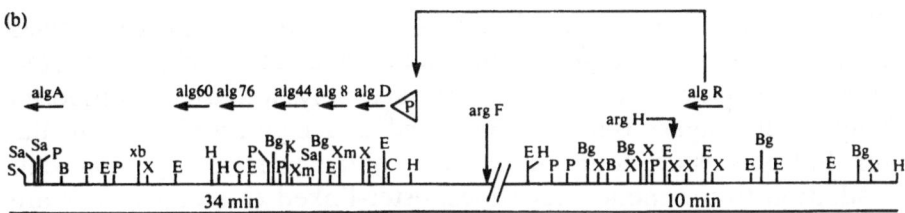

Fig. 8 The genetic map of alginate biosynthesis in *Pseudomonas aeruginosa. a*, Pathway. *b*, Restriction map showing the alginate gene cluster at 34 min and the *algR* gene located at 10 min.

possess features which may make them industrially useful biopolymers in their own right. A number of different Gram-negative bacterial species produce polysaccharides which closely resemble alginate from the traditional algal sources (Table 5). These bacteria therefore provide reliable alternative sources of alginate. To examine the feasibility of production from bacteria, one must pose several different questions: (1) How closely do the bacterial polysaccharides resemble the algal products? (2) How easily can these polymers be produced? (3) Is the production of bacterial alginate cost effective? (4) Can the bacterial alginates replace algal products in specific applications? (5) Are there legislative or other barriers to the use of the bacterial products?

Table 5 Bacterial strains producing alginate-like polymers

Strain	Ratio of mannuronic acid to guluronic acid
Azotobacter vinelandii	Wide range
Azotobacter chroococcum	
Pseudomonas aeruginosa	From 1:0 to ~0.6:0.4
Pseudomonas cepacia	
Pseudomonas fluorescens	~0.6:0.4
Pseudomonas maltophila	
Pseudomonas mendocina	
Pseudomonas phaseolicola	0.95:0.05
Pseudomonas pisi	0.83:0.17
Pseudomonas putida	~0.6:0.4

Composition and structure

All the sources of bacterial alginate are Gram-negative species belonging either to the genus *Pseudomonas* or to the closely related genus *Azotobacter*. All the products are polysaccharides composed solely of the sugars D-mannuronic acid and L-guluronic acid in the same linkages as are found in alginates obtained from algal material. In addition to the monosaccharides, ester-linked *O*-acetyl groups are present to a varying degree, always associated with the D-mannuronosyl residues. These acyl adornments markedly affect certain of the properties of the bacterial alginates. This is of course also true of chemically acetylated algal alginates. The most marked effect of the presence of acetyl groups is on the ion-binding capacity and on the selectivity of ion binding of bacterial alginates when compared with the nonacylated algal products. They also have some effect on the viscosity of aqueous solutions of the bacterial products. A further difference between the two types of alginate resides in the so-called block structure – the presence of contiguous sequences of either mannuronic acid or guluronic acid sequences or mixed sequences. As discussed earlier, the presence of these sequences and in particular of the polyguluronic blocks greatly influences the polymer properties. The Azotobacter alginates possess sequences of L-guluronic acid residues (Fig. 1) but all the *Pseudomonas* polysaccharides only have single residues of guluronic acid, separated by D-mannuronic acid residues or sequences of poly-D-mannuronic acid. The elastic modulus of the chemically acetylated calcium alginate gel decreases as a function of the degree of acylation. However, acetylation also greatly increases the water uptake of dry gels.

Further differences can be seen in the M_r of some of the bacterial products. The M_r of material from some sources is very low, but in others it is comparable to that of algal products. Some bacterial alginates have an extremely high M_r. A problem associated with some bacterial strains is the production of alginases, enzymes which may degrade the alginate produced. The nature of these enzymes and their products has already been discussed. In most of the alginate-producing bacterial species which have been examined, the alginases are specific towards D-mannuronosyl linkages.

Synthesis

Production of bacterial alginates is a function of the strain used. In some bacteria, very large quantities of exopolysaccharide can be produced, whereas in others yields are low. Usually alginate is the only extracellular polysaccharide synthesized, but there are strains of *Azotobacter chroococcum* that can produce more than one polysaccharide. There may also be competition for substrate from synthesis of internal storage products such as poly-β-hydroxy butyric acid (PHB). A further problem may be the ability of the bacteria to use carbon substrates and to grow under nonexacting nutrient conditions. Few bacteria are as flexible in their use of substrates as is *Xanthomonas campestris*, the bacterial species used commercially for xanthan production. This is especially true of the ability to use a wide range of carbon sources, which are available as cheap waste products from other industries. The biosynthetic process in some bacteria is exceedingly complex and there are major differences from the synthesis of other commercial bacterial products such as xanthan. In at least two of the Gram-negative bacteria which synthesize alginate, the main route of polysaccharide synthesis is not directly from monosaccharides. The hexose skeleton of the substrate is first fragmented to 3-carbon intermediates via the Entner–Doudoroff pathway. Supporting evidence for this route was obtained by studying the labelling pattern of alginate from glucose as substrate[15]. Incorporation from C6 was ten times that from C1 or C2 in alginate from either *P. aeruginosa* or *A. vinelandii*. From the glyceraldehyde 3-phosphate so formed, fructose 6-phosphate is formed and is the precursor for the mannuronic residues.

The final stage in bacterial alginate synthesis involves an extracellular modification of the polysaccharide by the enzyme polymannuronic acid C-5 epimerase. While the enzyme in *A. vinelandii* yields contiguous sequences of L-guluronosyl residues as in algal materials, the corresponding enzyme in all the *Pseudomonas* species examined introduces only single guluronic acid residues. Associated with the ease (or otherwise) of production is the question of the cost. Can the bacterial products really be made as cheaply as the algal alginates? This depends on the efficiency of the producing microorganism in converting substrate to polysaccharide, on the cost of substrate, and on the cost of energy for the fermenter and for product recovery. Using the production of xanthan as a guideline, it is probable that bacterial alginates will prove to be considerably more expensive, as none of the microorganisms producing them converts carbohydrate substrate to polysaccharide as efficiently as does *Xanthomonas campestris*.

Limitations of bacterial alginates

Some bacterial alginates may show close chemical similarity to the algal materials that find widespread industrial applications, but in general the bacterial polysaccharides may be sufficiently different in their detailed structure or in their M_r to render them unsuitable for specific purposes. The presence of *O*-acetyl groups on the native products may also affect properties sufficiently to limit the potential applications of bacterial alginates without preliminary deacetylation.

There may of course be legislative and other ethical barriers to the use of the bacterial products. Undoubtedly one would not wish to use pathogenic species of bacteria such as *Pseudomonas aeruginosa* which gives high yields of high M_r polysaccharide without forming PHB and other 'waste' products. Almost all the *P. aeruginosa* strains forming alginate are derived from infections of cystic fibrosis patients by the opportunist bacterial pathogen. There are however alternative bacterial species which could be used to form similar products. These include both *Pseudomonas* and *Azotobacter* species. *Pseudomonas mendocina* is one species offering an alternative to *P. aeruginosa*. The main legislative problem occurs if the producer wishes to use the bacterial polysaccharide in foodstuffs. This would require a lengthy and expensive evaluation process to ensure that the product was free from toxicity, carcinogenicity, etc. Thus, any potential uses for alginates derived from microorganisms are most likely to be in non-food applications.

The final conclusion is probably that bacterial alginates do not provide an exact alternative to algal products. Rather, they should be considered as unique bacterial extracellular polymers in their own right and evaluated as potentially useful polysaccharide products of biotechnology to add to dextrans, gellan and xanthan and to the traditional biopolymers such as algal alginates.

Alginate Usage

Alginates in food

Polysaccharides are incorporated into foods for several different purposes. They can be used as thickening agents (viscosifiers), as gelling agents, and as colloids (stabilizers) (Table 6). Alginates and their derivatives are suitable for several of these roles and either as specific salts or as propylene glycol alginate are approved food additives with E numbers from E400 to E405, respectively.

Table 6   Role of alginates used in foods

Water retention	Thickening agent
Gelation	Film formation
Emulsification	Suspending agent
Stabilization	Syneresis inhibitor

Incorporation of alginate into ice-cream maintains the texture and consistency of the product. The polysaccharide, frequently in the form of propylene glycol alginate, is used as a stabilizer in concentrations of 0.1–0.5%. Alginate also prevents the formation of large ice-crystals and gives good flavour release. It is also used for viscosity control in dairy products. Small amounts of alginate are added to some other food products to prevent drying-out and to stop the product from becoming tough or rubbery. Propylene glycol alginate is incorporated as a suspending agent into fruit-flavoured drinks and is also used as a foam stabilizer in beer.

The interaction between alginates and Ca^{2+} to form gels, is widely used in foodstuffs. The nature of the gels formed depends on the sequence structure of the polysaccharide. Alginates containing large proportions of polyguluronic acid sequences tend to form rigid brittle gels which are subject to syneresis. When polymannuronic acid sequences predominate, the gels formed are more elastic, less brittle and also less subject to syneresis. Thus, food use of alginates requires careful choice of the type of alginate to be utilized and the conditions under which the gel is formed. The soluble sodium alginate is reacted with a suitable calcium salt in order to produce a homogeneous gel. The gel formed differs from others used in food products in being thermally irreversible. It does not melt in the mouth but is broken up with relatively poor flavour release. The advantage of alginate gels is however in the formation of reconstituted fruit pieces, onion rings, pimento and other products. By using this technology, waste of the fruit or vegetable material is reduced and products of uniform size and shape are manufactured. Some of the many types of foodstuff in which alginates are used are listed in Table 7.

Table 7 Types of food and related products in which alginate is used

Dairy products	Canned foods
Bakery products	Salad dressings
Frozen foods	Beverages
Beer	
Pharmaceuticals	

As well as alginates in the salt form, derivatives such as propylene glycol alginate are used by the food industry. Such compounds are less susceptible to precipitation by acids or metallic cations and are thus suitable for use in foods and beverages in which the pH is low. These include drinks based on fruit juices. The hydrolytic stability of the propylene glycol alginate to weak acids increases with the level of substitution. In terms of food usage, alginates may well be the fourth most important non-starch polysaccharide, after carboxymethylcellulose, gum arabic and carrageenan.

Non-food applications

Apart from their use in food and related products, alginates find a number of applications in other industries and in medical care. They are used to control the rheology of fibre-reactive dyes in printing, as well as in paper coatings. In this latter application the main functions are film formation and binding at the paper surface where low viscosity alginates are used to improve the surface properties of paper. Unlike other sizes, paper penetration is very low, but there is increased strength and stiffness of the paper. The thin dense film formed at the surface of the paper resists penetration by oil, wax, and other agents. Alginates can also be applied as thickening agents to adhesives used in the manufacture of corrugated boards. In this role, bonding is also improved. Alginate is also incorporated into wallpaper adhesives.

As sodium alginate differs from other natural gums in being inert to reactive dyes, it is used as a thickener for printing pastes containing these dyes. Alginates are also used for machine and silk-screen printing.

Alginates have long been used as the major components of dental impression material where the ability to form a rapidly setting and strong gell is of importance. The advantage of alginate over other materials for this role is the ability to set at body temperature and yield a mould with good reproducibility. Other pharmaceutical applications of alginates include use as suspending agents and as lubricants. In tablets, it may be used to assist in their disintegration on the addition of water, or incorporated as a binding agent.

A recent application of alginates in wound care makes use of alginate fibres. These can be formed from calcium alginate and were initially used in the textile industry. They are now manufactured in the form of wound dressings which have excellent interaction properties with wounds, thus leading to improved healing and minimal risk of contamination. The wounds heal better if kept moist rather than being allowed to dry out; the alginate fibres retain wound exudate and maintain it in contact with the wound. The dressings have the additional advantage that they can be changed without further trauma to the patient. Some of the Ca^{2+} ions are exchanged with other ions at the wound surface to form a xerogel, but residual material can be dissolved with a saline solution. The exact nature of the actions of alginate wound dressings is still incompletely understood[16].

Biotechnology Applications

Cell and enzyme immobilization

Alginates have found many applications within the general field of biotechnology. Within these applications, the polymers must have specific structures and resultant well-defined physical properties. Alginates, in the form of calcium alginate gels, are used for the immobilization of cells and enzyme systems. In this role, they have many advantages over chemical methods in which there is likely to be considerable inactivation of enzymes or cells. The preparation of the alginate gels is fairly similar to the technology used for some alginate applications in the food industry.

Table 8 Immobilization of cells with alginate

Cell type	Product
Bacteria	
Anabena sp.	Ammonia
Erwinia sp.	Isomaltulose
Lactobacillus bulgaricus	Lactic acid
Synechococcus sp.	Glutamate
Streptomyces spp.	Antibiotics
Zymomonas mobilis	Ethanol
Fungi	
Aspergillus niger	Citric acid
Aureobasidium pullulans	Glucoamylase
Claviceps purpurea	Alkaloids
Saccharomyces cerevisiae	Ethanol, glycerol
Algae	
Dunalliella tertiolecta	Glycerol
Plant cells	Alkaloids, digitoxins
Animal cells	Monoclonal antibodies
	Insulin

Further examples are to be found in Skjaek-Braek & Martinsen (1991).

Several methods have been developed for the *in situ* gelation of alginate for the immobilization of living prokaryotic or eukaryotic cells. The species immobilized include eubacteria, cyanobacteria, and fungi as well as plant and animal cells and plant protoplasts (Table 8). Sodium alginate can be mixed with Ca^{2+} ions in the presence of EDTA and D-glucono-δ-lactone. The lactone hydrolyses slowly, gradually lowering the pH and releasing Ca^{2+} ions. This method is not entirely suitable for many applications in biotechnology because of the low pH and the toxicity of the reagents. But it does provide an extremely gentle procedure in which destruction or inactivation of the enzymes or viable cells is minimized. Use of Ca^{2+} is preferable to other cations causing gelation as there is lower toxicity than is the case for Ba^{2+}, Sr^{2+} or Al^{3+}. The size of the gel beads formed can be controlled through careful choice of alginate, polysaccharide concentration and solution viscosity. It is possible to produce beads ranging from 200 to 5000 μm in diameter. The physical properties of the alginate beads depend on the composition, the sequential structure and the molecular size of the polysaccharides used in their production. Martinsen *et al.*[17] noted that beads formed from alginate with 70% or more L-guluronic acid content, in which the average length of the polyguluronic acid sequences was greater than 15, showed the highest mechanical strength. Such beads also had the lowest shrinkage and the highest porosity, as well as optimal stability in the presence of monovalent cations.

The disadvantages of the alginate gels include their low stability. As the calcium ions are washed from the gels, crosslinking diminishes and the gels are destabilized leading to leakage and loss of the entrapped material. Such loss of integrity of the gels is also effected by chelating agents and by high concentrations of ions such as Na^+ or Mg^{2+} which do not themselves promote gelation. Unless the gels are very carefully prepared, they are liable to be relatively porous, again leading to leakage of the immobilized materials. Alginates from seaweeds may contain small quantities of toxic compounds such as polyphenols. These can injure sensitive cells. There may also be immunogenic and pyrogenic material present.

Recently, Draget *et al.*[18] have developed techniques for producing homogeneous alginate gels made by the internal liberation of Ca^{2+}. At neutral pH, gels were produced by mixing alginate solutions with particulate $CaCO_3$ and the slowly hydrolysing proton donor D-glucono-δ-lactone. One problem associated with production of these gels on a larger scale, was the entrapment of gas bubbles. This was overcome by degassing the alginate solution and yielded improved clarity of the gels. Alginates of high or low polyguluronic acid content gelled with the same speed, and initially, gel strength was independent of the L-guluronic acid content. Later, gels rich in guluronic acid increased in strength in proportion to the average length of the polyguluronic acid sequences in the alginate used for gel preparation.

The future usage of algal alginates seems likely to show a continuing increase. In part, this may result from the ingenious applications to which alginate is being put in biotechnological processes. There is also a continuing increase in the use of alginates in the pharmaceutical industry. Whether the bacterial alginates will be developed remains less certain. Much will depend on the cost of any product and on the capital cost of its development. What we may perhaps see is the preparation and use on a relatively large scale of a bacterial extracellular enzyme, the polymannuronic acid 5-epimerase from *A. vinelandii* to 'improve' algal alginates with naturally high mannuronosyl content. The enzyme can increase the content of L-guluronosyl residues and thus yield a polysaccharide with altered and improved properties for a number of applications.

References

1. Craigie, J.S., Morris, E.R., Rees, D.A. & Thom, D. (1984) Alginate block structure in Phaeophyceae from Nova Scotia: Variation with species, environment and tissue type. *Carbohyd. Polymers* **4**: 237–252.

2. Morris, E.R., Rees, D.A. & Thom, D. (1980) Characterisation of alginate composition and block structure by circular dichroism. *Carbohyd. Res.* **81**: 305–314.

3. Grasdalen, H. (1983) High field ^{1}H-n.m.r. spectroscopy of alginate sequential structure and linkage conformations. *Carbohyd. Res.* **118**: 255–260.

4. Martinsen, A., Skjaek-Braek, G., Smidsrod, O., Zanetti, F. & Paoletti, S. (1991) Comparison of different methods for determination of molecular weight and molecular weight distribution of alginates. *Carbohyd. Polymers* **15**: 171–193.

5. Smidsrod, O. & Haug, A. (1968) Dependence upon uronic acid composition of some ion-exchange properties of alginates. *Acta Chem. Scand.* **22**: 1989–1997.

6. Rees, D.A. (1981) Polysaccharide shapes and interactions – some recent advances. *Pure Appl. Chem.* **53**: 1–14.

7. Chitnis, C.E. & Ohman, D. (1990) Cloning of *Pseudomonas aeruginosa algG* which controls alginate structure. *J. Bact.* **172**: 2894–2900.

8. Indergaard, M. & Skjaek-Braek, G. (1987) Characteristics of alginate from *Laminaria digitata* cultivated in a high phosphate environment. *Hydrobiologia* **151/152**: 541–549.

9. Quatrano, R.S. (1982) Cell wall formation in *Fucus* zygotes: a model system to study the assembly and localization of wall polymers. In *Cellulose and Other Natural Polymer Systems* (ed. Brown, R.M.), pp. 45–49. Plenum.

10. Malmstrom, A. (1984) Biosynthesis of dermatan sulphate. II. Substrate specificity of the C-5 uronosyl epimerase. *J. Biol. Chem.* **259**: 161–165.

11. Feingold, D.S. & Bentley, R. (1987) Conformational aspects of the reaction mechanisms of polysaccharide lyases and epimerases. *FEBS Lett.* **223**: 207–211.

12. Skjaek-Braek, G., Smidsrod, O. & Larsen, B. (1986) Tailoring of alginates by

enzymatic modification *in vitro. Int. J. Biol. Macromol.* **8**: 330–336.

13. Larsen, B., Skjaek-Braek, G. & Painter, T. (1986) Action pattern of mannuronan C-5-epimerase: generation of block-copolymeric structures in alginates by a multiple attack mechanism. *Carbohyd. Res.* **146**: 342–345.

14. Darzins, A. & Chakrabarty, A.M. (1984) Cloning of genes controlling alginate biosynthesis from a mucoid cystic fibrosis isolate of *Pseudomonas aeruginosa. J. Bact.* **159**: 9–18.

15. Lynn, A.R. & Sokatch, J.R. (1984) *J. Bact.* **158**: 1161–1162.

16. Blair, S.D., Jarvis, P. & McCollum, C. (1990) Clinical trial of calcium alginate haemostatic swabs. *Br. J. Surg.* **77**: 568–570.

17. Martinsen, A., Skjaek-Braek, G. & Smidsrod, O. (1989) Alginate as immobilization material. I. Correlation between chemical and physical properties of alginate gel beads. *Biotechnol. Bioengng.* **33**: 79–89.

18. Draget, K.I., Ostgaard, K. & Smidsrod, O. (1991) Homogeneous alginate gels: a technical approach. *Carbohyd. Polymers* **14**: 159–178.

Other general sources

Aspinall, G.O. (1983) *The Polysaccharides*, Vol. 2. Academic.

Berry, A. *et al.* (1988) *Pseudomonas aeruginosa* infection in Cystic Fibrosis: molecular approaches to a medical problem. *Chimicaoggi*, 13–19.

Hoiby, N. *et al.* (1989) *Pseudomonas aeruginosa* Infection. *Antibiotics Chemother.* **42**.

Skjaek-Braek, G. (1988) Biosynthesis and structure-function relationships in alginates. Thesis, University of Trondheim NTH.

Skjaek-Braek, G. & Martinsen, A. (1991) Applications of some algal polysaccharides in biotechnology. In *Seaweed Resources in Europe: Uses and Potential.* (eds Guiry, M.D. & Blunden, G.), Ch. 9, pp. 219–257. Wiley.

Sutherland, I.W. (1990) *Biotechnology of Microbial Exopolysaccharides*. Cambridge University Press.

Yalpani, M. (1987) *Industrial Polysaccharides*. Elsevier.

8 Miscellaneous biomaterials

D. Byrom

ICI BioProducts and Fine Chemicals, PO Box 1, Billingham, Cleveland, TS23 1LB, UK

Introduction

The objective of this chapter is to review briefly a range of miscellaneous materials from biological sources and to cover aspects of some materials that have not been discussed elsewhere.

Poly-β-hydroxybutyrate

Synthesis

In recent years polyhydroxyalkanoates (PHAs) have attracted significant interest in both academic and industrial research laboratories. Poly-β-hydroxybutyrate (PHB) and its copolymer with poly-β-hydroxyvalerate (PHB/HV) are the only members of the series to be produced on a large scale. PHB/HV has been launched as a commercial product fabricated into bottles, and developments in other applications are well advanced[1].

A wide range of microorganisms will synthesize PHB and other PHAs from a number of substrates (see the chapter on PHAs in this volume), but in reality only a limited number of both organisms and feedstocks are practical for PHB production. Substrates derived from renewable resources have obvious attractions provided that the economics of use for large-scale production of polymer are favourable. It is likely, therefore, that for the immediate future, carbohydrates will be the substrate of choice for commercial production of PHB and PHB/HV.

Other substrates have been investigated for polymer synthesis and have, for various reasons, been rejected for large-scale use. A summary of the conclusions from these studies is given below. For more detail see ref. 2.

METHANOL

The primary reasons for rejecting methanol as a substrate for PHB synthesis were that the organisms that grew on methanol-made polymer that had low relative molecular mass (M_r), was difficult to extract from cells and was present at relatively low concentration. Some of those factors have not changed despite advances in the understanding of the fermentation physiology of methylotrophs.

Among the most recent workers to investigate the synthesis of PHB using methanol as substrate were Suzuki *et al.*[3] who grew *Pseudomonas* K to the remarkable dry weight of 160 g/l on a methanol/mineral salts medium. A computer-controlled fed-batch technique was employed to achieve tight control of dissolved oxygen and substrate concentration in the fermentation broth. Once the desired dry weight had been reached, nitrogen limitation was imposed by stopping the ammonia feed. Methanol feeding was continued and PHB accumulated to a concentration of 136 g/l. At that point the total biomas concentration was 206 g/l, with a PHB content in the cells of 66%. The M_r of the PHB was 3×10^5.

Investigation of the effects of nitrogen feeding during the PHB accumulation phase of the culture[4] enabled the total biomass content of the fermentation to be increased to 233 g/l, with a PHB concentration of 149 g/l. Fermentation time was shortened and the PHB yield from methanol was increased from 0.18 g PHB/g methanol to 0.20 (ref. 5). The average M_r of the polymer was influenced by the concentration of methanol in the medium. A methanol concentration of 0.05 g/l gave PHB of M_r 8×10^5, whereas it dropped to 0.5×10^5 at a methanol concentration of 32 g/l (ref. 6).

METHANE

Methane conversion to PHB has been noted by a number of authors over the years[7-9] and relatively high concentrations of PHB in cells have been recorded. There has, however, been little quantitative work using this substrate. Asenjo and Suk have demonstrated very significant levels of PHB accumulation (70% of dry weight) in a type II methanotroph *Methylocystis parvus* OBBP[10] and have developed a kinetic model for bioconversion of methane to PHB by that organism[11]. Industrial use of methane as a substrate is likely to be limited by gas–liquid mass transfer and the safety issues surrounding the potentially explosive mixtures of methane and oxygen required in these fermentations.

HYDROGEN/CARBON DIOXIDE/OXYGEN

Hydrogen is available industrially in large quantities in 'synthesis gas' used to make ammonia and methanol.

Hydrogen-oxidizing bacteria such as *Alcaligenes eutrophus* were considered for single-cell protein production (SCP) in the 1970s. It was known that they were capable of synthesizing PHB[12]. This capacity was recognized as a disadvantage for SCP production since protein yield was reduced. Subsequently, it was discovered that the organism would accumulate up to 80% of its dry weight as the polymer, and hydrogen was investigated as a substrate for PHB synthesis[13].

Yields of PHB are adequate, and basic raw-materials cost could be low using these substrates, but again safety considerations make the process unrealistic[2].

ORGANIC ACIDS

Acetate is an obvious choice as a feedstock for PHB synthesis since its metabolism leads directly to the key intermediate for the polymer. However, cell yield using acetate for growth is not good and consequently the overall yield of cells and polymer from the substrate is not economic. Acetate is a relatively toxic substance for microorganisms, even those capable of using it as a substrate. At large scale, good process control is required to prevent inadvertent build-up of acetate in the fermentation broth with the consequent death of the culture. Other organic acids are more expensive than acetate and therefore would be even more uneconomic for PHB production.

Industrial production of PHB and PHB/HV

W. R. GRACE

Industrial involvement with PHB goes back as far as the 1960s when the American company W. R. Grace carried out work on the polymer[14]. Ultimately development was curtailed because at that time the techniques available for extraction were not able to provide a product thermally stable in processing[15].

ICI

The development of the PHB process was begun by ICI in 1975–6 as a response to projected increases in oil prices. Although the oil-price rises have not materialized, the original attraction of a polymer based on renewable resources still remains. Biodegradability and biocompatibility are features that allow the product to be competitive in particular market sectors.

The issues involved in choice of substrate and organism have been mentioned above and are also discussed by Collins[2] and Byrom[16].

Biomaterials

The organism selected by ICI for PHB production was *Alcaligenes eutrophus*. A glucose-using mutant of the organism is grown in a two-stage fed-batch fermentation to a total biomass concentration in excess of 100 g/l (Fig. 1). Polymer accumulation is induced when the organism is nutrient-limited but supplied with excess carbon source (glucose). PHB concentration is between 70% and 80% of dry weight. In fact the polymer manufactured in the commercial process is almost exclusively the copolymer PHB/HV. The physical and mechanical properties of PHB/HV are superior to the homopolymer and allow it to be used for a wider range of applications. To produce the copolymer, a mixed substrate feed of glucose plus propionic acid is used. The HV content of the polymer can be controlled between 0–30 mole% by variation of the ratio of glucose:propionate in the feed to the fermentation[16,17]. The M_r of the polymer is $1.0–1.3 \times 10^6$.

Polymer separation and purification is accomplished using a proprietary aqueous wash process followed by drying. The resultant white powder is usually melt-extruded and comminuted to produce polymer pellets, the product form most familiar to plastics-fabricating

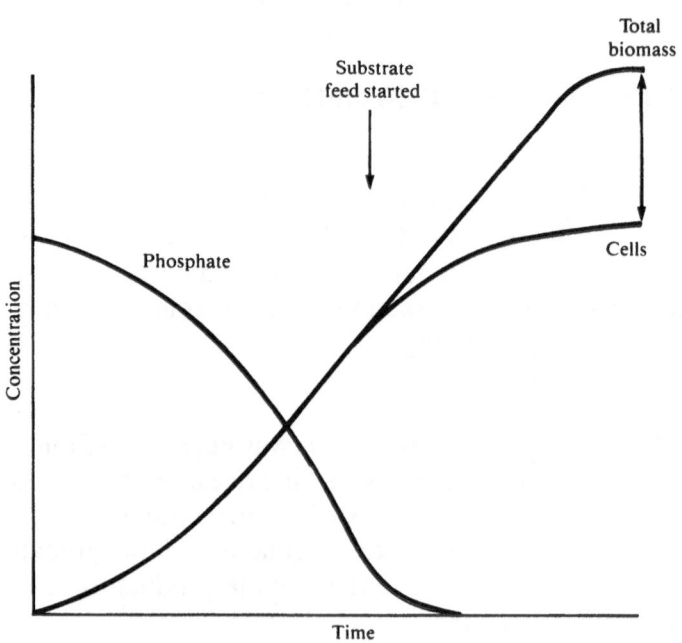

Fig. 1 Time course of PHB and PHB/HV fermentations using *Alcaligenes eutrophus*.

338

companies. Current scale of production is at the rate of some hundreds of tonnes per annum and is projected to rise to thousands of tonnes per annum by the mid to late 1990s.

'Biopol' is the trademark used by ICI for this PHB/HV family of PHAs. It is envisaged that applications for the polymers will be found in packaging, household items, slow-release encapsulation of pharmaceuticals and agrochemicals, personal hygiene products and biomedical devices. When in large scale production the price of the product is likely to be £3 to £5 per kilogramme.

CHEMI LINZ

In the 1980s, Chemi Linz AG began development of a process to produce PHB by fermentation. The polymer group became Petrochemia Danubia (PCD) and the fermentation and process development was carried out for that organization by btF, a biotechnological research unit affiliated to a number of Austrian companies. Aspects of the process have been patented[18,19].

The PCD process is different from that operated by ICI in several respects. (1) The production organism is *Alcaligenes latus* and sucrose is the substrate. (2) PHB formation is growth-associated. Nutrient limitation is not used to induce polymer accumulation. (3) Solvent extraction is used to separate and purify the polymer. (4) The process is used to make PHB homopolymer only. An outline of the process is given below.

The production organism is a mutant of *A. latus* DSM 1124 which is grown in a $15 M^3$ fermenter using sucrose as the substrate. PHB accumulation parallels cell growth and is reported as attaining over 60 g/l. At the end of the fermentation, the cells are killed by heating to 80°C and the biomass is separated from the fermentation medium by centrifugation. After a water wash, the cells are extracted with methylene chloride under pressure and at elevated temperature in a proprietary process. PHB is precipitated from the extract, centrifuged to remove excess solvent and dried to produce a white powder[20]. The M_r of the product is about 750,000. The production capacity of the plant is quoted as 20 tonnes per year rising to 1 tonne per week in the near future.

The product is blended with nucleating agents and plasticizers if required, and extruded into pellets. The pellets can be injection-moulded on conventional plastics-processing machinery. No applications for the material have been given.

Biodegradability

Biodegradability is one of the key features that has made PHAs an attractive commercial proposition, allowing an extension of the options available for the disposal of postconsumer plastics waste. Currently there are no generally agreed tests to determine biodegradability, but national bodies such as the ASTM in the USA and others in Europe and Japan are designing standard procedures.

PHB, PHB/HV and other PHAs containing short-carbon-chain-length monomers have been demonstrated by many investigators to be biodegradable in laboratory and field experiments[21-23]. Clearly the composition of the PHA is important in determining the rate of biodegradation – polymers containing, for example, 3-hydroxyoctanoate and longer chain-length monomers degrade very slowly compared to PHB and PHV[24].

A number of microorganisms have been isolated that secrete extracellular depolymerases active on PHB generating carbon substrates for growth. The organisms include *Alcaligenes faecalis*[25], *Pseudomonas lemoignei*[26,27], *Penicillium simplicissimum* and *Eupenicillium* sp.[28].

Hydrolytic degradation, especially at pH values away from neutrality, is also a mechanism of breakdown for PHB and its copolymers[29-31].

Properties

The properties of PHAs have been reviewed by Holmes[32] and will not be elaborated here.

Starch-Based Plastics

There are a number of applications where starch has been used either to produce a plastic material or to blend with a petroleum derived polymer such as polyethylene. The objective seems usually to have been to impart biodegradability to the resultant material.

Starch/polyethylene biodisintegrable blends

Starch/polyethylene blends will not be discussed in any detail here, because it is considered that since starch is the minor component in those materials, they are not biopolymers. The major uses of these products to date have been in packaging, for example, carrier bags, and agricultural mulch films. Methods of production can be found in ref. 33.

Starch-based biodegradable materials

Two product classes based on destructured starch have been developed by different companies.

Fertec (Ferruzzi, Ricerca e Tecnologia) have combined starch and a synthetic polymer to produce a material from which both films and rigid articles such as pots, bowls and plates can be made. The product is an alloy, since the starch granules have been disintegrated and form a continuous phase in which the other polymer is distributed. It is therefore different from the starch/polyethylene blends where the starch is dispersed in the basic polymer background forming discrete particles. Starch forms up to 50% by weight of the Fertec material. The second component, the synthetic polymer, has not been named but has an M_r between 5000 and 50,000 depending on the application of the final material. It is hydrophilic in nature and can form strong interactions with the starch. The composite is not water-soluble but will swell by about 4% when exposed to water.

The material can be processed using conventional polymer-processing technology such as moulding and film blowing. Different compositions are required depending on the processing technology to be used and the end product application.

In biodegradability trials using a modified Warburg test, the material shows results somewhere between those of nondegradable polyethylene and fully degradable paper[34].

Warner Lambert have developed the other starch-based plastic, called Novon. It differs from the Fertec product in that the starch content is higher. Some versions of Novon contain over 80% starch. Additives are used to improve processing. Other formulations also contain polymers such as ethylene/vinyl acetate copolymer or ethylene/acrylic acid copolymer to improve properties. The material is made by treating starch under conditions of controlled high pressure and temperature in an extruder or injection-moulding machine. Once the starch has been destructured in the process, articles can be moulded from the product.

The basic method for destructuring the starch is to heat the material under pressure in defined conditions of temperature and water content[35]. Reduction of M_r of the starch improves the mouldability of the material. Whilst this will take place without addition of a catalyst, a substantially reduced frequency of rejection of mouldings is seen if a chain-scission catalyst such as HCl or H_2SO_4 is used[36].

Control of the calcium and magnesium content of the starch has also been found to be important. Starch contains phosphate groups which are bridged by those metal ions. If washed with low pH water, the calcium and magnesium are removed from the starch. The resulting material can be processed at a lower temperature and with a shorter residence time in the equipment[37].

Development of the moulding technology has demonstrated that it is better to separate the destructuring step from the moulding step[38].

Melts of destructured material are compatible with melts of thermoplastic synthetic polymers. The resultant blends showed better stability than the starch-alone materials; for example, a starch rod lost up to 40% of its length in a few hours in humid air, whereas the thermoplastics blend shrank less than 4% in two days[39].

The basic material is brittle and plasticizers are added to make it tougher. The material is moisture-sensitive, becoming sticky on exposure to water, a feature which may limit its range of applications. Novon is biodegradable in accelerated landfill, controlled compost, field soil burial, aerobic and anaerobic aqueous environments and accelerated marine tests.

Warner Lambert envisage that the uses of the material include capsules for drugs, and disposable single-use products such as cups and food trays. The company has announced that a 25,000-tonnes-per-annum plant will be on stream in 1992.

Polylactide and Polyglycolide

In 1962, the American Cyanamid Corporation developed the first synthetic absorbable suture material. The product, Dexon, was a poly(glycolic acid) homopolymer[40–42]. Vicryl was developed by Dupont as a competitive material containing a 92:8 glycolic acid:lactic acid copolymer[43,44]. Since then, a range of applications have been developed for polylactides (PLA) and polyglycolides (PGA).

PLA and PGA are thermoplastic, biodegradable polyesters. They are produced from lactic and glycolic acids. Lactic acid is a fermentation product, and although the process used to polymerize it is chemical, it has been included here as a biopolymer. Glycolic acid is produced chemically as an intermediate in ethylene-glycol manufacture and is also polymerized chemically. Its use is inextricably involved with that of polylactic acid and so has been included for completeness.

Manufacture

Low M_r polylactic acid and polyglycolic acid are made by direct polymerization of the respective acids. The high-M_r PLA and PGA are made by ring-opening polymerization of 'lactide' and 'glycolide'[45,46]. These compounds are the cyclic diesters of the respective acids (Fig. 2) and are themselves made by controlled pyrolysis of low-M_r PLA and PGA. Normally the poly-L-lactide is synthesized. The racemic poly-D-L is amorphous and has inferior mechanical properties.

A variety of polymers can be produced: polyglycolic acid; poly-L-lactic acid; poly-D-lactic acid; polyglycolide; poly-L-lactide; poly-D-L-lactide; copolymers of glycolide and L-lactide; copolymers of D-L-lactide; and copolymers of L-lactide and D-L-lactide.

The polyglycolic and polylactic acids are low-M_r products and the polyglycolide and polylactide are of high-M_r.

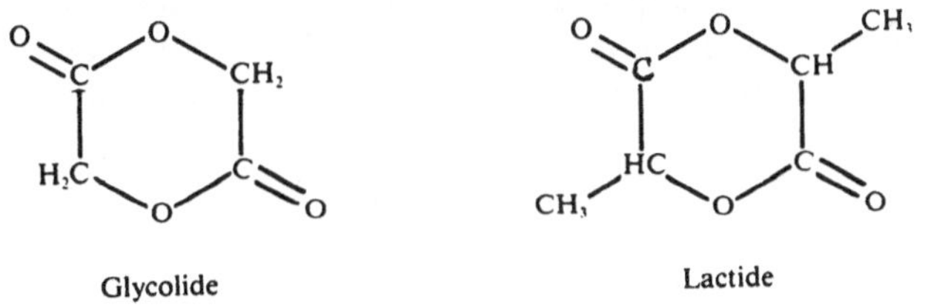

Glycolide   Lactide

Fig. 2 Cyclic diesters of glycolic and lactic acids.

Biodegradability

The PLA/PGA family of polymers are hydrolytically degraded in both physiological and environmental conditions to the component acids. It has not been definitely ascertained that microbial or enzymatic action contributes to their breakdown. The rate of degradation is dependent on the composition of the polymer, surface quality and the M_r (Table 1). *In vitro* tests have been used to estimate the rate of breakdown of these materials.

Table 1 Biodegradation times of polylactides and polyglycolides

Polymer	Degradation time
Poly-L-lactide	Months to years
Polyglycolide	Months
Poly-D-L-lactide	Weeks to months
Poly-L-lactic acid	Few weeks
Polyglycolic acid	Few days

The polymers are sensitive to humidity and must be stored in dry air.

In vitro they are hydrolysed nonspecifically to the component acids. Lactic acid is converted to pyruvate by normal intermediary metabolism and thence to carbon dioxide via the tricarboxylic acid cycle. Glycolate is excreted in urine or is converted first to glyoxylate then pyruvate, which is metabolized as above[47].

Properties

Poly-L-lactide and polyglycolide are highly crystalline with degrees of crystallinity of over 80%. Poly-D-L-lactide is amorphous. In many areas of application (see below), strength is an important attribute of the material. Tensile strength and modulus of elasticity depend on the conditions under which the product is processed. Increase in M_r gives stronger material in terms of its bending strength, elasticity and tensile strength.

Applications

The uses to date for PLA and PGA have been in medical applications.

SUTURES AND LIGAMENT REPLACEMENTS

Homopolymers of PGA as well as copolymers of PGA/PLA (90:10) are used in suture materials. The homopolymer of poly-L-lactide is not used despite its high initial strength and good strength retention owing to its low degradation rate. Sutures are manufactured by a melt-extrusion process yielding a multifilament which is stretched, braided, stretched again and annealed. PGA sutures are said to give high strengths, minimal tissue reactivity and a similar but more reproducible absorption rate to cat-gut[45].

Ligament replacement has been described as a potential use for PLA. In these applications, carbon fibres are coated with PLA which is eventually resorbed and replaced by collagen[48]. These applications require crystalline, oriented polymer filaments.

ORTHOPAEDIC REPAIRS

In many instances, metal plates, screws and wires are used in the treatment of fractures. They have to be removed after the healing process, necessitating a second operation. Replacement of these metal parts by a resorbable material is advantageous for patients and saves cost, since the second operation is then no longer necessary.

PLA and PGA homopolymers and copolymers have been used as resorbable plates and screws for fracture fixation[49–51]. These products are normally compression-moulded or fashioned from blocks. Applications such as these need significant mechanical integrity in the part and usually require partially crystalline compositions.

CONTROLLED DRUG RELEASE

PLA and PGA homopolymers are fairly crystalline, resorbed slowly and are not normally used for controlled-release applications. The crystallinity of PLA/PGA copolymers is determined by their composition. Copolymers containing 25–65 mole% glycolic acid are amorphous and are much more readily hydrolysed. The copolymer composition can be selected to give a material that degrades within a specified time.

The active substance can be micro-encapsulated, embedded in the polymer matrix or bound to the terminal groups of the polymer. Different mechanisms of release are seen depending on the nature of the formulation used. In the case of microencapsulation, drug release is by diffusion through micropores in the casing. Dispersion of the drug in a compact polymer gives release of the active substance by diffusion through the polymer matrix, by hydrolytic erosion of the matrix or by a combination of both these effects[52–54].

GRAFTS

Arterial grafts use PLA/PGA fibres woven into a mesh[55]. Artificial skin grafts can be formed from porous mixtures of polyurethane and PGA[56].

345

NONMEDICAL APPLICATIONS

It is understood that a number of companies, for example DuPont and Cargill, are developing processes to make commodity items from PLA. Little is known of the projects but it would appear that bio-degradability is the feature that has attracted the interest of these manufacturers.

Workers at Argonne National Laboratory, Illinois, are promoting a project that will convert a starch stream such as waste from a potato-processing plant to lactic acid in a single-stage bioreactor. The lactic acid will be used for PLA production[57]. They have forecast large potential markets for PLA plastics and coatings (Table 2).

Table 2 Potential markets for PLA plastics and coatings

>500,000 tonnes: degradable, controlled-release coatings for fertilizer and pesticide applications

250,000 tonnes: marine plastic applications

>100,000 tonnes: degradable conditioner coatings for paperboard stock

Tens of thousands of tonnes: speciality bags/sacks, e.g. compost waste bags

75,000 tonnes: agricultural mulch film

Other Biodegradable Plastics

Researchers at the University of Iowa are investigating the synthesis of biodegradable plastic using enzymes in organic solvents. The enzyme is being used to link a sugar and diacid into copolymer chains. Sucrose and adipic acid are the substrates for the reaction which is catalysed by lipases or proteases. It is predicted that the resulting polymers will be water-absorbent, which could enable them to be used in applications such as disposable nappies. The material will be biodegradable in both aerobic and anaerobic environments[58].

Polyglutamic Acid

Polyglutamate (PGLU) is a water-soluble polymer elaborated by several species of *Bacillus*. In some species such as *B. anthracis*, the polymer forms part of the capsular material of the organism, but in others, for example *B. subtilis*, PGLU is released into the growth medium.

The glutamate monomers in PGLU are linked through the α-amino of one residue to the γ-carboxyl of the second residue as shown in Fig. 3[59]. Both D and L isomers of glutamic acid were found in the polymer. Early workers in the field found that many strains of *B. subtilis* formed PGLU from free glutamic acid in the culture medium[60,61], but that only *B. subtilis* NRRL B-2612 was capable of using more complex sources of glutamate, such as wheat gluten, to synthesize the polymer[59]. Synthesis of the polymer has been reviewed by Housewright[62] and by Troy[63].

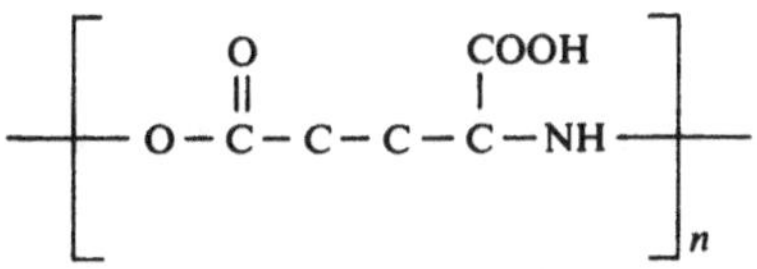

Fig. 3 γ-linked glutamic acid in polyglutamate.

Natto, a traditional fermented food produced in Japan from soya beans, contains a viscous material, one component of which is PGLU. The organism used to produce natto is *B. subtilis* (natto). Hara *et al.*[64] showed that transfer of genetic material from natto-producing *B. subtilis* strains to *B. subtilis* Marburg 168 conferred the ability to synthesize PGLU on the recipient organism. The recipient also gained γ-glutamyltranspeptidase activity at the same time, suggesting that this enzyme was responsible for PGLU synthesis. Subsequently a plasmid, pUH1, was found in *B. subtilis* (natto) which carried a gene encoding the γ-glutamyltranspeptidase activity[65].

A screen of *Bacillus* isolates from natto and from 'thau nao', the Siamese equivalent product, showed that all contained a cross-hybridizing plasmid when probed with fragments from pUH1[66]. The conclusion was that the plasmid sequences required for PGLU synthesis are highly conserved.

Production of PGLU

Fermentations to produce PGA were patented some years ago[67,68], but commercial manufacture does not appear to have taken place. In more recent times, a fed-batch fermentation has been described by Giannos *et al.* yielding approximately 40 g/l of PGLU in 5 days. They reported that this was the highest yield recorded in the literature for PGLU[69].

The author is not aware of any industrial application of the microbially produced polymer, although it is reported to have good metal-binding properties[70]. Chemically synthesized PGLU has been patented for use as a humectant in hair-care products[71], and when produced as a block copolymer with a polyester it has been patented as a synthetic leather coating[72].

Bioadhesives – Mollusc Glue

The interest in this material derives from the fact that it is an effective adhesive in an aqueous environment. Many conventional adhesives that exhibit excellent properties in dry conditions lose their adhesive power in wet environments. Furthermore, they cannot be cured in wet situations. Curing is the process whereby adhesivity is generated from the constituent chemicals of a formulation by chemical or enzymatic means. It has been very difficult to develop conventional adhesives for use in a wet environment such as marine situations or for use in medical and dental applications.

Marine mussels of the genus *Mytlius* deposit a substance which becomes cured and forms the attachment of the mussel to its substratum. A major component of the adhesive is a hydroxylated protein of M_r about 130,000 (ref. 73). Analysis has demonstrated that the protein contains a high proportion of lysine residues (20 per 100) and hydroxylated amino acids (60 residues per 100). Many of the hydroxylated residues are 3,4-dihydroxyphenylalanine (DOPA) and hydroxyproline which are formed by post-translational hydroxylation of tyrosine and proline. It is believed that these reactions are important in developing the adhesive properties of the protein[74,75].

Production of the bioadhesive

Two methods have been used to make the bioadhesive substance. In the first, decapeptides of the general formula shown in Fig. 4 are isolated from the protein extracted from the phenol gland of the mussel, or similar decapeptides are chemically synthesized. The decapeptides are then polymerized by chemical means using glutaraldehyde, oligopeptides or other bifunctional agents to yield materials of up to 1000-decapeptide repeating units[76]. Bio-Polymers Inc. have pioneered this methodology, but it is understood that it is not used commercially.

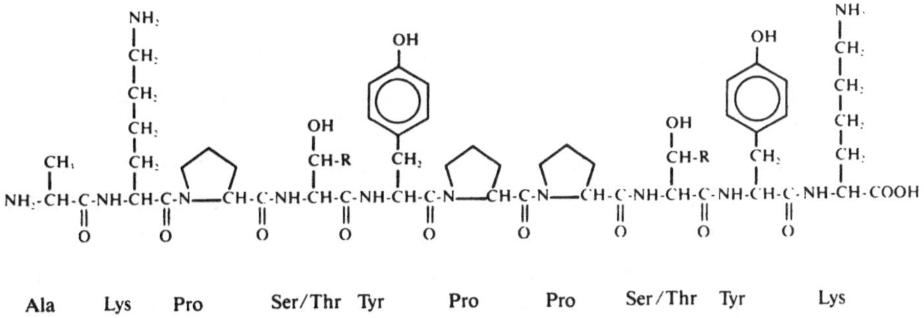

Fig. 4 Decapeptides from mollusc glue.

The second, and more practical, method used to produce the adhesive involves molecular genetics[77]. An oligonucleotide sequence was synthesized chemically that encoded a 'precursor protein', essentially repeating blocks of the decapeptide Ala-Lys-Pro-Ser-Tyr-Pro-Pro-Thr-Tyr-Lys. The oligonucleotide encoding the decapeptide was cloned into a plasmid immediately downstream of a fragment of the *trpB* gene. A fragment of the bovine chymosin gene was attached to the 3′ end of the synthetic sequence. This enabled efficient transcription and translation of repeating decapeptide units. Several contiguous copies of the oligonucleotide were cloned into this construct. Codons for methionine were inserted at the junctions with the *trpB* fragment and the chymosin gene fragment so that cleavage with cyanogen bromide would release the adhesive precursor protein (Fig. 5).

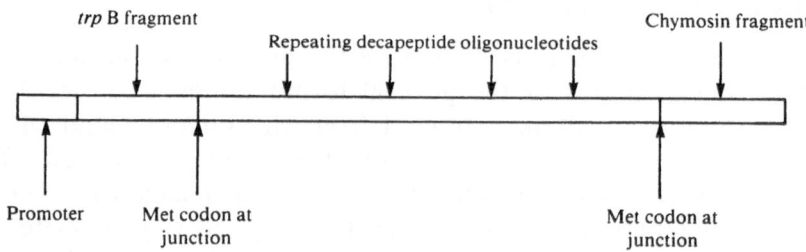

Fig. 5 Genetic construct for precursor-protein production of bioadhesive.

A family of decapeptides is found in the proteins isolated from mussels, and this situation was emulated by synthesis of a number of constructs where the oligonucleotide sequence of the precursor protein was varied. Precusor proteins were also made with additional amino acids added outside the decapeptide block, again to emulate the natural product. This technology has been developed by the Genex Corporation.

The material made both by extraction from mussels and by molecular genetics is cured to become adhesive by treatment with a monophenol oxidase such as tyrosinase, which converts tyrosine residues first to DOPA and then further oxidizes DOPA to the quinone. The quinone is capable of spontaneously forming crosslinks with pendant amino groups in the adhesive protein. The ability of the bioadhesive to function in wet environments is related to: (1) the manner in which it is cured, the hydroxylation reaction being thought to displace water molecules from the substrate surface; (2) the fact that it is a large molecule with low solubility in the cured state and therefore has a reduced tendency to diffuse from the surface being treated; and (3) the large number of reactive sites in the molecule (amine, hydroxyl, catechol) which give the structure the ability to participate in a number of interactions, including covalent, ionic and hydrogen-bonding, as well as Van der Waals' bonding, all of which contribute to cohesive and adhesive strength.

It is reported that W. R. Grace and Genex are developing the bioadhesive for medical and dental applications. It could be used in suture replacement, bone and wound repair and the bonding of caps and crowns.

Alternative Protein-Based Glues

Kaleem et al.[78] produced an adhesive derived from gelatin. To produce a curable substrate, the number of phenolic groups in the molecule was increased by reductive alkylation using phenolic aldehydes. The protein could then be crosslinked using polyphenol oxidase to produce an effective adhesive.

Chitin and Chitosan

Chitin is the major polysaccharide of the shells of crustaceans and exoskeletons of insects. Its also found in the cell walls of many fungi, yeast and algae.

Chitosan is a nonacetylated or partially deacetylated chitin derivative. It is found naturally in fungal cell walls but can also be produced by alkaline treatment of chitin.

Chitin is a linear homopolymer of $\beta(1-4)$-linked N-acetylglucosamine (Fig. 6). Three different conformations of chitin sheets are found: α, antiparallel packed sheets; β, parallel packed sheets; and γ, two parallel sheets separated by an antiparallel sheet. All three conformations are found in nature.

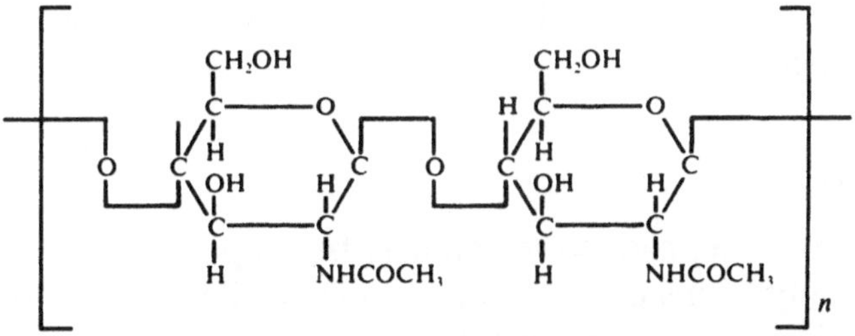

Fig. 6 Structure of chitin. Chitosan is the deacetylated derivative of this structure.

Biosynthesis

In the brine shrimp, *Artemia salina*, the immediate precursor of chitin is UDP-N-acetylglucosamine (UDPG). UDPG is transferred to a lipid acceptor and polymerized to form a 2–8-residue oligosaccharide. The oligosaccharide is in turn transferred to the growing chitin chain by a pronase-sensitive component[79].

Chitin synthesis in fungi also proceeds via UDPG. Chitin synthetase, which when isolated requires protease treatment for activation, is stimulated by divalent cations such as Mn^{2+} or Mg^{2+} (ref. 80). The site of chitin synthesis is at the growing tips of hyphae in filamentous fungi[81]. In yeast, chitin synthesis is the initial event in cell division[82]. Since chitin is an essential component of the cell wall in many fungi, it is not surprising that its synthesis is closely linked with growth processes.

Chitosan is produced from chitin by several fungi. Chitin deacetylase is a soluble, cytoplasmic enzyme that produces chitosan and acetate from chitin[83]. The enzyme cannot act on mature chitin, probably because of the crystalline nature of the material, and is thought to act on nascent chitin chains as they are formed.

The proportion of chitin and chitosan in an organism can vary with physiological and morphological state[84]. The microbial synthesis of chitin and chitosan has been reviewed by Gooday[85].

Manufacture

Chitin is extracted from shells that are the waste products from the crab- and prawn-processing industry. Japan and the USA are the major producers. Chitosan is produced from chitin by treatment with alkali and can be derivatized to give a wide range of materials with different properties.

Applications

New chitin/chitosan production capacity has appeared in the USA and Japan and many forms of chitosan are now available. New derivatives of chitin and chitosan continue to appear, providing solubility and useful properties. Commercial products are available as powders, solutions, gels, fibres and beads. The level of interest in these materials can be judged by the large number of publications and patents in the literature. Indeed a sourcebook of references to research on chitins has been published[86]. Table 3 lists some of the applications for which these versatile products have been used.

Table 3 Uses for chitin, chitosan and their derivatives

Chelating agent	Adhesive
Floccculant	Paper and textile processing aid
Coagulant	Medical applications
Drug carrier	Cosmetics
Membranes	Fungicide
Encapsulation	Photographic products
Water and waste treatment	

The more important of these applications will be briefly discussed and a few recent references cited. The area has been reviewed by Muzzarelli[87].

BIOMEDICAL APPLICATIONS

A wide range of biomedical applications for chitin and its derivatives have been considered. Many of the uses are listed in Table 4. Physical form of the materials is important in determining its application.

Chitin and chitosan were also among the materials chosen for study as artificial skin substitutes[104].

Further information on biomedical applications can be obtained in the review by Olsen *et al.*[105].

Table 4 Biomedical applications and physical form

Physical form	Application
Solution	Bacteriostatic agent[88]
	Homeostatic agent[89]
	Cosmetics[90]
Gel	Delivery vehicle[91]
	Spermicide[92]
Powder	Surgical glove powder[93]
	Enzyme immobilization[94]
Film/membrane	Dialysis membrane[95]
	Contact lens[89]
	Wound dressing[96]
Sponge	Mucosal haemostatic dressing[97]
	Wound dressing[98]
Miscellaneous	Anticholesteraemic materials[99]
	Antigastritis agents[100]
	Antisordes compositions[101]
	Antibilirubidaemia agent[102]
	Anticoagulants[103]

WASTE TREATMENT

The food industry has a requirement for reduction of solids in effluents, and chitosan has been used in this application. Combination with polyelectrolytes such as polyacrylic acid is also effective[106]. Chitosan has been studied as a flocculant to remove cell debris in an enzyme purification process[107].

COSMETICS

Chitosan and its derivatives have attracted increasing interest in the cosmetics industry. Lang and Clausen[108] have reviewed applications for these materials in that industry.

CELL AND ENZYME IMMOBILIZATION

Chitosan has been used to immobilize whole cells for use as biocatalysts. Fungi as well as bacteria have been immobilized. *Pleurobes ostreaties* was immobilized on chitosan by crosslinking with polyphosphate or glutaraldehyde[109]. The catalytic half-life of the cells producing 6-amino-penicillin acylase was extended from 2.5 days for the free cells to 25 days for the immobilized cells.

Escherichia coli with tryptophan synthetase activity was immobilized using chitosan with multivalent anionic counterions used to crosslink the gel matrix in a process called ionitrophic gelation. The beads formed contained *E. coli* trapped in micropores. The treatment is mild, inexpensive and simple[110].

A chitosan–collagen composite has been used to provide a substratum for eukaryotic cell culture. The cells adhere to the material and are then capable of successful proliferation[111].

Proteases[112,113], glucoamylase[114], lyzozyme[115,116] and β-glucosidase[117] have been immobilized on chitin, chitosan or their derivatives. More recently, studies have been carried out on the immobilization of pullulanase[118] and urease[119].

MISCELLANEOUS

A fabric has been made from shellfish chitin by the Asahi Chemical Industry and Asahi Chemical Textiles. It consists of a chemically treated chitin bound to a porous polyurethane resin coated onto a nylon sheet. The novel fabric is breathable, since sweat is absorbed by the chitin layer and then diffuses through the polyurethane and evaporates[120].

Researchers at Auburn University in the United States are planning to use a two-stage fermentation to convert chitin to alcohol or protein. In the first stage, chitin is hydrolysed by microbial action, and in the second phase, either *Zymomonas mobilis* or *Pachysolen tannophilus* is used to produce alcohol or protein[121].

Skin Substitutes

Several companies have developed living skin substitutes that are being used both in skin grafting in, for example, burns injuries, and as test systems by cosmetic and consumer-product companies to replace animals.

Organogenesis Inc. have used dermal cells cultured in a collagen matrix to give a layer on which epidermal cells taken from a patient may be grown to produce a skin substitute that can be used for grafting. The dermal fibroblasts are screened to eliminate immunogenic cells and to ensure that they are free of viral and other infections. After growth of the fibroblasts in a protein and collagen nutrient medium to produce a dermal layer, the epidermal cells are placed on top. The eventual result is a rejection-free graft.

It is reported that Avon Products will use the 'skin' to study skin absorption of cosmetics and skin-care products, and that Mary Kay Cosmetics and Estée Lauder are investigating the material as a replacement for animal testing[122,123].

Marrow-Tech has produced a full-thickness skin substitute based on growth of cells on a biodegradable polymer mesh developed by the American Cyanamid Corporation. The three-dimensional skin structure, which includes multilayer dermis, epidermis and pigment cells, is implanted into a wound such as a burn, and once the graft has taken, the mesh biodegrades. This product is also under consideration by cosmetic, chemical, drug and toxicology-testing companies as a possible replacement for some animal tests[124].

BioSurface Technology is developing a skin substitute based on cell tissue culture of human epidermal cells. It is understood that human epidermal growth factor is involved in the procedure.

Marion Laboratories were reported to be marketing a skin substitute that is an artificial dermal layer grown on a collagen–glycosaminoglycan sponge coated with silicone. The dermal layer is incorporated subsequently into the patient's own collagen network, and minimal scarring after wound healing is reported.

Clonetics Corporation is also reported to be active in this field, as are several academic groups.

References

1. Barstow, J.D. (personal communication).
2. Collins, S.H. (1987) In *Carbon Substrates in Biotechnology. Spec. Publ. Soc. Gen. Microbiol.* **21**: 161–169 (eds Stowell, J.D., Beardsmore, A.J., Keevil, C.W. & Woodward, J.R.). IRL Press.
3. Suzuki, T., Yamane, T. & Shimizu, S. (1986) *Appl. Microbiol. Biotechnol.* **23**: 322–329.
4. Suzuki, T., Yamane, T. & Shimizu, S. (1986) *Appl. Microbiol. Biotechnol.* **24**: 366–369.
5. Suzuki, T., Yamane, T. & Shimizu, S. (1986) *Appl. Microbiol. Biotechnol.* **24**: 370–374.
6. Suzuki, T., Deguchi, H., Yamane, T., Shimizu, S. & Gekko, K. (1988) *Appl. Microbiol. Biotechnol.* **27**: 487–491.
7. Kallio, R.E. & Harrington, A.A. (1960) *J. Bacteriol.* **80**: 321–324.
8. Whittenbury, R., Phillips, K.C. & Wilkinson, J.F. (1970) *J. Gen. Microbiol.* **61**: 205–218.
9. Higgins, I.J., Best, D.J., Hammond, R.C. and Scott, D. (1981) *Microbiol. Rev.* **45**: 556–590.
10. Asenjo, J.A. & Suk, J.S. (1986) *J. Ferment. Technol.* **64**: 271–278.
11. Asenjo, J.A. & Suk, J.S. (1985) *Biotech. Bioengng. Symp.* **15**: 225–234.
12. Schlegel, H.G., Gottschalk, G. & von Bartha, R. (1961) *Nature* **191**: 463–465.
13. Sonnleitner, B., Heinzle, E., Braunegg, G. & Lafferty, R.M. (1979) *Eur. J. Appl. Microbiol. Biotechnol.* **7**: 1–10.
14. US Patent US 3036959, 1962.
15. Griffin, M. & Magor, A.M. (1987) *Microbiol. Sci.* **4**: 357–361.
16. Byrom, D. (1987) *Trends Biotechnol.* **5**: 246–250.
17. European Patent EP 0052459, 1980.
18. European Patent EP 144017, 1984.
19. European Patent EP 149744, 1984.
20. Hanggii, U.J. (1990) In *Novel Biodegradable Microbial Polymers* (ed. Dawes, E.A.), pp. 65–70. Kluwer.
21. Chowdhury, A.A. (1963) *Arch. Microbiol.* **47**: 167–200.
22. Winton, J.M. (1985) *Chem. Week.* August 28, 55.
23. Uttley, N.L. (1986) *Appl. Biotechnol. Proc. BIOTECH '86*, Vol. 1, 171.
24. Brandl, H., Gross, R.A., Lenz, R.W. & Fuller, R.C. (1990) *Adv. Biochem. Eng./Biotechnol.* **41**: 78–91.
25. Tanio, T., Fukui, T., Shirakura, Y., Saito, T., Tomita, K., Kaiho, T. & Masamune, S. (1982) *Eur. J. Biochem.* **124**: 71–77.
26. Lusty, C.J. & Doudoroff, M. (1966) *Proc. Natl. Acad. Sci. U.S.A.* **56**: 960–965.
27. Nakayama, K., Saito, T., Fukui, T., Shirakura, Y. & Tomita, T. (1985) *Biochim. Biophys. Acta.* **827**: 63–72.
28. McLellan, D.W. & Halling, P.J. (1988) *FEMS Microbiol. Lett.* **52**: 215–218.

29. Holland, S.J., Jolly, A.M., Yasin, M. & Tighe, B.J. (1987) *Biomaterials* **8**: 289–295.
30. Miller, N.D. & Williams, D.F. (1987) *Biomaterials* **8**: 129–137.
31. Doi, Y., Kanesawa, Y., Kawaguchi, Y. & Kunioka, M. (1989) *Makromol. Chem. Rapid Commun.* **10**: 227–230.
32. Holmes, P.A. (1988) In *Developments in Crystalline Polymers* (ed. Bassett, D.C.), Vol. 2, pp. 1–65. Applied Science Publishers, London.
33. Griffin, G.J.L. (1974) *Adv. Chem. Ser.* **134**: 159–170.
34. Roper, H. & Koch, H. (1990) *Starch/Starke* **42**: 123–130.
35. European Patent EP 326517, 2 August 1989.
36. European Patent EP 282451, 14 September 1988.
37. European Patent EP 298920, 1 November 1989.
38. European Patent EP 327505, 9 August 1989.
39. European Patent EP 304401, 22 February 1989.
40. US Patent US 3297033, 1967.
41. US Patent US 3422871, 1969.
42. US Patent US 3626948, 1971.
43. US Patent US 1375008, 1971.
44. US Patent US 3839297, 1975.
45. Frazza, E.J. & Schmitt, E.E. (1971) *J. Biomed. Mater. Res. Symp.* **1**: 43.
46. Gilding, D.K. & Reed, A.M. (1979) *Polymer* **20**: 1459–1464.
47. Hollinger, J.O. & Schmitz, J.P. (1984) Report AD-A 146 141 US Army Institute of Dental Research, 26 July 1984.
48. Klein, A.J. (1986) *Adv. Mater. Processes* **5**: 18.
49. Vert, M., Christel, D., Chabot, F. & Lerav, J. (1984) In *Macromolecular Biomaterials* (eds Hastings, G.W. & Duchevne, P.). CRC Press.
50. Kronenthal, R.L. (1975) *Polym. Sci. Technol.* **8**: 119–137.
51. Cutright, D.E. & Hunsuck, E.E. (1972) *Oral Surg.* **33**: 28–34.
52. Wise, D.L., Fellman, Th.D., Sanderson, J.E. & Wentworth, R.L. (1979) In *Drug Carriers in Medicine*, pp. 237–270. Academic, London.
53. Wood, D.A. (1980) *Int. J. Pharm.* **7**: 1–18.
54. Heller, J. (1985) *CRC Crit. Rev., Therapeutic Drug Carrier Systems* **1**: 33–90.
55. Bowald, S., Busch, C. & Eriksson, I. (1978) *Lancet* **21**: 153.
56. Gogolewski, S. & Pennings, A.J. (1982) *Makromol. Chemie Rapid Comm.* **3**: 839–845.
57. Anonymous (1989) *Eur. Chem. News*, 20 November, p. 26.
58. Anonymous (1989) *Bioprocess. Technol.*, April, p. 6.
59. Ward, R.M., Anderson, R.F. & Dean, F.K. (1963) *Biotechnol. Bioengng.* **5**: 41–48.
60. Thorne, C.B., Gomez, C.G., Noyes, H.E. & Housewright, R.D. (1954) *J. Bacteriol.* **68**: 307–315.
61. Thorne, C.B. (1956) *Symp. Soc. Gen. Microbiol.* **6**: 68–84.
62. Housewright, R.D. (1962) In *The Bacteria: A Treatise on Structure and Function* (eds Gunsalas, I.C. & Stanier, R.Y.), Vol. 3, pp. 389–412. Academic, London.
63. Troy, F.A. (1982) In *Peptide Antibiotics: Biosynthesis and Functions* (eds Kleinkauf, H. & Dohren, H.V.), pp. 49–83. W. de Grutyer, New York.
64. Hara, T. & Ueda, S. (1982) *Agric. Biol. Chem.* **46**: 2275–2281.
65. Hara, T., Ueda, S. & Sakaki, Y. (1983) *Agric. Biol. Chem.* **47**: 1143–1144.
66. Hara, T., Chetanachit, C., Fujio, Y. & Ueda, S. (1986) *J. Gen. Appl. Microbiol.* **32**: 241–249.
67. US Patent US 28995882, 21 July 1959.
68. Japanese Patent JP 7724590, 1 July 1977.

69. Giannos, S.A., Shah, D., Gross, R.A., Kaplan, D.L. & Mayer, J.M. (1990) In *Novel Biodegradable Microbial Polymers* (ed. Dawes, E.A.), pp. 457–460. Kluwer, Netherlands.

70. McLean, R.J.C., Beauchemin, D., Clapham, L. & Beveridge, T.J. (1990) *Appl. Environ. Microbiol.* **56**: 3671–3677.

71. Japanese Patent JP 8375562, 28 April 1983.

72. Japanese Patent JP 694618, 23 January 1969.

73. Waite, H.J. (1983) *J. Biol. Chem.* **258**: 2911–2915.

74. Waite, H.J. (1983) *Mollusca* **1**: 467–504.

75. Pizzi, A. *et al.* (1982) *Ind. Eng. Chem. Prod. Res. Dev.* **21**: 309–369.

76. US Patent US 4585585.

77. Patent Cooperation Treaty (PCT) WO88/07076, 22 September 1988.

78. Kaleem, K., Chertok, F. & Erhan, S. (1987) *Angew. Makrmol. Chemie.* **155**: 31–43.

79. Horst, M.N. (1983) *Arch. Biochem. Biophys.* **223**: 254–263.

80. Cabib, E., Roberts, R. & Bowen, B. (1982) *Ann. Rev. Biochem.* **51**: 763–793.

81. Archer, D.B. (1977) *Biochem. J.* **164**: 653–658.

82. Molano, J., Bowen, B. & Cabib, E. (1980) *J. Cell Biol.* **85**: 199–212.

83. Bartnicki-Garcia, S. & Davis, L.L. (1982) *183rd ACS National Meeting, Abstr. Pap. Am. Chem. Soc.* **183**: Carb-21.

84. Domek, D.B. & Borgia, P.T. (1981) *J. Bacteriol.* **146**: 945–951.

85. Gooday, G.W. (1983) In *Prog. Ind. Microbiol.* (ed. Bushell, M.E.), Vol. 18. Elsevier, New York.

86. Pariser, E.R. & Lombardi, D.P. (1989) *Chitin Sourcebook: A Guide to the Chitin Research Literature.* Wiley, New York.

87. Muzzarelli, R.A.A. (1990) *Comm. Eur. Commun.* (Rep.) Eur 12757 Towards Carbohyd.-Based Chem., 77–97.

88. Malette, W.G., Quigley, H.J., Gaines, R.D., Johnson, N.D. & Rainer, W.G. (1983) *Ann. Thorac. Surg.* **36**: 55–58.

89. Allan, G.G., Altman, L.C., Bensinger, R.E., Ghosh, D.K., Hirabayashi, Y., Neogi, A.N. & Neogi, S. (1984) In *Chitin, Chitosan and Related Enzymes* (ed. Zikakis, J.P.), pp. 119–134. Academic.

90. US Patent US 4202881, 13 May 1980.

91. US Patent US 4659700, 21 April 1987.

92. US Patent US 4474769, 2 October 1984.

93. US Patent US 4064564, 27 December 1977.

94. Canadian Patent 1210717, 2 September 1986.

95. Japanese Patent JP 130870, 1975.

96. US Patent US 4570629, 18 February 1986.

97. Japanese Patent JP 60-142927, 29 July 1985.

98. US Patent US 4651725, 3 March 1987.

99. Nagyvary, J.J., Falk, J.D., Hill, M.L., Schmidt, M.L., Wilkins, A.K. & Bradbury, E.L. (1979) *Nutr. Rep. Int.* **20**: 677–684.

100. European Patent Application 219156, 22 September 1986.

101. Japanese Patent JP 61-151112, 9 July 1986.

102. US Patent US 4363801, 14 December 1982.

103. Hirano, S., Tanaka, Y., Hasegawa, M., Tobetto, K. & Nishioka, A. (1985) *Carb. Res.* **137**: 205–215.

104. Kim, K.Y. (1990) *Makromol. Chem. Macromol. Symp.* **33**: 311–318.

105. Olsen, R., Schwartzmiller, D., Weppner, W. & Winady, R. (1989) *Chitin Chitosan: Sources, Chem., Biochem., Phys. Prop. Appl. (Proc. 4th Int. Conf.),* 813–828.

106. Chavasit, V. & Torres, J.A. (1990) *Biotechnol. Prog.* **6**: 2–6.

107. Agerkvist, I., Eriksson, L. & Enfors, S.O. (1990) *Enzyme Microb. Technol.* **12**: 584–590.

108. Lang, G. & Clausen, T. (1989) *Chitin Chitosan: Sources, Chem., Biochem., Phys. Prop. Appl. (Proc. 4th Int. Conf.)*, 139–147.

109. Kluge, M., Klein, J. & Wagner, F. (1982) *Biotechnol. Lett.* **4**: 293–296.

110. Vorlop, K.D. & Klein, J. (1981) *Biotechnol. Lett.* **3**: 9–14.

111. European Patent Application 318286, 31 May 1989.

112. Lebua, J.-L. & Widmer, F. (1979) *Biotechnol. Lett.* **1**: 109–114.

113. Muzzarelli, R.A.A., Barontini, G. & Rocchetti, R. (1976) *Biotechnol. Bioengng.* **20**: 87–94.

114. Yamaguchi, R., Hirano, S., Arai, Y. & Itoh, T. (1978) *Agric. Biol. Chem.* **42**: 1981–1982.

115. Muzzarelli, R.A.A. (1973) *Natural Chelating Polymers*. Pergamon, New York.

116. Yamaguchi, R., Arai, Y., Kanebo, T. & Itoh, T. (1982) *Biotechnol. Bioengng.* **24**: 1081–1092.

117. Bissett, F. & Sternberg, D. (1978) *Appl. Envir. Microbiol.* **35**: 750–755.

118. Hisamatsu, M. & Yamada, T. (1989) *J. Ferment. Technol.* **67**: 219–220.

119. Wojcik, A., Lobarzewski, J. & Blaszczynska, T. (1990) *J. Chem. Technol. Biotechnol.* **48**: 287–301.

120. Anonymous (1989) *Bioprocess. Technol.* March, p. 8.

121. Anonymous (1989) *Bioprocess. Technol.* June, p. 28.

122. Anonymous. *Biotechnol. News* **9(27)**: 3.

123. Anonymous. *Biotechnol. News* **10(1)**: 7.

124. Anonymous. *Biotechnol. News* **9(25)**: 5.

Index